Fundamental Concepts in the Design of Experiments

Fundamental Concepts in the Design of Experiments

CHARLES R. HICKS
Director of Teacher Education
Purdue University

HOLT, RINEHART AND WINSTON
New York Chicago San Francisco Atlanta Dallas
Montreal Toronto London Sydney

Library of Congress Catalog Card Number: 72-90986

ISBN: 0-03-080132-X

Printed in the United States of America

3 4 5 6 038 1 2 3 4 5 6 7 8 9

Preface to the Second Edition

At first glance the reader may not notice much change in this edition from the original. An attempt has been made to keep chapter and section headings the same. It cannot be considered as a major revision although there are many new features. The new material adds approximately 25% to the size of the book and the number of problems has been increased by 70%.

Main features include a section on covariance analysis and analysis of attribute data in the 16th chapter, introduction of linear and curvilinear regression in Chapter 8, and an amplification of the concept of orthogonal contrasts and orthogonal polynomials. Some topics have been rearranged to appear earlier in the text and methods of testing means after analysis of variance have been changed and extended.

Several users of the original text have sent me helpful suggestions for this revision. Whenever a suggestion came from more than one person, I attempted to include it in this edition. Naturally, all suggestions were not followed and the responsibility for their adoption is mine alone. I am indebted to the following people who made detailed suggestions:

V. L. Anderson *Purdue University*
O. A. Beckendorf *General Motors Institute*
H. C. Charbomeau *General Motors Institute*
R. I. Fields *University of Louisville*
R. J. Fopman *University of Cincinnati*
B. Harshbarger *Virginia Polytechnic Inst.*
O. B. Moan *Arizona State University*
R. G. Morris *General Motors Institute*
V. V. Namilov *Moscow State University*
J. B. Neuhardt *The Ohio State University*
C. R. Peterson *General Motors Institute*
B. T. Rhodes, Jr. *University of Houston*
H. W. Stoker *The Florida State University*

B. J. Todd *Clemson University*
L. A. Weaver *University of South Florida*
I am also grateful to Purdue University for granting me sabbatical leave to complete this revised edition.

Lafayette, Indiana *Charles R. Hicks*
March, 1973

Preface to the First Edition

It is the primary purpose of this book to present the fundamental concepts in the design of experiments using simple numerical problems, many from actual research work. These problems are selected to emphasize the basic philosophy of design. Another purpose is to present a logical sequence of designs that fit into a consistent outline; for every type of experiment, the distinctions among the experiment, the design, and the analysis are emphasized. Since this theme of experiment-design-analysis is to be highlighted throughout the text, the first chapter presents and explains this theme. The second chapter reviews statistical inference, and the remaining chapters follow a general outline, with an experiment-design-analysis summary at the end of each chapter.

The book is written for anyone engaged in experimental work who has a good background in statistical inference. It will be most profitable reading to those with a background in statistical methods including analysis of variance.

Some of the latest techniques have not been included. The reason for this is again the purpose of presenting the *fundamental* concepts in design to provide a background for reading journals in the field and understanding some of the more recent techniques. Although it is not assumed that every reader of this book is familiar with the calculus, some theory is presented where calculus is used. The reader should be able to grasp the underlying ideas of each chapter even if he must omit these sections.

This book evolved from a course in Design of Experiments given at Purdue University over the past four or five years. The author is indebted to Professors Gayle McElrath of the University of Minnesota and Clyde Kramer of Virginia Polytechnic Institute whose lectures, along with the author's, form the basis of this book. Credit is also due Mr. James Dick and Mr. Roland Rockwell, who took notes of the author's lectures.

I shall be forever indebted to the University of Florida for granting me a visiting professorship during the academic year 1962–1963 with time to work on this book, and to Mrs. Alvarez of their Industrial Engineering Department for typing the manuscript.

Lafayette, Indiana *Charles R. Hicks*
November, 1963

Contents

4

Single-factor Experiments — Randomized Block Design 51

5

Single-factor Experiments — Latin and Other Squares 77

6

Factorial Experiments 86

7

2^n Factorial Experiments 108

8

Qualitative and Quantitative Factors 123

Fundamental Concepts in the Design of Experiments

1
The Experiment, the Design, and the Analysis

1.1 INTRODUCTION

In any experiment the experimenter is attempting to draw certain inferences or make a decision about some hypothesis or "hunch" that he has concerning the situation being studied. Life consists of a series of decision-making situations. As one tastes his morning cup of coffee, almost unconsciously he decides that it is better than, the same as, or worse than the coffee he has been drinking in the past. If, based on his idea of a "standard" cup of coffee, the new cup is sufficiently "bad," he may pour it out and make a new batch. If he believes it superior to his "standard," he may try and determine whether the brand is different than the one he is used to or the brewing is different, and so forth. In any event, it is the extreme differences which may cause him to decide to take action. Otherwise, he behaves as if nothing has changed. In any decision he runs the risk that the decision is wrong since it is based on a small amount of data.

In deciding whether or not to carry an umbrella on a given morning, one "collects" certain data: he taps the barometer, looks at the sky, reads the newspaper forecast, listens to the radio, and so on. After quickly assimilating all available data—including such predictions as "a 30-percent probability of rain today"—he makes a decision. Somehow a compromise is made between the inconvenience of carrying the umbrella and the possibility of having to spend money for a cleaning bill for one's clothes.

In these, and in most everyday events, decisions are made in the light of uncertainty. *Statistics* may be defined as a tool for decision making in the light of uncertainty. Uncertainty does not imply no knowledge, but only that the exact outcome is not completely predictable. If 10 coins are tossed, one knows that the number of heads will be some integer between 0 and 10. However, each specific integer has a certain chance of occurring and various results may be predicted in terms of their chance or probability of occurrence. When the results of an experiment which are observed could have occurred only 5 times in 100 by chance alone, most experimenters consider that this is a rare event and will state that the results are statistically significant at the 5-percent significance level. In such cases the hypothesis being tested is usually rejected as untenable. When statistical methods are used in experimentation, one can assess the magnitude of the risks taken in making a particular decision.

Since decision making based on the use of statistical tools almost always involves collection of data, the way in which the data are collected becomes extremely important. The design of an experiment has been defined very simply as the order in which an experiment is run. Too often in practice an experimenter is so anxious to collect data and stuff the results into a computer and make his decision that he neglects the important phase of the design of the experiment.

Experimental designs are used to help reduce the experimental error in the data collected. Randomization is employed to help average out the effect of many extraneous variables which may be present in an experiment. Other factors are purposefully varied during an experiment in order to make the results more valid for a large variety of situations which might occur in practice. By use of certain special designs, the effects of many important factors can be studied in one experiment as well as the interrelationship between these factors. In all such experiments, one seeks to obtain the maximum amount of reliable information at the minimum cost to the experimenter. This book will attempt to present many designs which have been found useful in accomplishing these purposes. Care will be taken to emphasize three important phases of every project: the *experiment*, the *design*, and the *analysis*.

1.2 THE EXPERIMENT

The experiment includes a statement of the problem to be solved. This sounds rather obvious, but in practice it often takes quite a while to get general agreement as to the statement of a problem. It is important to bring out all points of view to establish just what the experiment is intended to do. A careful statement of the problem goes a long way toward its solution.

Choice must also be made as to the dependent variable or variables to be studied. Are these measurable? How accurately can they be measured on the instruments available? If they are not measurable, what type of response can be expected? If the results are simply *yes* or *no*, *go* or *no-go*, what type of distribution of such results is reasonable?

It is also necessary to define the independent variables or factors which may affect the dependent or response variable. Are these factors to be held constant, to assume certain specified levels, or to be averaged out by a process of randomization? Are levels of the factors to be set at certain fixed values, such as temperature at 70°F, 90°F, and 110°F, or are such levels to be set at random among all possible levels? Are the factors to be varied quantitative (such as temperature) or qualitative (operators)? And how are the various factor levels to be combined?

All of the above considerations go into the definition of the experiment.

1.3 THE DESIGN

Of primary importance, since the remainder of this book will be devoted to it, is the design phase of a project. Many times an experiment is agreed upon, data are collected, and conclusions are drawn with little or no consideration given to *how* the data were collected. First, how many observations are to be taken? Considerations of how large a difference is to be detected, how much variation is present, and what size risks are to be tolerated are all important in deciding on the size of the sample to be taken for a given experiment. Without this information, the best alternative is to take as large a sample as possible. In practice this sample size is often quite arbitrary. However, as more and more tables become available, it should be possible to determine the sample size in a much more objective fashion (see Section 2.5).

Also of prime importance is the order in which the experiment is to be run, which should be random order. Once a decision has been made to control certain variables at specified levels, there are always a number of other variables which cannot be controlled. Randomization of the order of experimentation will tend to average out the effect of these uncontrolled variables.

For example, if an experimenter wishes to compare the average current flow through two types of computers and five of each type are to be tested, in what order are all ten to be tested? If the five of type I are tested, followed by the five of type II, and any general "drift" in line voltage occurs during the testing, it may appear that the current flow is greater on the first five (type I) than on the second five (type II), yet the real cause is the "drift" in line voltage. A random order for testing allows any time trends to average out. It is desirable to have the average current flow in the type I and type II computers equal, if the computer types do not differ in this respect. Randomization will help accomplish this. Randomization will also permit the experimenter to proceed as if the errors of measurement are independent, a common assumption in most statistical analyses.

What is meant by random order? Is the whole experiment to be completely randomized, with each observation made only after consulting a table of random numbers, or tossing dice, or flipping a coin? Or, possibly, once a temperature bath is prepared, is the randomization made only within this particular temperature and another randomization made at another temperature? In other words, what is the randomization procedure and how are the units arranged for testing? Once this step has been agreed upon, it is recommended that the experimenter keep a watchful eye on the experiment to see that it is actually conducted in the order prescribed.

Having agreed upon the experiment and the randomization procedure, a mathematical model can now be set up which should describe the experiment. This model will show the response variable as a function of all factors which are to be studied and any restrictions imposed on the experiment due to the method of randomization.

1.4 THE ANALYSIS

The final step, analysis, includes the procedure for data collection, data reduction, and the computation of certain test statistics to be used in making decisions about various aspects of an experiment. Analysis involves the computation of test statistics such as t, F, χ^2, and their corresponding decision rules for testing hypotheses about the mathematical model. Once the test statistics have been computed, decisions must be made. These decisions should be made in terms which are meaningful to the experimenter. They should not be couched in statistical jargon such as "the third-order $A \times B \times E$ interaction is significant at the 1-percent level," but instead should be expressed in graphical or tabular form, in order that they be clearly understood by the experimenter and by those persons who are to be "sold" by the experiment. The actual statistical tests should probably be included only in the appendix of a report on the experiment. These results should also be used as "feedback" to design a better experiment, once certain hypotheses seem tenable.

1.5 SUMMARY IN OUTLINE

I. *Experiment*
   - A. Statement of problem
   - B. Choice of response or dependent variable
   - C. Selection of factors to be varied
   - D. Choice of levels of these factors
     1. Quantitative or qualitative
     2. Fixed or random
   - E. How factor levels are to be combined.

II. *Design*
   - A. Number of observations to be taken
   - B. Order of experimentation
   - C. Method of randomization to be used
   - D. Mathematical model to describe the experiment.

III. *Analysis*
   A. Data collection and processing
   B. Computation of test statistics
   C. Interpretation of results for the experimenter.

Example 1.1 The personnel department of a plant was asked to conduct a special course in basic arithmetic to be given to all employees with only a high-school education. Since such a course could be taught by the usual lecture–recitation method or by use of computer aided instruction (CAI), it was decided to conduct a simple experiment to determine whether or not this CAI method produced as good or better results than the traditional teaching method. If the CAI method proved satisfactory, it would be much cheaper to administer to all new employees.

In this experiment, the response or dependent variable might be the score a person made on an achievement test in arithmetic after instruction by the CAI method. Realizing, however, that the trainees would enter the program with differing amounts of arithmetic ability, it is obvious that any comparison of final scores might reflect differences in initial ability as well as differences due to the method of instruction. Hence, it was decided that the best response variable would be the difference between a person's pretest and posttest scores as recorded before and after the instruction.

The chief factor affecting this "gain score" (difference between pretest and posttest scores) is assumed to be the method of instruction. Two methods are proposed, (1) the CAI method and (2) the lecture–recitation (traditional) method. Other factors which might affect this gain score include the instructor, the time the course is offered, the basic intelligence of the trainees, the motivation of the trainees, and so on. It was agreed to use the same instructor for the trainees taught by each of the methods. The experiment will be designed to help handle the other possible sources of variability.

The single factor under study—method of instruction—is at two fixed levels, and the levels are qualitative. Since only a single factor is involved, there is no problem of combining levels of factors.

The number of observations is somewhat restricted due to the number of trainees that can be effectively taught by one instructor in two groups. A class size of 25 was agreed upon with a total of 50 involved in the experiment. This number is also restricted by the number of computer terminals available for those taught by the CAI method.

Since the two groups of 25 trainees could differ considerably in basic intelligence, high-school backgrounds, culture, and so on, it was decided to assign the trainees to group 1 or 2 completely at random. A coin was tossed with a tail assigning the first trainee to group 1 and a head assigning him to group 2. The only restriction on this complete randomization of subjects (trainees) to treatments (methods) was that there be 25 in each group.

A mathematical model for this experimental design would be $Y_{ij} = \mu + M_j + \varepsilon_{i(j)}$, where Y_{ij} is the gain score for the ith trainee using method j. Here $j = 1, 2$ and $i = 1, 2, \cdots, 25$ for each group. μ represents a common effect in all observations (the true mean gain score of the population sampled), M_j the method effect, and $\varepsilon_{i(j)}$ the random error within each group of trainees.

The results of the experiment showed: $\bar{Y}_1 = 8.7$, $\bar{Y}_2 = 7.7$, $s_1^2 = 1.8$, and $s_2^2 = 1.6$, where subscripts 1 and 2 denote groups 1 and 2, respectively, and the statistics indicate mean and variance of the respective gain scores.

Since the groups are independent due to random selection and sample variances are quite similar, Student's t test can be applied, where

$$t = \frac{\bar{Y}_1 - \bar{Y}_2}{\sqrt{(s_1^2 + s_2^2)/n}}$$

with $2(n - 1)$ degrees of freedom and $n_1 = n_2 = n$.

Here, $t = 2.08$ which is significant at the 5-percent level for 48 degrees of freedom (see Appendix, Table B).

It is concluded that there is a real difference in the gain scores of the two groups, and those taught by the CAI method show a significantly greater improvement in basic arithmetic than those taught by the traditional method.

Example 1.2 The following example [14][1] is presented to show the three phases of the design of an experiment. It is not assumed that the reader is familiar with the design principles or analysis techniques in this problem. The remainder of the book is devoted to a discussion of many such principles and techniques, including those used in this problem.

An experiment was to be designed to study the effect of several factors on the power requirements for cutting metal with ceramic tools. The metal was cut on a lathe and the y or vertical component of a dynamometer reading was recorded. As this y component is proportional to the horsepower requirements in making the cut, it was taken as the measured variable. The y component is measured in millimeters of deflection on a recording instrument. Some of the factors which might affect this deflection are tool types, angle of tool edge bevel, type of cut, depth of cut, feed rate, and spindle speed. After much discussion, it was agreed to hold depth of cut constant at 0.100 in., feed rate constant at 0.012 in./min, and spindle speed constant at 1000 rpm. These levels were felt to represent typical operating conditions. The main objective of the study was to determine the effect of the other three factors (tool type, angle of edge bevel, and type of cut) on the power requirements. As only two ceramic tool types were available, this factor was considered at two levels. The angle of tool edge bevel was also set at two levels, 15° and 30°, representing the extremes for normal operation. The type of cut was either continuous or interrupted—again, two levels.

---

[1] Numbers in brackets refer to the references at the end of the book.

There are therefore two fixed levels for each of three factors, or eight experimental conditions (2^3) which may be set and which may affect the power requirements or y deflection on the dynamometer. This is called a 2^3 factorial experiment, since both levels of each of the three factors are to be combined with both levels of all other factors. The levels of two factors (type of tool and type of cut) are qualitative, whereas the angle of edge bevel (15° and 30°) is a quantitative factor.

The question of design for this experiment involves the number of tests to be made under each of the eight experimental conditions. After some preliminary discussion of expected variability under the same set of conditions and the costs of wrong decisions, it was decided to take four observations under each of the eight conditions, making a total of 32 runs. The order in which these 32 units are to be put in a lathe and cut was to be completely randomized.

In order to completely randomize the 32 readings, the experimenter decided on the order of experimentation from the results of three coin tossings. A penny was used to represent the tool type T: heads for one type, tails for the other; a nickel represented the angle of bevel B: heads 30°, tails 15°; and a dime represented the type of cut C: heads interrupted, tails continuous.

Thus if the first set of tosses came up *THT*, it would mean that the first tool to be used in the lathe would be of tool type 1, bevel 2, and subjected to the continuous type of cut. In Figure 1.1 a data layout is given, showing each of the 32 experimental conditions. The numbers 1, 2, 3, 4, and 5 indicate the first five conditions to be run on the lathe, assuming the coins came up *THT*, *TTH*, *HHT*, *HHT*, *HTH*.

In this layout it should be noted that the same set of conditions may be repeated (for example, runs 3 and 4) before all eight conditions are run once. The only restriction on complete randomization here is that once four repeated measures have occurred in the same cell, no more will be run using those same conditions.

	Tool Type T: 1		Tool Type T: 2	
Type Cut of C	Bevel B: 1	Bevel B: 2	Bevel B: 1	Bevel B: 2
1		1		3 4
2	2		5	

Figure 1.1 Data layout of power consumption for ceramic tools.

The coin flipping continues until the order of all 32 runs has been decided upon. This is a 2^3 factorial experiment with four observations per cell, run in a completely randomized manner. Complete randomization ensures the averaging out of any effects which might be correlated with the time of the experiment. If the lathe-spindle speed should vary and all of the type I tools were run through first and the type 2 tools followed, it might be that this extraneous effect of lathe-spindle speed would appear as a difference between tool types if the speed were faster at first and slower near the end of the experiment.

The mathematical model for this experiment and design would be $Y_{ijkm} = \mu + T_i + B_j + TB_{ij} + C_k + TC_{ik} + BC_{jk} + TBC_{ijk} + \varepsilon_{m(ijk)}$, where Y_{ijkm} represents the measured variable, μ a common effect in all observations (the true mean of the population from which all the data came), T_i the tool type effect where $i = 1, 2$, B_j the angle of bevel where $j = 1, 2$, and C_k the type of cut where $k = 1, 2$. $\varepsilon_{m(ijk)}$ represents the random error in the experiment where $m = 1, 2, 3, 4$. The other terms stand for interactions between the main factors T, B, and C.

The analysis of this experiment consists of collecting 32 items of data in the spaces indicated in Figure 1.1 in a completely randomized manner. The results in millimeter deflection are given in Table 1.1.

Table 1.1 Data for Power Requirement Example

	Tool Type T			
	1		2	
	Bevel Angle B		Bevel Angle B	
Type of Cut C	15°	30°	15°	30°
Continuous	29.0	28.5	28.0	29.5
	26.5	28.5	28.5	32.0
	30.5	30.0	28.0	29.0
	27.0	32.5	25.0	28.0
Interrupted	28.0	27.0	24.5	27.5
	25.0	29.0	25.0	28.0
	26.5	27.5	28.0	27.0
	26.5	27.5	26.0	26.0

This experiment and the mathematical model suggest a three-way analysis of variance (ANOVA) which yields the results in Table 1.2.

In testing the hypotheses that there is no type of tool effect, no bevel effect, no type of cut effect, and no interactions, the EMS column indicates that all observed mean squares are to be tested against the error mean square of 2.23 with 24 degrees of freedom (df). The proper test statistic is the F

Table 1.2 ANOVA for Power Requirement Example

Source of Variation	Degrees of Freedom (dF)	Sum of Squares (SS)	Mean Square (MS)	Expected Mean Square (EMS)
Tool types T	1	2.82	2.82	$\sigma_e^2 + 16\phi_T$
Bevel B	1	20.32	20.32*	$\sigma_e^2 + 16\phi_B$
$T \times B$ interaction	1	0.20	0.20	$\sigma_e^2 + 8\phi_{TB}$
Type of cut C	1	31.01	31.01*	$\sigma_e^2 + 16\phi_C$
$T \times C$ interaction	1	0.01	0.01	$\sigma_e^2 + 8\phi_{TC}$
$B \times C$ interaction	1	0.94	0.94	$\sigma_e^2 + 8\phi_{BC}$
$T \times B \times C$ interaction	1	0.19	0.19	$\sigma_e^2 + 4\phi_{TBC}$
Error $\varepsilon_{m(ijk)}$	24	53.44	2.23	σ_e^2
Totals	31	108.93		

* One asterisk indicates significance at the 5-percent level; two, the 1-percent level; and three, the 0.1-percent level. This notation will be used throughout this book.

statistic (Appendix, Table D) with 1 and 24 df. At the 5-percent significance level ($\alpha = 0.05$), the critical region of F is $F \geq 4.26$. Comparing each mean square with the error mean square indicates that only two hypotheses can be rejected: bevel has no effect on deflection and type of cut has no effect on deflection. None of the other hypotheses can be rejected, and it is concluded that only the angle of bevel and type of cut affect power consumption as measured by the y deflection on the dynamometer. Tool type appears to have little effect on the y deflection, and all interactions are negligible. Calculations on the original data of Table 1.2 show the average y deflections given in Table 1.3.

Table 1.3 Average y Deflections

Tool type T	1	28.1
	2	27.5
Bevel angle B	15°	27.0
	30°	28.6
Type of cut C	1	28.8
	2	26.8

These averages seem to bear out the conclusions that bevel affects y deflection, with a 30° bevel requiring more power than the 15° bevel (note the difference in average y deflection of 1.6 mm), and that type of cut affects y deflection, with a continuous cut averaging 2.0 mm more deflection than an interrupted cut. The difference of 0.6 mm in average deflection due to tool type is not significant at the 5-percent level. Graphing all four B–C combinations in Figure 1.2 shows the meaning of no significant interaction.

A brief examination of this graph indicates that the y deflection is increased by an increase in degree of bevel. The fact that the line for the continuous cut C_1 is above the line for the interrupted cut C_2 shows that the continuous cut requires more power. The fact that the lines are nearly parallel is characteristic of no interaction between two factors. Or, it can easily be seen that an increase in the degree of bevel produced about the same average increase in y deflection regardless of which type of cut was made. This is another way to interpret the presence of no interaction.

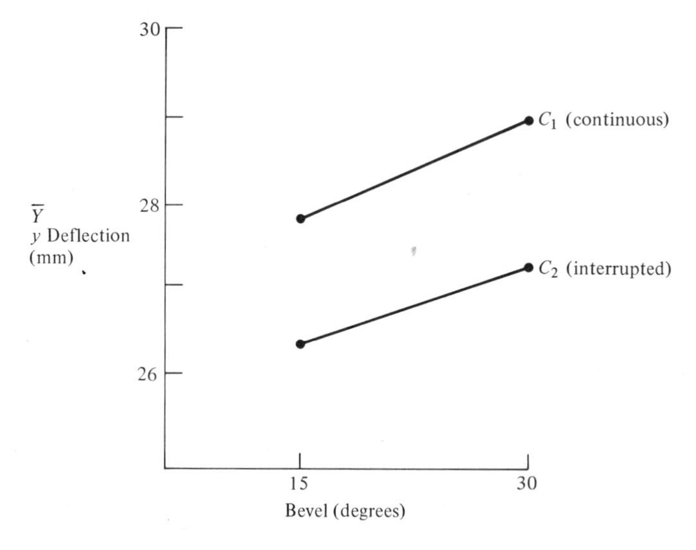

Figure 1.2 $B \times C$ interaction for power requirement example.

The experiment here was a three-factor experiment with two levels for each factor. The design was a completely randomized design and the analysis was a three-way ANOVA with four observations per cell. From the results of this experiment the experimenter not only found that two factors (angle of bevel and type of cut) affect the power requirements, but he has also determined that within the range of the experiment it makes little difference which ceramic tool type is used and that there are no significant interactions between the three factors.

These are but two examples which might have been used to illustrate the three phases of the project: experiment, design, and analysis.

2
Review of Statistical Inference

2.1 INTRODUCTION

The purpose of this chapter is to review basic concepts of statistical inference and also to introduce some of the definitions and notations which will be used throughout the book.

Present-day *statistics* may be defined as a tool for decision making in the light of uncertainty. Uncertainty does not imply ignorance, merely that an exact determination of the outcome of an experiment is not possible. A range of possible outcomes can often be determined on the basis of past experience or from observed sample data. Assuming the range or pattern of variability to be expected, a decision can be made as to whether or not observed data could reasonably have been taken from a population which has this assumed pattern. Since random variation is present in all measurements, the real variation in an experiment must be detected in the presence of this random variation or measurement error. Thus, statistics may also be defined as decision making in the light of random variation.

Statistical inference refers to the process of inferring something about a population from a sample drawn from that population. The population consists of all possible values of some random variable Y, where Y represents the response or dependent variable being measured. The response Y may represent tensile strength, weight, score, reaction time, or whatever criterion is being used to evaluate the experimental results. Characteristics of the population of this random variable are called *parameters*. In general θ will be used to designate any given population parameter. The average or expected value of the random variable is designated as $E(Y) = \mu$. If the probability function defining the random variable is known,

$$E(Y) = \sum Y_i p(Y_i)$$

where $p(Y_i)$ is a discrete probability function, and

$$E(Y) = \int Y f(Y)\, dY$$

where $f(Y)$ is a continuous probability density function.

The long-range average of squared deviations from the mean of a population is called the *population variance* σ_Y^2. Or

$$E[(Y - \mu_Y)^2] = \sigma_Y^2$$

The square root of this parameter is called *population standard deviation* σ_Y. Quantities computed from the sample values drawn from the population are called *sample statistics*, or simply, statistics. Examples include the sample mean

$$\overline{Y} = \sum_{i}^{n} Y_i/n$$

where n is the number of observations in the sample, and the sample variance

$$s^2 = \sum_{i=1}^{n} (Y_i - \overline{Y})^2/(n - 1)$$

Italic letters will be used to designate sample statistics. The symbol u will be used to designate a general statistic corresponding to the population parameter θ.

Most statistical theory is based on the assumption that samples drawn are *random samples*, that is, that each member of the population has an equal chance of being included in the sample and that the pattern of variation in the population is not changed by this deletion of the n members for the sample.

The notion of statistical inference may be divided into two parts: (1) estimation and (2) tests of hypotheses.

2.2 ESTIMATION

The objective of statistical estimation is to make an estimate of a population parameter based on a sample statistic drawn from this population. Two types of estimates are usually needed, point estimates and interval estimates.

A *point estimate* is a single statistic used to estimate a parameter. For example, the sample mean $\overline{Y}$ is a point estimate of the population mean μ. Point estimates are usually expected to have certain desirable characteristics. They should be unbiased, consistent, and have minimum variance.

An *unbiased statistic* is one whose expected or average value taken over an infinite number of similar samples equals the population parameter being estimated. Symbolically, $E(u) = \theta$, where $E(\cdot)$ means the expected value of the statistic in $(\cdot)$. The sample mean is an unbiased statistic, since it can be proved that

$$E(\overline{Y}) = \mu_Y \qquad \text{(or simply, } \mu)$$

Likewise, the sample variance as defined above is unbiased, since

$$E(s^2) = E\left[\sum_{i} (Y_i - \overline{Y})^2/(n - 1)\right] = \sigma_Y^2 \qquad (\text{or } \sigma^2)$$

Note that the sum of squares $\sum_{i=1}^{n} (Y_i - \bar{Y})^2$ must be divided by $n - 1$ and not by n if s^2 is to be unbiased. The sample standard deviation

$$s = \sqrt{\sum_{i=1}^{n} (Y_i - \bar{Y})^2/(n - 1)}$$

is not unbiased, since it can be shown that

$$E(s) \neq \sigma$$

This somewhat subtle point is proved in Burr [6, pp. 174–176].

A few basic theorems involving expected values that will be useful later in this book include

$$E(k) = k \qquad \text{(where } k \text{ is a constant)}$$
$$E(kY) = kE(Y)$$
$$E(Y_1 + Y_2) = E(Y_1) + E(Y_2)$$
$$E(Y - \mu_Y)^2 = E(Y^2) - \mu_Y^2 = \sigma_Y^2$$

The statistic $\sum_i (Y_i - \bar{Y})^2$ is used so frequently that it is often referred to as the *sum of squares*, or SS_Y. It is actually the sum of the squares of deviations of the sample values from their sample mean.

From the above it is easily seen that

$$E\left[\sum_i (Y_i - \bar{Y})^2\right] = (n - 1)\sigma_Y^2$$

or

$$E(SS_Y) = (n - 1)\sigma_Y^2$$

where $n - 1$ represents the degrees of freedom associated with the sum of squares on the left. In general, for any statistic u,

$$E\left[\sum_i (u_i - \bar{u})^2\right] = (\text{df on SS}_u)\sigma_u^2$$

or

$$E(\text{SS}_u) = (\text{df on SS}_u)\sigma_u^2$$

When a sum of squares is divided by its degrees of freedom, it indicates an averaging of the squared deviations from the mean, and this statistic, SS/df, is called a *mean square*, or MS. Hence

$$E(\text{MS}) = E(\text{SS/df}) = \sigma^2$$

of the variable or statistic being studied.

A *consistent statistic* is one whose value comes closer and closer to the parameter as the sample size is increased. Symbolically,

$$\lim_{n \to \infty} \Pr(|u_n - \theta| < \varepsilon) \to 1$$

which expresses the notion that as n increases, the probability approaches certainty (or 1) that the statistic u_n, which depends on n, will be within a distance ε, however small, of the true parameter θ.

Minimum variance applies where two or more statistics are being compared. If u_1 and u_2 are two estimates of the same parameter θ, the estimate having the smaller standard deviation is called the minimum-variance estimate. This is shown diagrammatically in Figure 2.1. In Figure 2.1, $\sigma_{u_1} < \sigma_{u_2}$, and u_1 is then said to be the *minimum-variance estimate* as variance is the square of the standard deviation.

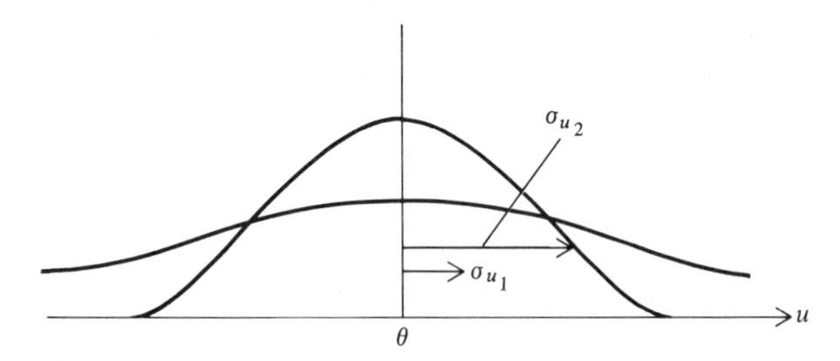

Figure 2.1 Minimum-variance estimates.

An *interval estimate* consists of the interval between two values of a sample statistic which is asserted to include the parameter in question. The band of values between these two limits is called a *confidence* interval for the parameter, since its width may be determined from the degree of confidence that is assumed when we say that the parameter lies in this band. Confidence limits (the end points of the confidence interval) are determined from the observed sample statistic, the sample size, and the degree of confidence desired. A 95-percent confidence interval on μ is given by

$$\bar{Y} \pm 1.96 \frac{\sigma}{\sqrt{n}}$$

where 1.96 is taken from normal distribution values (see Appendix, Table A), n is the sample size, and σ is the population standard deviation. If 100 sample means based on n observations each are computed, and 100 confidence intervals are set up using the above formula, we expect that about 95 of the 100 intervals will include μ. If only one interval is set up based on one sample of n, as is usually the case, we can state that we have 95-percent confidence that this interval includes μ. If σ is unknown, the Student's t distribution (Appendix, Table B) is used, and confidence intervals are given by

$$\bar{Y} \pm t_{1-\alpha/2} \frac{s}{\sqrt{n}}$$

where s is the sample standard deviation and t has $n - 1$ df. $100(1 - \alpha)$ percent gives the degree of confidence desired.

2.3 TESTS OF HYPOTHESES

A *statistical hypothesis* is an assumption about the population being sampled. It usually consists of assigning a value to one or more parameters of the population. For example, it may be hypothesized that the average number of miles per gallon obtained with a certain carburetor is 19.5. This is expressed as $H_0: \mu = 19.5$ mi/gal. The basis for the assignment of this value to μ usually rests on past experience with similar carburetors. Another example would be to hypothesize that the variance in weight of filled vials for the week is 40 grams^2 or $H_0: \sigma^2 = 40$ g^2. When such hypotheses are to be tested, the other parameters of the population are either assumed or estimated from data taken on a random sample from this population.

A *test of a hypothesis* is simply a rule by which a hypothesis is either accepted or rejected. Such a rule is usually based on sample statistics, called *test statistics*, when they are used to test hypotheses. For example, the rule might be to reject $H_0: \mu = 19.5$ mi/gal if a sample of 25 carburetors averaged 18.0 mi/gal ($\overline{Y}$) or less when tested. The *critical region* of a test statistic consists of all values of the test statistic where the decision is made to reject H_0. In the example above, the critical region for the test statistic $\overline{Y}$ is where $\overline{Y} \leq 18.0$ mi/gal.

Since hypothesis testing is based on observed sample statistics computed on n observations, the decision is always subject to possible errors. If the hypothesis is really true and it is rejected by the sample, a *type I error* is committed. The probability of a type I error is designated as α. If the hypothesis is accepted when it is not true, that is, some alternative hypothesis is true, a *type II error* has been made and its probability is designated as β. These α and β error probabilities are often referred to as the risks of making incorrect decisions, and one of the objectives in hypothesis testing is to design a test whose α and β risks are both small. In most such test procedures α is set at some predetermined level, and the decision rule is then formulated in such a way as to minimize the other risk, β. In quality control work, α is the producer's risk and β the consumer's risk.

In order to review hypothesis testing, a series of steps can be taken which will apply to most types of hypotheses and test statistics. To help clarify these steps and to illustrate the procedure, a simple example will be given parallel to the steps.

Steps in Hypothesis Testing	*Examples*
1. Set up the hypothesis and its alternative.	1. $H_0: \mu = 19.5$ mi/gal. $H_1: \mu < 19.5$ mi/gal.
2. Set the significance level of the test, α. (Size of the type I error.)	2. $\alpha = 0.05$.
3. Choose a test statistic to test H_0.	3. Test statistic $\overline{Y}$ or standardized $\overline{Y}: Z = \dfrac{\overline{Y} - \mu}{\sigma/\sqrt{n}}$ (assume $\sigma = 2$).

4. Determine the sampling distribution of this test statistic when H_0 is true.

4. $\overline{Y}$ is normally distributed with mean μ and standard deviation $\sigma/\sqrt{n}$. Or Z is $N(0, 1)$.

5. Set up a critical region on this test statistic where H_0 will be rejected in $(100)\alpha$ percent of the samples when H_0 is true.

5.

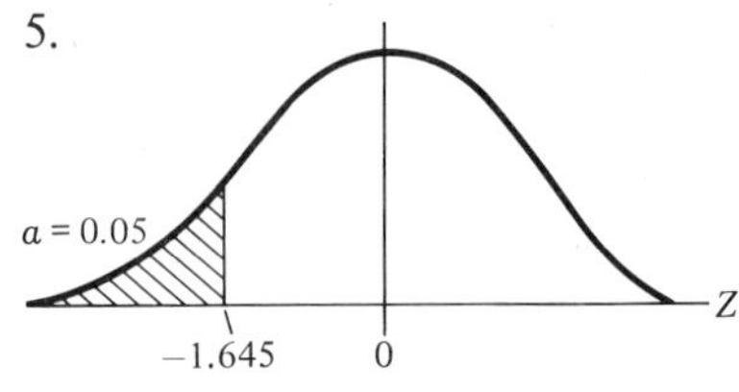

Critical region: $Z \leq -1.645$

6. Choose a random sample of n observations, compute the test statistic, and make a decision on H_0.

6. If $n = 25$ and $\overline{Y} = 18.9$ mi/gal,

$$Z = \frac{18.9 - 19.5}{2/\sqrt{25}} = -1.5.$$

As $-1.5 > -1.645$, do not reject H_0.

In the example above a one-sided or one-tailed test was used. This is dictated by the alternative hypothesis, since we only wish to reject H_0 when low values of $\overline{Y}$ are observed. The size of the significance level α is often set in an arbitrary fashion such as 0.05 or 0.01. It should reflect the seriousness of rejecting many carburetors when they are really satisfactory, or when the actual mean of the lot (population) is 19.5 mi/gal or better. In using the normal variate Z, σ is assumed known; a different test statistic would be used if σ were unknown, namely, Student's t. The critical region may also be expressed in terms of $\overline{Y}$ using the critical Z value of -1.645:

$$-1.645 = \frac{\overline{Y}_c - 19.5}{2/\sqrt{25}}$$

or $\overline{Y}_c = 18.8$, and the decision rule can be expressed as: Reject H_0 if $\overline{Y} \leq 18.8$. Here $\overline{Y} = 18.9$ and the hypothesis is not rejected.

The procedure outlined above may be used to test many different hypotheses. The nature of the problem will indicate what test statistic is to be used, and proper tables can be found to set up the required critical region. Well known are tests such as those on a single mean, two means with various assumptions about the corresponding variances, one variance, and two variances. A good discussion of various tests may be found in Dixon and Massey [8, pp. 88–138].

2.4 POWER OF A TEST

In the example given above no mention was made of the type II error of size β. In order to compute the probability β of a type II error, the hypothesis must be assumed untrue and some specific alternative assumed true. Then a value can be determined for the probability of accepting H_0 when the specified alternative H_1 is really true. Thus β is a function of the specific alternative, and several values of β will be obtained as μ is specified at various points below 19.5.

For example, when $\mu = 19.0$, the distribution of sample means around 19.0 would appear as in Figure 2.2. Here the critical value of $\bar{Y}$, $\bar{Y}_c$, is still 18.8 as determined above, since the hypothesis under test is $\mu = 19.5$ with an $\alpha = 0.05$. The probability of accepting H_0: $\mu = 19.5$ is indicated by the shaded region above 18.8, when actually $\mu = 19.0$. This probability is β and can be computed using normal distribution areas:

$$Z = \frac{18.8 - 19.0}{2/\sqrt{25}} = \frac{-0.2}{0.4} = -0.5$$

$$\beta = 0.6915 \text{ or } 0.69 \text{ (from Appendix Table A)}$$

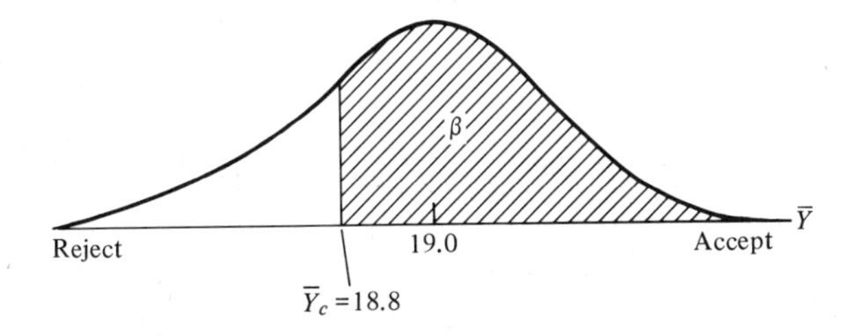

Figure 2.2

If μ is assumed to be at 18.5 (see Figure 2.3), the results are

$$Z = \frac{18.8 - 18.5}{2/\sqrt{25}} = 0.75$$

$$\beta = 0.2266 \text{ or } 0.23 \text{ (Appendix Table A)}$$

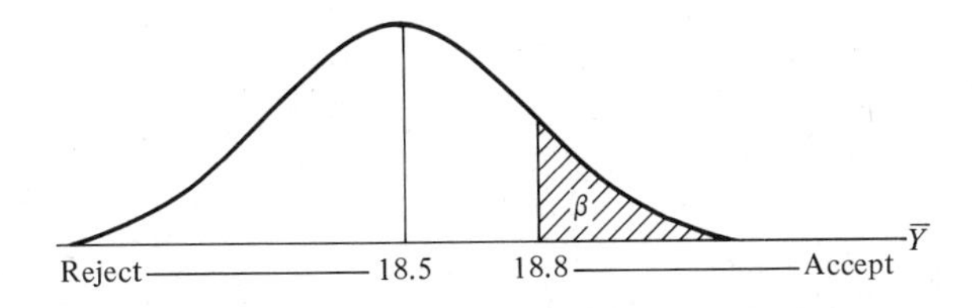

Figure 2.3

Thus it is observed that β varies with the true value of μ, becoming smaller when μ is farther to the left of 19.5 mi/gal. Since β does vary with μ, a curve can be plotted of μ versus β. This is called an *operating characteristic curve*, used in testing

$$H_0: \mu = 19.5$$

with $\alpha = 0.05$, $\sigma = 2$, $n = 25$, versus

$$H_1: \mu < 19.5$$

Sometimes the complement of β is plotted, namely, the probability of rejecting H_0 for various values of μ. This plot of $1 - \beta$ versus μ is called the *power curve* for this test (see Figure 2.4).

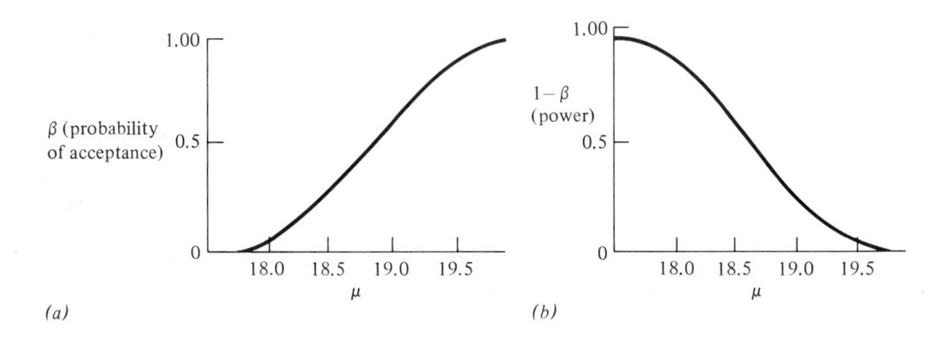

Figure 2.4 (*a*) Operating characteristic curve. (*b*) Power curve.

A table such as Table 2.1 makes these curves easy to plot using Appendix Table A.

Table 2.1 Data for Operating Characteristic Curve

μ	$Z = \dfrac{18.8 - \mu}{2/\sqrt{25}}$	β	$1 - \beta$
18.0	+2.00	0.02	0.98
18.3	+1.25	0.11	0.89
18.5	+0.75	0.23	0.77
19.0	−0.50	0.69	0.31
19.2	−1.00	0.84	0.16
19.5	−1.75	0.96	0.04*

*(approximately $= \alpha$)

It should be noted that when μ is quite a distance from the hypothesized value of 19.5, the test does a fairly good job of detecting this shift, that is, if $\mu = 18.5$, the probability of rejecting 19.5 is 0.77, which is fairly high. On the other hand, if μ has shifted only slightly from 19.5, say 19.2, the probability of detection is only 0.16. The power of a test may be increased by increasing the sample size or increasing the risk α.

2.5 HOW LARGE A SAMPLE?

The question of how large a sample to take from a population for making a test is one often asked of a statistician. This question can be answered provided the experimenter can answer each of the following questions.

1. How large a shift (from 19.5 to 19.0) in a parameter do you wish to detect?
2. How much variability is present in the population? (Based on past experience, $\sigma = 2$ mi/gal.)
3. What size risks are you willing to take? ($\alpha = 0.05$ and $\beta = 0.10$.)

If numerical values can be at least estimated in answering the above questions, the sample size may be determined.

Set up two sampling distributions $\overline{Y}$ one when H_0 is true, $\mu = 19.5$, and the other when the alternative to be detected is true, $\mu = 19.0$ (Figure 2.5).

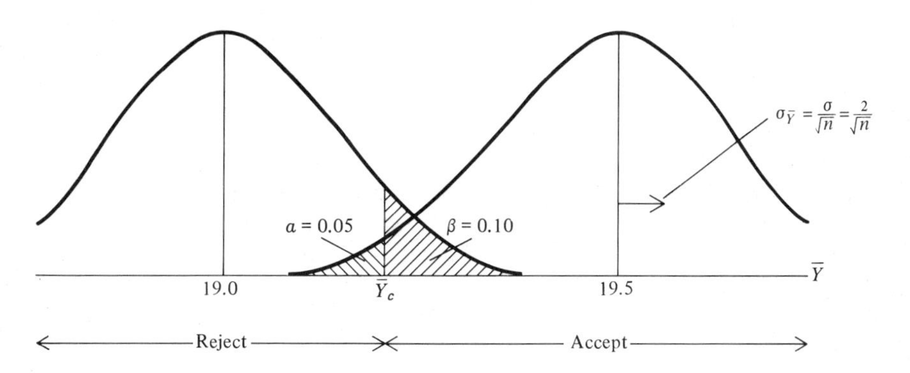

Figure 2.5 Determining sample size.

Indicate by $\overline{Y}_c$ a value between the two μ's which will become a critical point, rejecting H_0 for observed values of $\overline{Y}$ below it and accepting H_0 for $\overline{Y}$ values above it. Indicate the α and β risks on the diagram. Set up two simultaneous equations, standardizing $\overline{Y}_c$ first with respect to a μ of 19.5 (α equation) and second with respect to a μ of 19.0 (β equation). Solve these equations for n and $\overline{Y}_c$:

$$\alpha \text{ equation: } \quad \frac{\overline{Y}_c - 19.5}{2/\sqrt{n}} = -1.645 \quad (\text{based on } \alpha = 0.05)$$

$$\beta \text{ equation: } \quad \frac{\overline{Y}_c - 19.0}{2/\sqrt{n}} = +1.282 \quad (\text{based on } \beta = 0.10)$$

Subtracting the second equation from the first and multiplying both sides of each equation by $2/\sqrt{n}$ gives

$$-0.5 = -2.927\left(\frac{2}{\sqrt{n}}\right)$$

$$\sqrt{n} = \frac{5.854}{0.5} = 11.71$$

$$n = (11.71)^2 = 136.9 \text{ or } 137$$

Keeping $\alpha = 0.05$ gives

$$\overline{Y}_c = 19.5 - 1.645\left(\frac{2}{\sqrt{137}}\right) = 19.22 \text{ mi/gal}$$

The decision rule is then: Choose a random sample of 137 carburetors, and if the mean mi/gal of these is less than 19.22, reject H_0; otherwise, accept H_0.

Excellent tables are available for determining n for several tests of hypotheses, such as those in Davies [7, pp. 606–615] or Owen [16, pp. 19, 23, 36, 41, ff.].

Example 2.1 (*Variance*) As another example of a test of a hypothesis, consider testing whether or not the variances of two normal populations are equal. This example is included here as the test statistic involved will have many applications in later chapters.

In accordance with the steps outlined in Section 2.3, we have the following.

1. $H_0: \sigma_1^2 = \sigma_2^2 \qquad H_1: \sigma_1^2 > \sigma_2^2$.
2. $\alpha = 0.05$.
3. Test statistic $F = s_1^2/s_2^2$ (often called the variance ratio, where s_1^2 is based on $n_1 - 1$ df and s_2^2 is based on $n_2 - 1$ df).
4. If the two samples are independently chosen from normal populations and H_0 is true, the F statistic follows a skewed distribution, formed as the ratio of two independent chi-square distributions. A table giving a few percentiles of this F distribution appears as Appendix Table D. This table is entered with $n_1 - 1$ df for the numerator and $n_2 - 1$ df for the denominator.

5. The critical region is set at $F \geq F_{0.95}$ in this example ($\alpha = 0.05$) with the degrees of freedom dependent on the size of the two samples.
6. If the first sample results are

$$n_1 = 8 \qquad s_1^2 = 156$$

and the second are

$$n_2 = 10 \qquad s_2^2 = 100$$

then

$$F = \frac{156}{100} = 1.56$$

and the critical region is $F \geq 3.29$ for 7 and 9 df. Hence, the hypothesis is not rejected.

PROBLEMS

2.1 For the following data on tensile strength in psi (pounds per square inch), determine the mean and variance of this sample:

Tensile Strength (psi)	Frequency
18,461	2
18,466	12
18,471	15
18,476	10
18,481	8
18,486	3
Total	50

2.2 Using the results of Problem 2.1, test the hypothesis that the mean tensile strength of the population sampled is 18,470 psi (assume that $\sigma^2 = 40$, $\alpha = 0.05$).

2.3 Plot the operating characteristic curve for Problem 2.2.

2.4 Determine how large a sample would be needed to detect a 10-psi increase in the mean tensile strength of Problem 2.2 if $\alpha = 0.05$, $\beta = 0.02$, and $\sigma^2 = 40$.

2.5 Repeat Problem 2.4 for the detection of a 10-psi shift in the mean in either direction.

2.6 For a sample mean of 124 g (grams) based on 16 observations from a population whose variance is 25 g^2, set up 90-percent confidence limits on the mean of the population.

The following problems may require test statistics not reviewed in this chapter but assumed to be known to the reader. The tables at the end of this book should be adequate for their solution.

2.7 For a sample variance of 62 based on 12 observations, test the hypothesis that the population variance is 40. Use a one-sided test and a 1-percent significance level.

2.8 For Problem 2.7, set up 95-percent confidence limits (two-sided) on σ^2.

2.9 Two samples are taken, one from each of two machines. For this process

$$\begin{aligned} n_1 &= 8, & \bar{Y}_1 &= 42.2 \text{ g}, & s_1^2 &= 10 \text{ g}^2 \\ n_2 &= 15, & \bar{Y}_2 &= 44.5 \text{ g}, & s_2^2 &= 18 \text{ g}^2 \end{aligned}$$

Test these results for a significant difference in variances.

2.10 Test the results in Problem 2.9 for a significant difference in means.

2.11 Reflection light box readings before and after dichromating the interior of a metal cone were

Test Number	1	2	3	4	5	6	7	8
Before	6.5	6.0	7.0	6.8	6.5	6.8	6.2	6.5
After	4.4	4.2	5.0	5.0	4.8	4.6	5.2	4.9

Test for a significant difference in mean light-box readings.

2.12 To test the hypothesis that the defective fraction p of a process is 0.20, a sample of 100 pieces was drawn at random.

1. Use an $\alpha = 0.05$ and set up a critical region for the number of observed defectives to test this hypothesis against a two-sided alternative.
2. If the process now shifts to a 0.10 fraction defective, find the probability of an error of the second kind for this alternative.
3. Without repeating the work, would you expect the β error to be the same if the process shifted to 0.30? Explain.

2.13 Given the following sample data on a random variable, Y: 12, 8, 14, 20, 26, 26, 20, 21, 18, 24, 30, 21, 18, 16, 10, 20. Assuming $\sigma = 7$, and $\alpha = 0.05$, test each of the following hypotheses:

1. $H_0: \mu = 12$
   $H_1: \mu > 12$
2. $H_0: \mu = 16$
   $H_1: \mu \neq 16$
3. $H_0: \mu = 18$
   $H_1: \mu > 18$

2.14 For test 1 in Problem 2.13 evaluate the power of the test with respect to several alternative hypotheses from $\mu = 13$ to $\mu = 24$. Plot the power curve.

2.15 Repeat the results of Problem 2.14 with $\alpha = 0.01$. Plot on the same graph and compare the two power curves.

2.16 Given the following data on the current flow in amperes through a cereal forming machine: 8.2, 8.3, 8.2, 8.6, 8.0, 8.4, 8.5, 7.3. Test the hypothesis of the standard deviation $\sigma = 0.35$ ampere versus the alternative that $\sigma > 0.35$ ampere.

2.17 For the data in Problem 2.16 test the hypothesis that the true mean current flow is $\mu = 7.0$.

2.18 The percent moisture content in a puffed cereal where samples are taken from two different "guns" showed

Gun I: 3.6, 3.8, 3.6, 3.3, 3.7, 3.4
Gun II: 3.7, 3.9, 4.2, 4.2, 4.9, 3.6, 3.5, 4.0

Test the hypothesis of equal variances and equal means. Use any assumptions you believe appropriate.

2.19 Pretest data for experimental and control groups on course content in a special vocational–industrial course indicated:

Experimental: $\bar{Y}_1 = 9.333,\ s_1 = 4.945,\ n_1 = 12$
Control: $\bar{Y}_2 = 8.375,\ s_2 = 1.187,\ n_2 = 8$

Make any statistical tests which the data might suggest. Comment on the results.

2.20 Assuming that the pretest results in Problem 2.19 can be combined, set up a 95-percent confidence interval on the true pretest mean.

3
Single-Factor Experiments with No Restrictions on Randomization

3.1 INTRODUCTION

In this and several subsequent chapters single-factor experiments will be considered. In this chapter no restrictions will be placed on the randomization so that the design will be completely randomized. Many of the techniques of analysis for a completely randomized single-factor experiment can be applied with little alteration to more complex experiments.

For example, the single factor could be steel manufacturers, where the main interest of an analyst centers on the effect of several different manufacturers on the hardness of steel purchased from them. It could be temperature in an instance where the experimenter is concerned about the effect of temperature on penicillin yield. Whenever only one factor is varied, whether the levels be quantitative or qualitative, fixed or random, the experiment is referred to as a *single-factor experiment*, and the symbol τ_j will be used to indicate the effect of the jth level of the factor. τ_j suggests that the general factor may be thought of as a "treatment" effect.

If the order of experimentation applied to the several levels of the factor is completely random, so that any material to which the treatments might be applied is considered approximately homogeneous, the design is called a *completely randomized design*. The number of observations for each level of the treatment or factor will be determined from cost considerations and the power of the test. The model then becomes

$$Y_{ij} = \mu + \tau_j + \varepsilon_{ij}$$

where Y_{ij} represents the ith observation ($i = 1, 2, \cdots, n_j$) on the jth treatment ($j = 1, 2, \cdots, k$ levels). For example, Y_{23} represents the second observation using level 3 of the factor. μ is a common effect for the whole experiment, τ_j represents the effect of the jth treatment, and ε_{ij} represents the random error present in the ith observation on the jth treatment.

The error term ε_{ij} is usually considered a normally and independently distributed (NID) random effect whose mean value is zero and whose variance is the same for all treatments or levels. This is expressed as: ε_{ij}'s are NID $(0, \sigma_e^2)$ where σ_e^2 is the common variance within all treatments. μ is always a fixed parameter, and $\tau_1, \tau_2, \cdots, \tau_j, \cdots, \tau_k$ are considered to be fixed parameters if the levels of treatments are fixed. It is also assumed that

$$\sum_{j=1}^{k} \tau_j = 0$$

If the k levels of treatments are chosen at random, the τ_j's are assumed NID $(0, \sigma_\tau^2)$. Whether the levels are fixed or random depends upon how these levels are chosen in a given experiment.

The analysis of a single-factor completely randomized experiment usually consists of a one-way analysis of variance test where the hypothesis $H_0 : \tau_j = 0$ for all j, is tested. If this hypothesis is true, then no treatment effects exist and each observation Y_{ij} is made up of its population mean μ and a random error ε_{ij}. After an analysis of variance (ANOVA), many other tests may be made, and some of these will be shown in the example which follows.

Example 3.1 Interest is centered on the effect on tube conductivity of four different types of coating of TV tubes. As only four types of coating are used, the experiment is a single-factor experiment at four fixed levels, and these are also qualitative levels since no numerical value can be assigned to the four coatings. It is agreed that five observations per coating should be adequate and that the order for testing the 20 tubes can be completely randomized. A table of random numbers indicated the proper order for testing. A mathematical model might be

$$Y_{ij} = \mu + \tau_j + \varepsilon_{ij}$$

with

$$i = 1, 2, \cdots, 5 \qquad j = 1, 2, \cdots, 4$$

There are four treatments (coatings) and five observations per coating. The data are shown in Table 3.1.

Table 3.1 TV Tube Coating Data

Coating			
I	II	III	IV
56	64	45	42
55	61	46	39
62	50	45	45
59	55	39	43
60	56	43	41

An analysis of variance performed on these data gave the results in Table 3.2.

Table 3.2 TV Tube Coating ANOVA

Source	df	SS	MS
Between coatings τ_j	3	1135.0	378.3
Within coatings or error ε_{ij}	16	203.2	12.7
Totals	19	1338.2	

To test $H_0: \tau_j = 0$ for all $j = 1, 2, 3$, and 4, the test statistic is

$$F_{3,16} = \frac{378.3}{12.7} = 29.8$$

which is a highly significant value (see Appendix Table D). This indicates rejection of the hypothesis and a claim that there is a considerable difference in average conductivity between the four coatings. This result is no surprise when the data of Table 3.1 are examined.

3.2 ANALYSIS OF VARIANCE RATIONALE

To review the basis for the F test in a one-way analysis of variance, k populations, each representing one level of treatment, can be considered with observations as shown in Table 3.3.

Table 3.3 Population Layout for One-Way ANOVA

	Treatment					
	1	2	...	j	...	k
	Y_{11}	Y_{12}		Y_{1j}		Y_{1k}
	Y_{21}	Y_{22}		Y_{2j}		Y_{2k}
	Y_{31}	Y_{32}		—		—
	—	—		—		—
	Y_{i1}	Y_{i2}		Y_{ij}		Y_{ik}
Population means	$\mu_{.1}$	$\mu_{.2}$	...	$\mu_{.j}$	...	$\mu_{.k}$

Here the use of the "dot notation" indicates a summing over all observations in the population. Since each observation could be returned to the population and measured, there could be an infinite number of observations taken on each population, so the average or $E(Y_{i1}) = \mu_{.1}$, and so on. μ will represent the average Y_{ij} over all populations, or $E(Y_{ij}) = \mu$. In the model τ_j, the treatment effect can also be indicated by $\mu_{.j} - \mu$, and then the model is either

$$Y_{ij} = \mu + \tau_j + \varepsilon_{ij}$$

or

$$Y_{ij} \equiv \mu + (\mu_{.j} - \mu) + (Y_{ij} - \mu_{.j})$$

This last expression is seen to be an identity true for all values of X_{ij}. Expressed another way,

$$Y_{ij} - \mu \equiv (\mu_{.j} - \mu) + (Y_{ij} - \mu_{.j}) \tag{3.1}$$

Since these means are unknown, random samples are drawn from each population and estimates can be made of the treatment means and the grand mean. If n_j observations are taken for each treatment where the numbers need not be equal, a sample layout would be as shown in Table 3.4.

Table 3.4 Sample Layout for One-Way ANOVA

	Treatment						
	1	2	$\cdots$	j	$\cdots$	k	
	Y_{11}	Y_{12}	$\cdots$	Y_{1j}	$\cdots$	Y_{1k}	
	Y_{21}	Y_{22}	$\cdots$	Y_{2j}	$\cdots$	Y_{2k}	
	$\vdots$	$\vdots$		$\vdots$		$\vdots$	
	Y_{i1}	Y_{i2}	$\cdots$	Y_{ij}		Y_{ik}	
	$\vdots$	$\vdots$		$\vdots$		$\vdots$	
	$Y_{n_1 1}$	$\vdots$		$Y_{n_j j}$		$\vdots$	
		$Y_{n_2 2}$				$Y_{n_k k}$	
Totals	$T_{.1}$	$T_{.2}$	$\cdots$	$T_{.j}$	$\cdots$	$T_{.k}$	$T_{..}$
Number	n_1	n_2	$\cdots$	n_j	$\cdots$	n_k	N
Means	$\bar{Y}_{.1}$	$\bar{Y}_{.2}$	$\cdots$	$\bar{Y}_{.j}$	$\cdots$	$\bar{Y}_{.k}$	$\bar{Y}_{..}$

Here $T_{.j}$ represents the total of the observations taken under treatment j, n_j represents the number of observations taken for treatment j, and $\bar{Y}_{.j}$ is the observed mean for treatment j. $T_{..}$ represents the grand total of all observations taken where

$$T_{..} = \sum_{j=1}^{k} \sum_{i=1}^{n_j} Y_{ij} = \sum_{j=1}^{k} T_{.j}$$

and

$$N = \sum_{j=1}^{k} n_j$$

and $\bar{Y}_{..}$ is the mean of all N observations. Note too that

$$\bar{Y}_{..} = \sum_{j=1}^{k} n_j \bar{Y}_{.j}/N$$

If these sample statistics are substituted for their corresponding population parameters in Equation (3.1), we get a sample equation (also an identity) of the form

$$Y_{ij} - \bar{Y}_{..} \equiv (\bar{Y}_{.j} - \bar{Y}_{..}) + (Y_{ij} - \bar{Y}_{.j}) \tag{3.2}$$

This equation states that the deviation of any observation from the grand mean can be broken into two parts: the deviation of the observation from its own treatment mean plus the deviation of the treatment mean from the grand mean.

If both sides of Equation (3.2) are squared and then added over both i and j, we have

$$\begin{aligned}\sum_{j=1}^{k}\sum_{i=1}^{n_j}(Y_{ij} - \bar{Y}_{..})^2 &= \sum_{j=1}^{k}\sum_{i=1}^{n_j}(\bar{Y}_{.j} - \bar{Y}_{..})^2 + \sum_{j=1}^{k}\sum_{i=1}^{n_j}(Y_{ij} - \bar{Y}_{.j})^2 \\ &\quad + \sum_{j=1}^{k}\sum_{i=1}^{n_j}(\bar{Y}_{.j} - \bar{Y}_{..})(Y_{ij} - \bar{Y}_{.j})\end{aligned} \tag{3.3}$$

Examining the last expression on the right, we find that

$$\sum_{j=1}^{k}\sum_{i=1}^{n_j}(\bar{Y}_{.j} - \bar{Y}_{..})(Y_{ij} - \bar{Y}_{.j}) = \sum_{j=1}^{k}(\bar{Y}_{.j} - \bar{Y}_{..})\left[\sum_{i=1}^{n_j}(Y_{ij} - \bar{Y}_{.j})\right]$$

The term in brackets is seen to equal zero, as the sum of the deviations about the mean within a given treatment equals zero. Hence

$$\sum_{j=1}^{k}\sum_{i=1}^{n_j}(Y_{ij} - \bar{Y}_{..})^2 = \sum_{j=1}^{k}\sum_{i=1}^{n_j}(\bar{Y}_{.j} - \bar{Y}_{..})^2 + \sum_{j=1}^{k}\sum_{i=1}^{n_j}(Y_{ij} - \bar{Y}_{.j})^2 \tag{3.4}$$

This may be referred to as "the fundamental equation of analysis of variance," and it expresses the idea that the total sum of squares of deviations from the grand mean is equal to the sum of squares of deviations between treatment means and the grand mean plus the sum of squares of deviations within treatments. In Chapter 2 an unbiased estimate of population variance was determined by dividing the sum of squares $\sum_{i=1}^{n}(Y_i - \bar{Y})^2$ by the corresponding number of degrees of freedom, $n - 1$. If the hypothesis being tested in analysis of variance is true, namely, that $\tau_j = 0$ for all j, or that there is no treatment effect, then $\mu_{.1} = \mu_{.2} = \cdots = \mu_{.j} \cdots = \mu_{.k}$ and all there is in the model is the population mean μ and random error ε_{ij}. Then, any one of the three terms in Equation (3.4) may be used to give an unbiased estimate of this common population variance. For example, dividing the left-hand term by its degrees of freedom, $N - 1$ will yield an unbiased estimate of population variance σ^2. Within the jth treatment, $\sum_i (Y_{ij} - \bar{Y}_{.j})^2$ divided by $n_j - 1$ df would yield an unbiased estimate of the variance within the jth treatment. If the variances within the k treatments are really all alike, their estimates may be pooled to give $\sum_{j=1}^{k} \sum_{i=1}^{n_j} (Y_{ij} - \bar{Y}_{.j})^2$ with degrees of freedom

$$\sum_{j}^{k} (n_j - 1) = N - k$$

which will give another estimate of the population variance. Still another estimate can be made by first estimating the variance $\sigma_{\bar{Y}}^2$ between means drawn from a common population with $\sigma_Y^2 = n_j\sigma_{\bar{Y}_{.j}}^2$. An unbiased estimate for $\sigma_{\bar{Y}}^2$ is given by $\sum_{j=1}^{k} (\bar{Y}_{.j} - \bar{Y}_{..})^2/(k - 1)$ so that an unbiased estimate of

$$\sigma_Y^2 = \sum_{j=1}^{k} n_j(\bar{Y}_{.j} - \bar{Y}_{..})^2/(k - 1)$$

which is found by summing the first term on the right-hand side of Equation (3.4) over i and dividing by $k - 1$ df. Thus there are three unbiased estimates of σ^2 possible from the data in a one-way ANOVA if the hypothesis is true. Now all three are not independent since the sum of squares is additive in Equation (3.4). However, it can be shown that if each of the terms (sums of squares) on the right of Equation (3.4) is divided by its proper degrees of freedom, it will yield two independent chi-square distributed unbiased estimates of σ^2 when H_0 is true. If two such independent unbiased estimates of the same variance are compared, their ratio can be shown to be distributed as F with $k - 1$, $N - k$ df. If, then, H_0 is true, the test of the hypothesis can be made using a critical region of the F distribution with the observed F at $k - 1$ and $N -$ k df given by

$$F_{k-1,N-k} = \frac{\sum_{j=1}^{k} n_j(\bar{Y}_{.j} - \bar{Y}_{..})^2/(k - 1)}{\sum_{j=1}^{k} \sum_{i=1}^{n_j} (Y_{ij} - \bar{Y}_{.j})^2/(N - k)} \tag{3.5}$$

The critical region is usually taken as the upper tail of the F distribution, rejecting H_0 if $F \geq F_{1-\alpha}$ where α is the area above $F_{1-\alpha}$. In this F ratio, the sum of squares between treatments is always put into the numerator, and then a significant F will indicate that the differences between means has something in it besides the estimate of variance. It probably indicates that there is a real difference in treatment means $(\mu_{.1}, \mu_{.2}, \cdots)$ and that H_0 should be rejected. These unbiased estimates of population variance, sums of squares divided by df, are also referred to as *mean squares*.

The actual computing of sums of squares indicated in Equation (3.5) is much easier if they are first expanded and rewritten in terms of treatment totals. These computing formulas are given in Table 3.5. Applying the formulas in Table 3.5 to the problem data in Table 3.1, the data may first be coded by subtracting 50 from all readings (coding will leave the F statistic unchanged in value). Table 3.6 shows the coded data and useful statistics computed from these data.

Table 3.5 One-Way ANOVA

Source	df	SS	MS
Between treatments τ_j	$k-1$	$\sum_{j=1}^{k} n_j(\bar{Y}_{.j} - \bar{Y}_{..})^2 = \sum_{j=1}^{k} \frac{T_{.j}^2}{n_j} - \frac{T_{..}^2}{N}$	$\text{SS}_{\text{treatment}}/(k-1)$
Within treatments or error ε_{ij}	$N-k$	$\sum_{j=1}^{k}\sum_{i=1}^{n_j} (Y_{ij} - \bar{Y}_{.j})^2 = \sum_{j=1}^{k}\sum_{i=1}^{n_j} Y_{ij}^2 - \sum_{j=1}^{k} \frac{T_{.j}^2}{n_j}$	$\text{SS}_{\text{error}}/(N-k)$
Totals	$N-1$	$\sum_{j=1}^{k}\sum_{i=1}^{n_j} (Y_{ij} - \bar{Y}_{..})^2 = \sum_{j=1}^{k}\sum_{i=1}^{n_j} Y_{ij}^2 - \frac{T_{..}^2}{N}$	

Table 3.6 Coded TV Tube Coating Data

	Treatment 1	2	3	4	
	6	14	−5	−8	
	5	11	−4	−11	
	12	0	−5	−5	
	9	5	−11	−7	
	10	6	−7	−9	
$T_{.j}$	42	36	−32	−40	$T_{..} = 6$
n_j	5	5	5	5	$N = 20$
$\sum_{i=1}^{n_j} Y_{ij}^2$	386	378	236	340	$\sum_{j=1}^{k}\sum_{i=1}^{n_j} Y_{ij}^2 = 1340$

The sums of squares can be computed quite easily from this table. The total sum of squares states, "square each observation, add over all observations, and subtract the correction term." This latter is the grand total squared and divided by the total number of observations

$$\text{SS}_{\text{total}} = \sum_{j=1}^{k}\sum_{i=1}^{n_j} Y_{ij}^2 - \frac{T_{..}^2}{N} = 1340 - \frac{(6)^2}{20} = 1338.2$$

The sum of squares between treatments is found by totaling n_j observations for each treatment, squaring this total, dividing by the number of observations, adding for all treatments, and then subtracting the correction term:

$$SS_{\text{treatment}} = \sum_{j=1}^{k} \frac{T_{.j}^2}{n_j} - \frac{T_{..}^2}{N}$$

$$= \frac{(42)^2}{5} + \frac{(36)^2}{5} + \frac{(-32)^2}{5} + \frac{(-40)^2}{5} - \frac{(6)^2}{20} = 1135.0$$

The sum of squares for error is then determined by subtraction

$$SS_{\text{error}} = SS_{\text{total}} - SS_{\text{treatment}} = 1338.2 - 1135.0 = 203.2$$

These results are then displayed as in Table 3.2, and the F test is run on the $H_0\colon \tau_j = 0$ as shown before.

3.3 AFTER ANOVA—TESTS ON MEANS

Having concluded, as in the problem above, that there is a significant difference in treatment means, questions naturally arise, "Which means differ?" "Does the mean of the first coating differ from the mean of the second?" "Does the average of 1 and 2 differ from the average of 3 and 4?" The answers to questions about means, after ANOVA, may be handled in two ways, depending upon when a selection is made of those contrasts among means that are to be of interest—before the experiment is performed or after the data are collected.

Tests on Means Set prior to Experimentation—Orthogonal Contrasts

If the above decision is made prior to the running of an experiment, such comparisons can usually be set without disturbing the risk α of the original ANOVA. This means that the contrasts must be chosen with care, and the number of such contrasts should not exceed the number of degrees of freedom between the treatment means. The method usually used here is called the *method of orthogonal contrasts.*

In comparing treatments 1 and 2, it should seem quite logical to examine the difference in their means or their totals since all treatment totals are based on the same sample size n. One such comparison might be $T_{.1} - T_{.2}$. If, on the other hand, one wished to compare the first and second treatment with the third, the third treatment total should be weighted by a factor of 2 since it is based on only five observations and the treatment totals of 1 and 2 together represent ten observations. Such a comparison would then be $T_{.1} + T_{.2} - 2T_{.3}$. Note that for each of the two comparisons or contrasts, the sum of the coefficients of the treatment totals always adds to zero:

$$T_{.1} - T_{.2} \text{ has coefficients } +1 - 1 = 0$$
$$T_{.1} + T_{.2} - 2T_{.3} \text{ has coefficients } +1 + 1 - 2 = 0$$

Hence, a *contrast* C_m is defined for any linear combination of treatment totals as follows:

$$C_m = c_{1m}T_{.1} + c_{2m}T_{.2} + \cdots + c_{jm}T_{.j} + \cdots + c_{km}T_{.k}$$

where

$$c_{1m} + c_{2m} + \cdots + c_{jm} + \cdots + c_{km} = 0$$

or, more compactly,

$$C_m = \sum_{j=1}^{k} c_{jm}T_{.j}$$

where

$$\sum_{j=1}^{k} c_{jm} = 0$$

In dealing with several contrasts, it is highly desirable to have the contrasts independent of each other. When the contrasts are independent—one having no projection on any of the others—they are said to be orthogonal contrasts, and independent tests of hypotheses can be made by comparing the mean square of each such contrast with the mean square of the error term in the experiment. Each contrast carries one degree of freedom.

It may be recalled that when two straight lines, say

$$c_{11}X + c_{21}Y = b_1$$
$$c_{12}X + c_{22}Y = b_2$$

are perpendicular, or orthogonal, to each other, the slope of one line is the negative reciprocal of the slope of the other line. Changing the two lines above into slope-intercept form,

$$Y = -c_{11}/c_{21}X + b_1/c_{21}$$
$$Y = -c_{12}/c_{22}X + b_2/c_{22}$$

To be orthogonal,

$$-c_{11}/c_{21} = -(-c_{22}/c_{12})$$

or,

$$c_{11}/c_{12} = -c_{21}c_{22}$$

or

$$c_{11}c_{12} + c_{21}c_{22} = 0$$

The sum of the products of the corresponding coefficients (on X and on Y) must therefore add to zero. This idea may be extended to a general definition.

Two contrasts C_m and C_q are said to be *orthogonal contrasts*, provided

$$\sum_{j=1}^{k} c_{jm}c_{jq} = 0$$

for equal n's.

The sums of squares for a contrast are given by

$$SS_{C_m} = \frac{C_m^2}{n \sum_{j=1}^{k} c_{jm}^2}$$

To apply this procedure to the problem above, three orthogonal contrasts may be set up since there are 3 df between treatments. One such set of three might be

$$\begin{aligned} C_1 &= T_{.1} \qquad\qquad\qquad\quad - T_{.4} \\ C_2 &= \qquad\quad T_{.2} - T_{.3} \\ C_3 &= T_{.1} - T_{.2} - T_{.3} + T_{.4} \end{aligned} \tag{3.6}$$

C_1 is a contrast to compare the mean of the first treatment with the fourth, C_2 compares the second treatment with the third, and C_3 compares the average of treatments one and four with the average of two and three. The coefficients of the $T_{.j}$'s for the three contrasts are given in Table 3.7.

Table 3.7 Orthogonal Coefficients

	$T_{.1}$	$T_{.2}$	$T_{.3}$	$T_{.4}$
C_1	+1	0	0	−1
C_2	0	+1	−1	0
C_3	+1	−1	−1	+1

It can be seen from Table 3.7 that the sum of coefficients c_{jm} adds up to zero for each contrast, and that the sum of products of coefficients of each pair of contrasts is also zero.

For the example given in Section 3.2, using the coded data in Table 3.6, the contrasts are

$$\begin{aligned} C_1 &= +1(42) + 0(36) + 0(-32) - 1(-40) = 82 \\ C_2 &= 0(42) + 1(36) - 1(-32) + 0(-40) = 68 \\ C_3 &= +1(42) - 1(36) - 1(-32) + 1(-40) = -2 \end{aligned}$$

The corresponding sums of squares are

$$SS_{C_1} = \frac{(82)^2}{5(2)} = \frac{6724}{10} = 672.4$$

$$SS_{C_2} = \frac{(68)^2}{5(2)} = \frac{4624}{10} = 462.4$$

$$SS_{C_3} = \frac{(-2)^2}{5(4)} = \frac{4}{20} = \frac{0.2}{1135.0}$$

Each sum of squares for a contrast has 1 df, and the total sum of squares for the treatments is 1135.0 as in Table 3.2. Each of these sums of squares may be tested against the error mean square with 1 and 16 df as follows:

$$H_1: \tau_1 = \tau_4 \qquad F_{1,16} = \frac{672.4/1}{12.7} = 52.9$$

$$H_2: \tau_2 = \tau_3 \qquad F_{1,16} = \frac{462.4/1}{12.7} = 36.4$$

$$H_3: \tau_1 + \tau_4 = \tau_2 + \tau_3 \qquad F_{1,16} = \frac{0.2/1}{12.7} = 0.016.$$

Comparing these with $F_{1,16}$ at the 5-percent significance level, which is 4.49 (Appendix Table D), the first two hypotheses are rejected and the third one is not. We can therefore conclude that there is a significant difference in the mean conductivity between the first and fourth coatings and between the second and third coatings. However, there is no significant difference between the average of coatings one and four and two and three. These results can be assessed only if the decision is made beforehand as to the contrasts to be performed.

As this method of orthogonal contrasts is used quite often in experimental design work, the definitions and formulas are given below for the case of unequal numbers of observations per treatment.

C_m is a *contrast* if

$$\sum_{j=1}^{k} n_j c_{jm} = 0$$

and C_m and C_q are *orthogonal contrasts* if

$$\sum_{j=1}^{k} n_j c_{jm} c_{jq} = 0$$

The *sum of squares for such contrasts* is given by

$$SS_{C_m} = \frac{C_m^2}{\sum_{j=1}^{k} n_j c_{jm}^2}$$

Tests on Means after Experimentation

If the decision on what comparisons to make is withheld until after the data are examined, comparisons may still be made, but the α level is altered because such decisions are not taken at random but are based on observed results. Several methods have been introduced to handle such situations [13], but only the Newman–Keuls range test [12] and Scheffé's test [17] will be described here.

Range Test After the data have been compiled the following steps are taken.

1. Arrange the k means in order from low to high.
2. Enter the ANOVA table and take the error mean square with its degrees of freedom.
3. Obtain the standard error of the mean for each treatment

$$s_{\bar{Y}_{.j}} = \sqrt{\frac{\text{error mean square}}{\text{number of observations in } \bar{Y}_{.j}}}$$

   where the error mean square is the one used as the denominator in the F test on means $\bar{Y}_{.j}$.
4. Enter a Studentized range table (Appendix, Table E) of significant ranges at the α level desired, using n_2 = degrees of freedom for error mean square and $p = 2, 3, \cdots, k$, and list these $k - 1$ ranges.
5. Multiply these ranges by $s_{\bar{Y}_{.j}}$ to form a group of $k - 1$ least significant ranges.
6. Test the observed ranges between means, beginning with largest versus smallest, which is compared with the least significant range for $p = k$; then test largest versus second smallest with the least significant range for $p = k - 1$; and so on. Continue this for second largest versus smallest, and so forth, until all $k(k - 1)/2$ possible pairs have been tested. The sole exception to this rule is that no difference between two means can be declared significant if the two means concerned are both contained in a subset with a nonsignificant range.

To see how this works, consider the coded data of the problem in Tables 3.1, 3.2, and 3.6. Here the means for treatments 1, 2, 3, 4 are

$$\bar{Y}_{.j}: 8.4, 7.2, -6.4, -8.0$$

Following the steps given above,

1. $k = 4$ means are $-8.0 \quad -6.4 \quad 7.2 \quad 8.4$
   for treatments $\quad 4 \quad 3 \quad 2 \quad 1$.

2. From Table 3.2, error mean square = 12.7 with 16 df.

3. Standard error of a mean is

$$s_{\bar{Y}_{.j}} = \sqrt{\frac{12.7}{5}} = 1.59$$

4. From Appendix Table E, at the 5-percent level, the significant ranges are, for $n_2 = 16$,

$$\begin{array}{rccc} p = & 2 & 3 & 4 \\ \text{ranges} = & 3.00 & 3.65 & 4.05 \end{array}$$

5. Multiplying by the standard error of 1.59, the least significant ranges (LSR) are

$$\begin{array}{rccc} p = & 2 & 3 & 4 \\ \text{LSR} = & 4.77 & 5.80 & 6.44 \end{array}$$

6.

Largest versus smallest:	1 versus 4 =	16.4 > 6.44
Largest versus next smallest:	1 versus 3 =	14.8 > 5.80
Largest versus next largest:	1 versus 2 =	1.2 < 4.77
Second largest versus smallest:	2 versus 4 =	15.2 > 5.80
Second largest versus next largest:	2 versus 3 =	13.6 > 4.77
Third largest versus smallest:	3 versus 4 =	1.6 < 4.77.

Here there is a significant difference between treatments 1 and 4, 1 and 3, 2 and 4, and 2 and 3, but not between 1 and 2 or 3 and 4. This may be shown by underlining means which are not significant and could therefore have come from a common population. These appear on a one-dimensional scale as shown in Figure 3.1. Here any two means *not* underscored by the same line are significantly different, and any two means underscored by the same line are *not* significantly different.

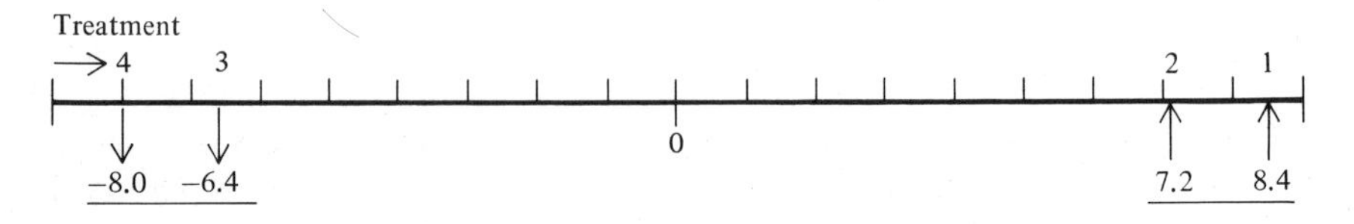

Figure 3.1

Scheffé's Test Since the Newman–Keuls test is restricted to comparing pairs of means and it is often desirable to examine other contrasts which represent combinations of treatments, many experimenters prefer a test devised by Scheffé [17]. Scheffé's method uses the concept of contrasts presented earlier but the contrasts need not be orthogonal. In fact, any and all conceivable contrasts may be tested for significance. Since comparing means in pairs is a special case of contrasts, Scheffé's scheme is more general than the Newman–Keuls. However, since the Scheffé method must be valid for such a large set of possible contrasts, it requires larger observed differences to be significant than some of the other schemes. To see how Scheffé's method applies, the following steps are taken after the data have been compiled.

1. Set up all contrasts of interest to the experimenter and compute their numerical values.
2. Determine the significant F statistic for the ANOVA just performed based on α and degrees of freedom $k - 1, N - k$.
3. Compute $A = \sqrt{(k - 1)F}$, using the F from step 2.
4. Compute the standard error of each contrast to be tested. This standard error is given by

$$s_{C_m} = \sqrt{(\text{error mean square}) \textstyle\sum_j n_j c_{jm}^2}$$

5. If a contrast C_m is numerically larger than A times s_{C_m}, it is declared significant. Or, if $|C_m| > As_{C_m}$, reject the hypothesis that the true contrast among means is zero.

Applying this technique to the coded data of Tables 3.1, 3.2, and 3.6, proceed as follows.

1. Consider two contrasts

$$\begin{aligned} C_1 &= T_{.1} - T_{.2} &&= 42 - 36 &&= 6 \\ C_2 &= 3T_{.1} - T_{.2} - T_{.3} - T_{.4} &&= 126 - (36) - (-32) - (-40) &&= 162 \end{aligned}$$

   (Note that these are not orthogonal. The first compares treatment 1 with treatment 2 and the second compares treatment 1 with all the other 3 treatments.)
2. Since $\alpha = 0.05$, $k - 1 = 3$, $N - k = 16$, $F = 3.24$.
3. $A = \sqrt{(3)(3.24)} = 3.12$.
4. $s_{C_1} = \sqrt{12.7[5(1)^2 + 5(-1)^2]} = 11.3$

   $s_{C_2} = \sqrt{12.7[5(3)^2 + 5(-1)^2 + 5(-1)^2 + 5(-1)^2]} = 27.6.$
5. Since $C_1 = 6$ is <35.3, do not reject—this contrast is not significant. Since $C_2 = 162$ is >86.1, reject—this contrast is significant and a real difference between the conductivity of coating I and the conductivity of the other three coatings is indicated.

3.4 CONFIDENCE LIMITS ON MEANS

After an analysis of variance, it is often desirable to set confidence limits on a treatment mean. The $100(1 - \alpha)$ percent confidence limits on $\mu_{.j}$ are given by

$$\bar{Y}_{.j} \pm t_{1-\alpha/2} \sqrt{\frac{\text{mean square used to test treatment mean square}}{n_j}} \tag{3.6}$$

where the mean square used to test treatment mean square is the error mean square in a one-way ANOVA but may be a different mean square in more complex analyses. The degrees of freedom used with the Student's t statistic are the degrees of freedom that correspond to the mean square used for testing the treatment mean.

In Example 3.1, 95-percent confidence limits on the mean conductivity for the first coating $\mu_{.1}$ would be $\overline{Y}_{.1} \pm t_{0.975}\sqrt{12.7/5}$ with 16 df on t. On coded data, $8.4 \pm 2.12\sqrt{12.7/5}$ (t from Appendix Table B), or 8.4 ± 3.37, or from 5.03 to 11.77. Decoding by adding 50 gives 55.03 to 61.77 for 95-percent confidence limits on $\mu_{.1}$.

3.5 COMPONENTS OF VARIANCE

In Example 3.1 the levels of the factor were considered as fixed, since only four coatings were available and a decision was desired on the effect of these four coatings only. If, however, the levels of the factor were random (such as operators, days, or samples where the levels in the experiment might have been chosen at random from a large number of possible levels), the model is then called a *random model*, and inferences are to be extended to all levels of the population (of which the observed four levels are random samples). In a random model, the experimenter is not usually interested in testing hypotheses, setting confidence limits, or making contrasts in means; rather, he is interested in estimating components of variance. How much of the variance in the experiment might be considered as due to true differences in treatment means, and how much might be due to random error about these means?

Example 3.2 In order to see how to analyze and interpret a component of variance or random model, consider the following problem.

A company supplies a customer with several hundred batches of a raw material every year. The customer is interested in a high yield from the raw material in terms of percent usable chemical. He usually makes three sample determinations of yield from each batch in order to control the quality of the incoming material. He expects and gets some variation between determinations on a given batch, but he suspects that there may be significant batch-to-batch variation as well.

In order to check this, he selects five batches at random from several batches available and runs three yield determinations per batch. His 15 yield determinations are completely randomized. The mathematical model is again

$$Y_{ij} = \mu + \tau_j + \varepsilon_{i(j)}$$

except that in this experiment the k levels of the treatment (batches) are chosen at random, rather than being fixed levels. The data are shown in Table 3.8.

Table 3.8 Chemical Yield by Batch Data

Batch				
1	2	3	4	5
74	68	75	72	79
76	71	77	74	81
75	72	77	73	79

An ANOVA performed on these data gave the results in Table 3.9.

Table 3.9 Chemical Yield ANOVA

Source	df	SS	MS	EMS
Between batches τ_j	4	147.74	36.94	$\sigma_\varepsilon^2 + 3\sigma_\tau^2$
Error $\varepsilon_{i(j)}$	10	17.99	1.80	σ_ε^2
Totals	14	165.73		

The F test gives $F_{4,10} = 36.94/1.80 = 20.5$ which is highly significant. Since these batches are but a random sample of batches, we may be interested in how much of the variance in the experiment might be attributed to batch differences and how much to random error. To help answer these questions, another column has been added to the ANOVA table (Table 3.9). This is the expected mean square (EMS) column. It can be derived by inserting the mathematical model into the operational formulas for mean squares in Table 3.5 and then computing the expected values of these mean squares. The results will depend upon whether a fixed or random model is used as will be discussed in greater length in the next part of this section.

The interpretation of Table 3.9 then is that the error variance σ_ε^2 is best estimated as 1.80, its corresponding mean square. Also the mean square 36.94 is an estimate of $\sigma_\varepsilon^2 + 3\sigma_\tau^2$. If the numerical mean squares are set equal to the variance components that they are estimating

$$s_\varepsilon^2 = 1.80 \qquad s_\tau^2 + 3s_\varepsilon^2 = 36.94$$

where s^2 is used since these are estimates of the corresponding σ^2's.

Solving these expressions,

$$s_\tau^2 = \frac{36.94 - 1.80}{3} = 11.71$$

The total variance can be estimated as

$$s_{\text{total}}^2 = s_\tau^2 + s_\varepsilon^2 = 11.71 + 1.80 = 13.51$$

and 11.71/13.51 or 86.7 percent of the total variance is attributable to batch differences and only 1.80/13.51 or 13.3 percent is attributable to errors within batches.

It is interesting to note that the above estimate of total variance of 13.51 would give a standard deviation of 3.68. One might therefore expect all the data in such a small experiment ($N = 15$) to fall within four standard deviations or within $4(3.68) = 14.72$. The actual range is 13 (high, 81–low, 68). This breakdown of total variance into its two components then seems quite reasonable.

Expected Mean Square (EMS) Derivation

For a single factor experiment

$$Y_{ij} = \mu + \tau_j + \varepsilon_{ij} \tag{3.7}$$

From Table 3.5, the sum of squares for treatments with equal n's is

$$\text{SS}_{\text{treatment}} = \sum_{j=1}^{k} n(\bar{Y}_{.j} - \bar{Y}_{..})^2$$

From model Equation (3.7),

$$\begin{aligned}
\bar{Y}_{.j} &= \sum_{i=1}^{n} Y_{ij}/n = \sum_{i=1}^{n} (\mu + \tau_j + \varepsilon_{ij})/n \\
&= \frac{n\mu}{n} + \frac{n\tau_j}{n} + \sum_{i=1}^{n} \varepsilon_{ij}/n \\
&= \mu + \tau_j + \sum_{i=1}^{n} \varepsilon_{ij}/n
\end{aligned} \tag{3.8}$$

Also

$$\begin{aligned}
\bar{Y}_{..} &= \sum_{j=1}^{k} \sum_{i=1}^{n} Y_{ij}/nk = \sum_{j}^{k} \sum_{i}^{n} (\mu + \tau_j + \varepsilon_{ij})/nk \\
&= \frac{nk\mu}{nk} + n \sum_{j}^{k} \tau_j/nk + \sum_{j}^{k} \sum_{i}^{k} \varepsilon_{ij}/nk \\
&= \mu + \sum_{j}^{k} \tau_j/k + \sum_{j}^{k} \sum_{i}^{n} \varepsilon_{ij}/nk
\end{aligned} \tag{3.9}$$

Subtracting Equation (3.9) from Equation (3.8) gives

$$\bar{Y}_{.j} - \bar{Y}_{..} = \tau_j - \sum_{j=1}^{k} \tau_j/k + \sum_{i=1}^{n} \varepsilon_{ij}/n - \sum_{i}^{n} \sum_{j}^{k} \varepsilon_{ij}/nk$$

Squaring gives

$$(\bar{Y}_{.j} - \bar{Y}_{..})^2 = \left(\tau_j - \sum_{j=1}^{k} \tau_j/k\right)^2 + \frac{1}{n^2}\left(\sum_{i}^{n} \varepsilon_{ij} - \sum_{i}^{n} \sum_{j}^{k} \varepsilon_{ij}/k\right)^2$$

(+ cross products)

Multiplying by n and summing over j gives

$$\text{SS}_{\text{treatment}} = \sum_{j=1}^{k} n(\bar{Y}_{.j} - \bar{Y}_{..})^2 = n \sum_{j}^{k} \left(\tau_j - \sum_{j}^{k} \tau_j/k\right)^2$$

$$+ \frac{n}{n^2} \sum_{j} \left(\sum_{i}^{n} \varepsilon_{ij} - \sum_{i}^{n} \sum_{j}^{k} \varepsilon_{ij}/k\right)^2 + n \sum_{j} (\text{cross product})$$

The expected value operator may now be applied to this $\text{SS}_{\text{treatment}}$

$$E(\text{SS}_{\text{treatment}}) = nE\left[\sum_{j}^{k} \left(\tau_j - \sum_{j}^{k} \tau_j/k\right)^2\right]$$

$$+ \frac{1}{n} E\left[\sum_{j}^{k} \left(\sum_{i}^{n} \varepsilon_{ij} - \sum_{i}^{n} \sum_{j}^{k} \varepsilon_{ij}/k\right)^2\right]$$

as it can be shown that the expected value of the cross-product term equals zero.

If the treatment levels are fixed,

$$\sum_{j=1}^{k} \tau_j = \sum_{j=1}^{k} (\mu_{.j} - \mu) = 0$$

and the $E(\text{SS}_{\text{treatment}})$ becomes

$$E(\text{SS}_{\text{treatment}}) = n \sum_{j=1}^{k} \tau_j^2 + \frac{1}{n} (nk - n)\sigma_\varepsilon^2$$

since errors are random and $\sum_{j=1}^{k} \tau_j^2$ is a constant. The $E(\text{MS}_{\text{treatment}}) = E[\text{SS}_{\text{treatment}}/(k - 1)]$, so

$$E(\text{MS}_{\text{treatment}}) = \left[n \sum_{j=1}^{k} \tau_j^2/(k - 1)\right] + \frac{n(k - 1)}{n(k - 1)} \sigma_\varepsilon^2$$

If, on the other hand, treatment levels are random and their variance is σ_τ^2

$$E(\text{MS}_{\text{treatment}}) = \frac{n(k - 1)\sigma_\tau^2}{(k - 1)} + \sigma_\varepsilon^2$$

the EMS term corresponding to the treatments is then either

$$\sigma_\varepsilon^2 + n\left[\sum_{j} \tau_j^2/(k - 1)\right]$$

or

$$\sigma_\varepsilon^2 + n\sigma_\tau^2$$

depending upon whether treatments are fixed or random. When fixed, we shall designate $\sum_j \tau_j^2/(k - 1)$ as ϕ_τ, so the fixed treatment EMS is $\sigma_\varepsilon^2 + n\phi_\tau$.

For the error mean square,

$$SS_{error} = \sum_{j=1}^{k} \sum_{i=1}^{n} (Y_{ij} - \bar{Y}_{.j})^2$$

Subtracting Equation (3.8) from Equation (3.7) gives

$$Y_{ij} - \bar{Y}_{.j} = \varepsilon_{ij} - \sum_{i=1}^{n} \varepsilon_{ij}/n$$

Squaring and adding gives

$$\sum_{j}^{k} \sum_{i}^{n} (Y_{ij} - \bar{Y}_{.j})^2 = \sum_{j}^{k} \sum_{i}^{n} \left(\varepsilon_{ij} - \sum_{i} \varepsilon_{ij}/n\right)^2$$

Taking the expected value, we have

$$\begin{aligned} E(SS_{error}) &= E\left[\sum_{j}^{k} \sum_{i}^{n} \left(\varepsilon_{ij} - \sum_{i} \varepsilon_{ij}/n\right)^2\right] \\ &= \sum_{j}^{k} E\left[\sum_{i} \left(\varepsilon_{ij} - \sum_{i} \varepsilon_{ij}/n\right)^2\right] \\ &= \sum_{j}^{k} (n - 1)\sigma_{\varepsilon}^2 \\ &= k(n - 1)\sigma_{\varepsilon}^2 \end{aligned}$$

and

$$E(MS_{error}) = E[SS_{error}/k(n - 1)] = \sigma_{\varepsilon}^2$$

as shown in Table 3.9.

3.6 GENERAL REGRESSION SIGNIFICANCE TEST

A method will be presented in this section that can be used on any ANOVA problem but is much more general than the method of Section 3.2. The results of this general regression significance test will be the same as given above when applied to the example, but the method is important as it can easily be extended to more complicated problems. This method will be given in three parts: (1) the steps in a general regression significance test, (2) these steps applied to the example of this chapter, (3) some of the theory behind this method.

Steps in General Regression Significance Test

In the model, $Y_{ij} = \mu + \tau_j + \varepsilon_{ij}$, let m be the least squares best estimate of μ, t_j a similar estimate of τ_j, and e_{ij} that of ε_{ij}. Proceed with the following steps.

1. Obtain totals for each term in the model which is a distinct classification.
2. Obtain the grand total of the experiment.
3. Obtain normal equations, one equation for each total in steps 1 and 2. These equations are in terms of the estimates of the parameters.
4. Solve these equations for m and the t_j's.
5. Obtain regression sum of squares due to all estimates.
6. Rewrite the model, omitting the parameters assumed to be zero when the hypothesis under test is true.
7. Determine normal equations for this reduced model.
8. Solve these equations and determine the regression sum of squares due to the estimates left. Call these estimates primed from original, for example, m'.
9. Obtain between-treatments sum of squares:

$$SS_{\text{between treatments}} = SS_{\text{regression}}(m, t) - SS_{\text{regression}}(m')$$

10. Obtain

$$SS_{\text{error}} = \sum_{j=1}^{k} \sum_{i=1}^{n_j} Y_{ij}^2 - SS_{\text{regression}}(m, t)$$

11. Make F test where

$$F = \frac{SS_{\text{treatment}}/\text{df}}{SS_{\text{error}}/\text{df}}$$

to test hypothesis that treatment effects are zero.

Example 3.3 (*Based on Coded Data in Table 3.6*) Following the above steps with the data below, we have

1. Distinct classification—4 treatments, $T_{.j}$: 42, 36, -32, -40.
2. $T_{..} = 6$.
3. Normal equations:

$$\begin{aligned} 6 &= 20m + 5t_1 + 5t_2 + 5t_3 + 5t_4 \\ 42 &= 5m + 5t_1 \\ 36 &= 5m + 5t_2 \\ -32 &= 5m + 5t_3 \\ -40 &= 5m + 5t_4 \end{aligned}$$

Note that the coefficients of m and the t_j's are the number of times that estimate appears in the total on the left-hand side of the equation.

4. Since

$$\sum_{j=1}^{k} t_j = 0$$

the first equation becomes

$$6 = 20m \quad \text{and} \quad m = \tfrac{6}{20} = 0.3$$

The others give

$$t_1 = \frac{42 - 5m}{5} = 8.1$$

$$t_2 = \frac{36 - 5m}{5} = 6.9$$

$$t_3 = \frac{-32 - 5m}{5} = -6.7$$

$$t_4 = \frac{-40 - 5m}{5} = -8.3$$

$$\sum_{j=1}^{4} t_j = 0$$

5. Regression sum of squares is found by multiplying each term on the left of the equations (the totals) by the corresponding estimates obtained in step 4:

$$\begin{aligned} SS_{\text{regression}}(m, t_j) &= 6(0.3) + 42(8.1) + 36(6.9) \\ &\quad + (-32)(-6.7) + (-40)(-8.3) \\ &= 1.8 + 340.2 + 248.4 + 214.4 + 332.0 \\ &= 1136.8 \end{aligned}$$

6. $Y_{ij} = \mu + \varepsilon_{ij}$.
7. $6 = 20m'$.
8. $m' = \dfrac{6}{20} = 0.3$ (as before)

$$SS_{\text{regression}}(m') = 6(.3) = 1.8$$

9. $$\begin{aligned} SS_{\text{between treatments}} &= SS_{\text{regression}}(m, t) - SS_{\text{regression}}(m') \\ &= 1136.8 - 1.8 = 1135.0. \end{aligned}$$

10. $$\begin{aligned} SS_{\text{error}} &= \sum_j \sum_i Y_{ij}^2 - SS_{\text{regression}}(m, t) \\ &= 1340 - 1136.8 = 203.2. \end{aligned}$$

11. $$F = \frac{SS_{\text{treatment}}/\text{df}}{SS_{\text{error}}/\text{df}} = \frac{1135.0/3}{203.2/16} = \frac{378.3}{12.7} = 29.8 \text{ (as before)}.$$

Rationale for General Regression Significance Test

Consider the model

$$Y_{ij} = \mu + \tau_j + \varepsilon_{ij}$$

To find the least squares best estimates of the parameters $\mu, \tau_1, \tau_2, \cdots, \tau_k$, form the sum of squares of the errors

$$\sum_j \sum_i \varepsilon_{ij}^2 = \sum_j \sum_i (Y_{ij} - \mu - \tau_j)^2 \tag{3.10}$$

The object is to find estimates of $\mu = m$, $\tau_j = t_j$ which will minimize the sum of squares of these errors. To minimize, differentiate partially with respect to μ, take $j = 1, 2, \cdots, k$, and differentiate partially with respect to $\tau_1, \tau_2, \cdots, \tau_k$:

$$\text{for } \mu: \frac{\partial}{\partial \mu}\left(\sum_i \sum_j \varepsilon_{ij}^2\right) = 2 \sum_i \sum_j (Y_{ij} - \mu - \tau_j)(-1) = 0$$

$$\text{for } \tau_1: \quad \frac{\partial}{\partial \tau_1}\left(\sum_i e_{i1}^2\right) = 2 \sum_i (Y_{i1} - \mu - \tau_1)(-1) = 0$$

$$\text{for } \tau_2: \quad \frac{\partial}{\partial \tau_2}\left(\sum_i \varepsilon_{i2}^2\right) = 2 \sum_i (Y_{i2} - \mu - \tau_2)(-1) = 0$$

$$\vdots$$

$$\text{for } \tau_k: \quad \frac{\partial}{\partial \tau_k}\left(\sum_i \varepsilon_{ik}^2\right) = 2 \sum_i (Y_{ik} - \mu - \tau_k)(-1) = 0$$

From these equations the least squares normal equations are written

$$\begin{aligned}
\sum_{i=1}^{n} \sum_{j=1}^{k} Y_{ij} &= \sum_i \sum_j \mu + \sum_i \sum_i \tau_j \\
\sum_i Y_{i1} &= \sum_i \mu + \sum_i \tau_1 \\
\sum_i Y_{i2} &= \sum_i \mu + \sum_i \tau_2 \\
&\vdots \\
\sum_i Y_{ik} &= \sum_i \mu + \sum_i \tau_k
\end{aligned}$$

When they are added as indicated these become

$$\begin{aligned}
T_{..} &= Nm + n \sum_{j=1}^{k} t_j \\
T_{.1} &= nm + nt_1 \\
T_{.2} &= nm \qquad\qquad + nt_2 \\
&\vdots \\
T_{.k} &= nm \qquad\qquad\qquad + nt_k
\end{aligned} \tag{3.11}$$

where μ and τ have been replaced by m and t_j as they are estimates of the parameters. These equations are the normal equations of step 3. Solving these for m and the t_j's gives the following.

Since $\sum_{j=1}^{k} t_j = 0$,

$$m = \frac{T_{..}}{N} = \bar{Y}_{..}$$

$$t_1 = \frac{T_{.1} - nm}{n} = \frac{T_{.1}}{n} - \frac{T_{..}}{N} = \bar{Y}_{.1} - \bar{Y}_{..}$$

$$t_2 = \frac{T_{.2} - nm}{n} = \frac{T_{.2}}{n} - \frac{T_{..}}{N} = \bar{Y}_{.2} - \bar{Y}_{..}$$

$$\vdots \qquad\qquad \vdots \qquad\qquad \vdots$$

$$t_k = \frac{T_{.k} - nm}{n} = \frac{T_{.k}}{n} - \frac{T_{..}}{N} = \bar{Y}_{.k} - \bar{Y}_{..}$$

To obtain the regression sum of squares due to this regression of Y_{ij} on m and the t_j's, consider the error sum of squares, Equation (3.10) and substitute the best estimates

$$\sum_i \sum_j e_{ij}^2 = \sum_i \sum_j (Y_{ij} - m - t_j)^2$$

$$\begin{aligned}
SS_{\text{error}} &= \sum_i \sum_j (Y_{ij} - [m + t_j])^2 \\
&= \sum_i \sum_j Y_{ij}^2 - 2 \sum_i \sum_j Y_{ij}(m + t_j) + \sum_i \sum_j (m + t_j)^2 \\
&= \sum_i \sum_j Y_{ij}^2 - 2m \sum_i \sum_j Y_{ij} - 2 \sum_i \sum_j Y_{ij}t_j + m^2N \\
&\quad + 2m \sum_i \sum_j t_j + \sum_i \sum_j t_j^2 \\
&= \sum_i \sum_j Y_{ij}^2 - 2m \sum_i \sum_j Y_{ij} - 2 \sum_i \sum_j Y_{ij}t_j \\
&\quad + m\left[Nm + n \sum_j t_j\right] + \sum_j t_j[nm + nt_j]
\end{aligned}$$

where the fifth term ($2m \sum_i \sum_j t_j$) is divided in half in the two bracketed expressions. From the least squares Equation (3.11) these terms in brackets may be replaced by $T_{..}$ and $T_{.j}$, respectively, and the resulting SS_{error} is

$$SS_{\text{error}} = \sum_i \sum_j Y_{ij}^2 - 2mT_{..} - 2 \sum_i \sum_j Y_{ij}t_j + mT_{..} + \sum_j T_{.j}t_j$$

or

$$SS_{\text{error}} = \sum_i \sum_j Y_{ij}^2 - mT_{..} - \sum_j T_{.j}t_j$$

The regression sum of squares is equal to the total sum of squares less the error sum of squares. Therefore,

$$\begin{aligned} SS_{\text{regression}}(m, t) &= \sum_i \sum_j Y_{ij}^2 - SS_{\text{error}} \\ &= mT_{..} + \sum_j T_{.j}t_j \end{aligned}$$

Note that this is each estimate $m, t_1, \cdots, t_k$, multiplied by the corresponding terms on the left of Equation (3.11) and added.

In general, then

$$\begin{aligned} SS_{\text{regression}}(m, t) &= mT_{..} + \sum_j T_{.j}t_j \\ &= (\bar{Y}_{..})(T_{..}) + T_{.1}(\bar{Y}_{.1} - \bar{Y}_{..}) + T_{.2}(\bar{Y}_{.2} - \bar{Y}_{..}) \\ &\quad + \cdots + T_{.k}(\bar{Y}_{.k} - \bar{Y}_{..}) \end{aligned}$$

If H_0 is assumed true, so that $\tau_j = 0$ for all j, the model becomes

$$Y_{ij} = \mu + \varepsilon'_{ij}$$

Again, applying least squares to this error

$$\sum_i \sum_j \varepsilon'^2_{ij} = \sum_i \sum_j (Y_{ij} - \mu)^2$$

and the estimate of $\mu = m' = \bar{Y}_{..}$

$$SS_{\text{regression}}(m' \text{ only}) = m'T_{..} = (\bar{Y}_{..})(T_{..})$$

then

$$\begin{aligned} SS_{\text{treatment}} &= SS_{\text{regression}}(m, t) - SS_{\text{regression}}(m') \\ &= (\bar{Y}_{..})(T_{..}) + \sum_j T_{.j}(\bar{Y}_{.j} - \bar{Y}_{..}) - (\bar{Y}_{..})(T_{..}) \\ &= \sum_j T_{.j}(\bar{Y}_{.j} - \bar{Y}_{..}) = \sum_j T_{.j}\left(\frac{T_{.j}}{n} - \frac{T_{..}}{N}\right) \\ &= \sum_j \frac{T_{.j}^2}{n} - \frac{T_{..}}{N}\sum_j T_{.j} \\ &= \sum_j \frac{T_{.j}^2}{n} - \frac{T_{..}^2}{N} \quad \text{as} \quad \sum_j^k T_{.j} = T_{..} \end{aligned}$$

and

$$\begin{aligned} SS_{\text{error}} &= \sum_i \sum_j Y_{ij}^2 - SS_{\text{regression}}(m, t) \\ &= \sum_i \sum_j Y_{ij}^2 - (\bar{Y}_{..})(T_{..}) - \sum_j T_{.j}(\bar{Y}_{.j} - \bar{Y}_{..}) \\ &= \sum_i \sum_j Y_{ij}^2 - \frac{T_{..}^2}{N} - \sum_j \frac{T_{.j}^2}{n} + \frac{T_{..}^2}{N} \\ &= \sum_i \sum_j Y_{ij}^2 - \sum_j \frac{T_{.j}^2}{n} \end{aligned}$$

and the F test becomes

$$F = \frac{\text{SS}_{\text{treatment}}/\text{df}}{\text{SS}_{\text{error}}/\text{df}} = \frac{\left[\sum_j \frac{T_{.j}^2}{n} - \frac{T_{..}^2}{N}\right]\Big/(k-1)}{\left[\sum_i \sum_j Y_{ij}^2 - \sum_j \frac{T_{.j}^2}{n}\right]\Big/(N-k)}$$

which is identical with the expression for F from Table 3.5 if the n_j's are all equal.

3.7 SUMMARY

In this chapter consideration has been given to

Experiment	*Design*	*Analysis*
I. Single factor	Completely randomized	One-way ANOVA

In applying the ANOVA techniques, certain assumptions should be kept in mind:

1. The process is in control, that is, it is repeatable.
2. The population distribution being sampled is normal.
3. The variance of the errors within all k levels of the factor are homogeneous.

Many texts [3], [7] discuss these assumptions and what may be done if they are not met in practice.

PROBLEMS

3.1 Assuming a completely randomized design, do a one-way analysis of variance on the following data in order to familiarize yourself with the technique:

	Factor A Level 1	2	3	4	5
Measurement	8	4	1	4	10
	6	−2	2	6	8
	7	0	0	5	7
	5	−2	−1	5	4
	8	3	−3	4	9

3.2 The cathode warm-up time in seconds was determined for three different tube types using eight observations on each type of tube. The order of experimentation was completely randomized. The results were

	Tube Type					
	A		*B*		*C*	
Warm-up time (seconds)	19	20	20	40	16	19
	23	20	20	24	15	17
	26	18	32	22	18	19
	18	35	27	18	26	18

Do an analysis of variance on these data and test the hypothesis that the three tube types require the same average warm-up time.

3.3 For Problem 3.2, set up orthogonal contrasts between the tube types and test your contrasts for significance.

3.4 Set up 95-percent confidence limits for the average warm-up time for tube type *C* in Problem 3.2.

3.5 Use the Newman–Keuls range method to test for differences between tube types.

3.6 The following data are on the pressure in a torsion spring for several settings of the angle between the legs of the spring in a free position:

	Angle of Legs of Spring (degrees)				
	67	71	75	79	83
Pressure (psi)	83	84	86	89	90
	85	85	87	90	92
		85	87	90	
		86	87	91	
		86	88		
		87	88		
			88		
			88		
			88		
			89		
			90		

Assuming a completely randomized design, complete a one-way analysis of variance for this experiment and state your conclusion concerning the effect of angle on the pressure in the spring.

3.7 Set up orthogonal contrasts for the angles in Problem 3.6.

3.8 Do Problem 3.1 using the general regression significance test.

3.9 Show that the expanded forms for the sums of squares in Table 3.5 are correct.

3.10 Assume that the levels of factor A in Problem 3.1 were chosen at random and determine the proportion of variance attributable to differences in level means and the proportion due to error.

3.11 Verify the results given in Table 3.9 from the data of Table 3.8.

3.12 Set up two or more contrasts for the data of Problem 3.2 and test their significance by the Scheffé method.

3.13 Since the Scheffé method is not restricted to equal sample sizes, set up several contrasts for Problem 3.6 and test for significance by the Scheffé method.

3.14 It is suspected that the environmental temperature in which batteries are activated affects their activated life. Thirty homogeneous batteries were tested, six at each of five temperatures, and the data shown below were obtained. Analyze and interpret the data.

	Temperature (°C)				
	0	25	50	75	100
Activated life (seconds)	55	60	70	72	65
	55	61	72	72	66
	57	60	73	72	60
	54	60	68	70	64
	54	60	77	68	65
	56	60	77	69	65

3.15 A highway research engineer wishes to determine the effect of four types of subgrade soil on the moisture content in the top soil. He takes five samples of each type of subgrade soil and the total sum of squares is computed as 280, whereas the sum of squares among the four types of subgrade soil is 120.

1. Set up an analysis-of-variance table for these results.
2. Set up a mathematical model to describe this problem, define each term in the model, and state the assumptions made on each term.
3. Set up a test of the hypothesis that the four types of subgrade soil have the same effect on moisture content in the top soil.
4. Set up a set of orthogonal contrasts for this problem.
5. Explain briefly how to set up a test on means after the analysis of variance for these data.
6. Set up an expression for 90-percent confidence limits on the mean of type 2 subgrade soil. Insert all numerical values that are known.

4

Single-Factor Experiments—Randomized Block Design

4.1 INTRODUCTION

Consider the problem of determining whether or not different brands of tires exhibit different amounts of tread loss after 20,000 miles of driving. A fleet manager wishes to consider four brands which are available and make some decision about which brand might show the least amount of tread wear after 20,000 miles. The brands to be considered are A, B, C, and D and although driving conditions might be simulated in the laboratory, he wants to try these four brands under actual driving conditions. The variable to be measured is the difference in maximum tread thickness on a tire between the time it is mounted on the wheel of a car and after it has completed 20,000 miles on this car. The measured variable Y_{ij} is this difference in thickness in mils (0.001 inch), and the only factor of interest is brands, say, τ_j where $j = 1$, 2, 3, and 4.

Since the tires must be tried on cars and since some measure of error is necessary, more than one tire of each brand must be used and a set of four of each brand would seem quite practical. This means 16 tires, four each of four different brands, and a reasonable experiment would involve at least four cars. Designating the cars as I, II, III, and IV, one might put brand A's four tires on car I, brand B's on car II, and so on, with a design as shown in Table 4.1.

Table 4.1 Design 1 for Tire Brand Test

	Car			
	I	II	III	IV
Brand distribution	A	B	C	D
	A	B	C	D
	A	B	C	D
	A	B	C	D

One look at this design shows its fallibility, since averages for brands are also averages for cars. If the cars travel over different terrains, using different drivers, any apparent brand differences are also car differences. This design is called completely confounded, since we cannot distinguish between brands and cars in the analysis.

A second attempt at design might be to try a completely randomized design, as given in Chapter 3. Assigning the 16 tires to the four cars in a completely random manner might give results as in Table 4.2. In Table 4.2 the

Table 4.2 Design 2 for Tire Brand Test

	Car			
	I	II	III	IV
Brand distribution and loss in thickness	$C(12)$	$A(14)$	$D(10)$	$A(13)$
	$A(17)$	$A(13)$	$C(11)$	$D(9)$
	$D(13)$	$B(14)$	$B(14)$	$B(8)$
	$D(11)$	$C(12)$	$B(13)$	$C(9)$

loss in thickness is given for each of the 16 tires. The purpose of complete randomization here is to average out any car differences which might affect the results. The model would be

$$Y_{ij} = \mu + \tau_j + \varepsilon_{ij}$$

with

$$j = 1, 2, 3, 4 \qquad i = 1, 2, 3, 4$$

An analysis of variance on these data gives the results in Table 4.3.

Table 4.3 ANOVA for Design 2

Source	df	SS	MS
Brands	3	30.6	10.2
Error	12	50.3	4.2
Totals	15	80.9	

The F test shows $F_{3,12} = 2.43$, and the 5-percent critical region for F is $F_{3,12} \geq 3.49$ (Appendix, Table D), so there is no reason to reject the hypothesis of equal average tread loss among the four brands.

4.2 RANDOMIZED COMPLETE BLOCK DESIGN

A more careful examination of design 2 in Table 4.2 will reveal some glaring disadvantages of the completely randomized design in this problem. One thing to be noted is that brand A is never used on car III nor brand B on car I. Also any variation within brand A may reflect variation between cars I, II, and IV. Thus, the random error may not be merely an experimental error but may include variation between cars. Since the chief objective of experimental design is to reduce the experimental error, a better design might be one in which car variation is removed from error variation. Although the completely randomized design averaged out the car effects, it did not eliminate the variance among cars. A design which requires that each brand be used once on each car is a *randomized complete block design*, given in Table 4.4.

Table 4.4 Design 3: Randomized Block Design for Tire Brand Test

	Car			
	I	II	III	IV
Brand distribution and loss in thickness	B(14)	D(11)	A(13)	C(9)
	C(12)	C(12)	B(13)	D(9)
	A(17)	B(14)	D(11)	B(8)
	D(13)	A(14)	C(10)	A(13)

In this design the order in which the four brands are placed on a car is random and each car gets one tire of each brand. In this way better comparisons can be made between brands since they are all driven over approximately the same terrain, and so on. This provides a more homogeneous environment in which to test the four brands. In general, these groupings for homogeneity are called *blocks* and randomization is now restricted within blocks. This design also allows the car (block) variation to be independently assessed and removed from the error term. The model for this design is

$$Y_{ij} = \mu + \beta_i + \tau_j + \varepsilon_{ij} \tag{4.1}$$

where β_i now represents the block effect (car effect) in the example above.

The analysis of this model is a two-way analysis of variance, since the block effect may now also be isolated. A slight rearrangement of the data in Table 4.4 and a coding by subtracting 13 from all readings gives Table 4.5.

The total sum of squares is computed as in Chapter 3:

$$SS_{\text{total}} = \sum_{j=1}^{4} \sum_{i=1}^{4} Y_{ij}^2 - \frac{T_{..}^2}{N}$$

$$= 95 - \frac{(-15)^2}{16} = 80.9$$

Table 4.5 Randomized Block Design Coded Data for Tire Brand Test

Car	Brand *A*	*B*	*C*	*D*	$T_{i.}$
I	4	1	−1	0	4
II	1	1	−1	−2	−1
III	0	0	−3	−2	−5
IV	0	−5	−4	−4	−13
$T_{.j}$	5	−3	−9	−8	$-15 = T_{..}$
$\sum_{i=1}^{4} Y_{ij}^2$	17	27	27	24	$95 = \sum_i^4 \sum_j^4 Y_{ij}^2$

The brand (treatment) sum of squares is computed as usual:

$$SS_{\text{brand}} = \sum_j \frac{T_{.j}^2}{n} - \frac{T_{..}^2}{N}$$

$$= \frac{(5)^2 + (-3)^2 + (-9)^2 + (-8)^2}{4} - \frac{(-15)^2}{16} = 30.6$$

Since the car (block) effect is similar to the brand effect but totaled across the rows of Table 4.5, the car sum of squares is computed exactly like the brand sum of squares, using row totals $T_{i.}$ instead of column totals. Calling the number of treatments (brands) in general k, then

$$SS_{\text{car}} = \sum_{i=1}^{n} \frac{T_{i.}^2}{k} - \frac{T_{..}^2}{N}$$

$$= \frac{(4)^2 + (-1)^2 + (-5)^2 + (-13)^2}{4} - \frac{(-15)^2}{16} = 38.6$$

The error sum of squares is now the remainder after subtracting both brand and car sum of squares from the total sum of squares:

$$SS_{\text{error}} = SS_{\text{total}} - SS_{\text{brand}} - SS_{\text{car}}$$

$$= 80.9 - 30.6 - 38.6 = 11.7$$

Table 4.6 is an ANOVA table of these data.

Table 4.6 ANOVA for Randomized Block Design of Tire Brand Test

Source	df	SS	MS	EMS
Brands	3	30.6	10.2	$\sigma_\varepsilon^2 + 4\phi_\tau$
Cars	3	38.6	12.9	$\sigma_\varepsilon^2 + 4\sigma_\beta^2$
Error	9	11.7	1.3	σ_ε^2
Totals	15	80.9		

To test the hypothesis, $H_0: \mu_{.1} = \mu_{.2} = \mu_{.3} = \mu_{.4}$, the ratio is

$$F_{3,9} = \frac{10.2}{1.3} = 7.8$$

which is significantly larger than the corresponding critical F even at the 1-percent level (Appendix Table D). The hypothesis of equal brand means is thus rejected. It is to be noted that this hypothesis could not be rejected using a completely randomized design. The randomized block design which allows for removal of the block (car) effect definitely reduced the error variance estimate from 4.2 to 1.3.

It is also possible, if desired, to test the hypothesis that the average tread loss of all four cars is the same. $H_1: \mu_{1.} = \mu_{2.} = \mu_{3.} = \mu_{4.}$, and $F_{3,9} = 12.9/1.3 = 9.9$, which is also significant at the 1-percent level (Appendix Table D). Here this hypothesis is rejected and a car-to-car variation is detected.

Even though an effect due to cars (blocks) has been isolated, the main objective is still to test brand differences. Thus it is still a single-factor experiment, the blocks representing only a restriction on complete randomization due to the environment in which the experiment was conducted. Other examples include testing differences in materials which are fed into several different machines, testing differences in fertilizers which must be spread on several different plots of ground, testing the effect of different teaching methods on several pupils. In these examples, the blocks are machines, plots, and pupils, respectively, and the levels of the factors of interest can be randomized within each block.

4.3 ANOVA RATIONALE

For this randomized complete block design, the model is

$$Y_{ij} = \mu + \beta_i + \tau_j + \varepsilon_{ij} \qquad (4.2)$$

or

$$Y_{ij} = \mu + (\mu_{i.} - \mu) + (\mu_{.j} - \mu) + (Y_{ij} - \mu_{i.} - \mu_{.j} + \mu) \qquad (4.3)$$

where $\mu_{i.}$ represents the true mean of block i. The last term can be obtained by subtracting the treatment and block deviations from the overall deviation as follows:

$$(Y_{ij} - \mu) - (\mu_{i.} - \mu) - (\mu_{.j} - \mu) \equiv Y_{ij} - \mu_{i.} - \mu_{.j} + \mu$$

Best estimates of the parameters in Equation (4.3) give the sample model (after moving $\bar{Y}_{..}$ to the left of the equation):

$$Y_{ij} - \bar{Y}_{..} = (\bar{Y}_{i.} - \bar{Y}_{..}) + (\bar{Y}_{.j} - \bar{Y}_{..}) + (\bar{Y}_{ij} - \bar{Y}_{i.} - \bar{Y}_{.j} + \bar{Y}_{..})$$

Squaring both sides and adding with $i = 1, 2, \cdots, n$, $j = 1, 2, \cdots, k$,

$$\sum_{i=1}^{n} \sum_{j=1}^{k} (Y_{ij} - \bar{Y}_{..})^2 = \sum_{i}^{n} \sum_{j}^{k} (\bar{Y}_{i.} - \bar{Y}_{..})^2 + \sum_{i}^{n} \sum_{j}^{k} (\bar{Y}_{.j} - \bar{Y}_{..})^2$$
$$+ \sum_{i}^{n} \sum_{j}^{k} (Y_{ij} - \bar{Y}_{i.} - \bar{Y}_{.j} + \bar{Y}_{..})^2 + 3 \text{ cross products} \qquad (4.4)$$

A little algebraic work on the sums of the cross products will show that they all reduce to zero and the remaining equation becomes the fundamental equation of a two-way analysis of variance. The equation states that

$$SS_{\text{total}} = SS_{\text{block}} + SS_{\text{treatment}} + SS_{\text{error}}$$

one sum of squares for each variable term in the model [Equation (4.2)]. Each sum of squares has associated with it its degrees of freedom, and dividing any sum of squares by its degrees of freedom will yield an unbiased estimate of population variance σ_ε^2 if the hypotheses under test are true.

The breakdown of degrees of freedom here is

total		blocks		treatments		error
$(nk - 1)$	=	$(n - 1)$	+	$(k - 1)$	+	$(n - 1)(k - 1)$

The error degrees are derived from the remainder

$$(nk - 1) - (n - 1) - (k - 1) = nk - n - k + 1 = (n - 1)(k - 1)$$

It can be shown that each sum of squares on the right of Equation (4.4) when divided by its degrees of freedom provides mean squares which are independently chi-square distributed, so that the ratio of any two of them is distributed as F.

The sums of squares formulas given in Equation (4.4) are usually expanded and rewritten to give formulas which are easier to apply. These are shown in Table 4.7. They are the formulas applied to the data of Table 4.5 where $nk = N$.

Table 4.7 ANOVA for Randomized Block Design

Source	df	SS	MS
Between blocks β_i	$n-1$	$\sum_i^n \frac{T_{i.}^2}{k} - \frac{T_{..}^2}{nk}$	$SS_{\text{block}}/(n-1)$
Between treatments τ_j	$k-1$	$\sum_j^k \frac{T_{.j}^2}{n} - \frac{T_{..}^2}{nk}$	$SS_{\text{treatment}}/(k-1)$
Error ε_{ij}	$(n-1)(k-1)$	$\sum_i^n \sum_j^k Y_{ij}^2 - \sum_{i=1}^n \frac{T_{i.}^2}{k} - \sum_{j=1}^k \frac{T_{.j}^2}{n} + \frac{T_{..}^2}{nk}$	$SS_{\text{error}}/(n-1)(k-1)$
Totals	$nk-1$	$\sum_i^n \sum_j^k Y_{ij}^2 - \frac{T_{..}^2}{nk}$	

4.4 INTERPRETATIONS

Since the above example shows significant differences in brand effects, it might be desirable to make further investigation of the brand means. Orthogonal contrasts could be made on these if set prior to experimentation, or tests like the Newman–Keuls range test could be made after experimentation. For the latter the means are arranged in order of magnitude as follows (coded):

	Brand			
	C	D	B	A
$T_{.j}$	−9	−8	−3	5
$\bar{Y}_{.j}$	−2.25	−2.00	−0.75	1.25

The standard error of a treatment (brand) mean is

$$s_{\bar{Y}_{.j}} = \sqrt{\frac{1.3}{4}} = 0.57$$

The tabled ranges are $Z_{p,9\text{ df}}(0.05)$

$$Z_{2,9}(0.05) = 3.20$$
$$Z_{3,9}(0.05) = 3.95$$
$$Z_{4,9}(0.05) = 4.42$$

and the least significant ranges for this example are

$$R_2 = (3.20)(0.57) = 1.82$$
$$R_3 = (3.34)(0.57) = 2.25$$
$$R_4 = (3.41)(0.57) = 2.51$$

Testing the brand averages,

$$A - C = 1.25 - (-2.25) = 3.5 > 2.52^*$$
$$A - D = 1.25 - (-2.00) = 3.25 > 2.25^*$$
$$A - B = 1.25 - (-0.75) = 2.00 > 1.82^*$$
$$B - C = -0.75 - (-2.25) = 1.50 < 2.25$$
$$B - D = -0.75 - (-2.00) = 1.25 < 1.82$$
$$D - C = -2.00 - (-2.25) = 0.25 < 1.82$$

where, again, one asterisk indicates significance at the 5-percent level. These results show A significantly greater than all other brands, but the other three do not differ significantly from each other. This is shown graphically in Figure 4.1. Since a high average means a large loss in tread, these results indicate that brand A is the poorest of the four tested, but there is little difference in the other three brands. Confidence limits might be set on brand A

$$\bar{Y}_{.1} \pm t_{1-(\alpha/2)} s_{\bar{Y}_{.j}} \qquad \text{with 9 df on } t$$
$$1.25 \pm (2.26)(0.57)$$
$$1.25 \pm 1.29$$

or

$$-0.04 \text{ to } 2.54$$

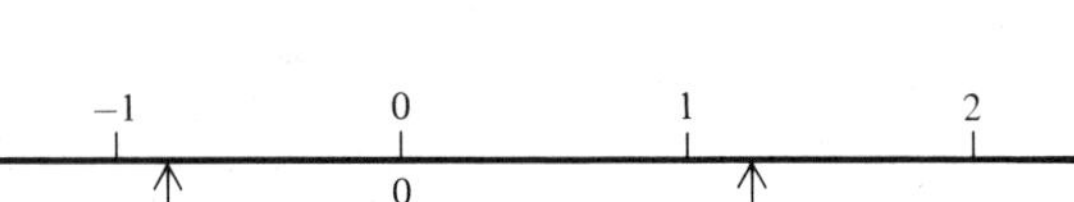

Figure 4.1

Decoded, we obtain 12.96 to 15.54 mm. Similar tests might be run on car effects if there is any concern about these four specific cars. Often the cars are selected at random and it might be desirable to estimate the variance in car means. From Table 4.6,

$$s_\varepsilon^2 = 1.3$$

$$s_\varepsilon^2 + 4s_\beta^2 = 12.9$$

$$s_\beta^2 = \frac{12.9 - 1.3}{4} = \frac{11.6}{4} = 2.9$$

4.5 GENERAL REGRESSION SIGNIFICANCE TEST APPROACH

A randomized block design may also be analyzed by the general regression significance test approach given in Chapter 3. The only essential difference is that there are n more totals taken across the k treatments, making $n + k + 1$ least squares normal equations in $n + k + 1$ unknowns. For the randomized block model,

$$Y_{ij} = \mu + \beta_i + \tau_j + \varepsilon_{ij}$$

The least squares best estimates give a model

$$Y_{ij} = m + b_i + t_j + e_{ij}$$

with

$$i = 1, 2, \cdots, n \qquad j = 1, 2, \cdots, k$$

Applying this approach to the coded data of Table 4.5 on brands and cars, the following normal equations result:

$$\begin{aligned}
-15 &= 16m + 4b_1 + 4b_2 + 4b_3 + 4b_4 + 4t_1 + 4t_2 + 4t_3 + 4t_4 \\
4 &= 4m + 4b_1 + t_1 + t_2 + t_3 + t_4 \\
-1 &= 4m + 4b_2 + t_1 + t_2 + t_3 + t_4 \\
-5 &= 4m + 4b_3 + t_1 + t_2 + t_3 + t_4 \\
-13 &= 4m + 4b_4 + t_1 + t_2 + t_3 + t_4 \\
5 &= 4m + b_1 + b_2 + b_3 + b_4 + 4t_1 \\
-3 &= 4m + b_1 + b_2 + b_3 + b_4 + 4t_2 \\
-9 &= 4m + b_1 + b_2 + b_3 + b_4 + 4t_3 \\
-8 &= 4m + b_1 + b_2 + b_3 + b_4 + 4t_4
\end{aligned} \tag{4.5}$$

Since

$$\sum_{i=1}^{4} b_i = \sum_{j=1}^{4} t_j = 0$$

these can be solved easily to give

$$m = -\frac{15}{16} = -0.94$$

$$\left.\begin{aligned}
b_1 &= \frac{4 - 4(-15/16)}{4} = 1.94 \\
b_2 &= \frac{-1 - 4(-15/16)}{4} = +0.69 \\
b_3 &= \frac{-5 - 4(-15/16)}{4} = -0.31 \\
b_4 &= \frac{-13 - 4(-15/16)}{4} = -2.31
\end{aligned}\right\} \sum_i b_i = 0.01 \approx 0$$

$$\left.\begin{aligned}
t_1 &= \frac{5 - 4(-15/16)}{4} = +2.19 \\
t_2 &= \frac{-3 - 4(-15/16)}{4} = 0.19 \\
t_3 &= \frac{-9 - 4(-15/16)}{4} = -1.31 \\
t_4 &= \frac{-8 - 4(-15/16)}{4} = -1.06
\end{aligned}\right\} \sum_j t_j = 0.01 \approx 0$$

$$\begin{aligned}
SS_{\text{regression}}(m, b_i, t_j) = {} & (-15)(-0.94) + 4(1.94) - 1(0.69) \\
& - 5(-0.31) - 13(-2.31) + 5(2.19) \\
& - 3(0.19) - 9(-1.31) - 8(-1.06) = 83.40
\end{aligned}$$

$$\begin{aligned}
SS_{\text{error}} &= \sum_i^4 \sum_j^4 Y_{ij}^2 - SS_{\text{regression}}(m, b_i, t_j) \\
&= 95 - 83.40 = 11.60
\end{aligned}$$

which agrees within rounding error with the value in ANOVA Table 4.6.

To determine the treatment sum of squares, assume no treatment effect. Therefore, the reduced model is

$$Y_{ij} = m' + b_i' + e_{ij}'$$

The normal equations for this model turn out to be the same as the first five equations in Equation (4.5), so that the estimates are

$$\begin{aligned} m' &= m = -0.94 \\ b_1' &= b_1 = 1.94 \\ b_2' &= b_2 = 0.69 \\ b_3' &= b_3 = -0.31 \\ b_4' &= b_4 = -2.31 \end{aligned}$$

and

$$SS_{\text{treatment}}(m, b_i) = (-15)(-0.94) + 4(1.94) - 1(0.69) - 5(-0.31) - 13(-2.31) = 52.75$$

Hence,

$$\begin{aligned} SS_{\text{regression}} &= SS_{\text{regression}}(m, b_i, t_j) - SS_{\text{regression}}(m, b_i) \\ &= 83.40 - 52.75 = 30.65 \end{aligned}$$

agreeing with 30.6 of Table 4.6. If block sums of squares are desired, assume block effect zero and use the reduced model

$$Y_{ij} = m'' + t_j'' + e_{ij}''$$

The normal equations are now equations 1, 6, 7, 8, and 9 in Equation (4.5) with the same solutions as before, and

$$SS(m, t_j) = (-15)(-0.94) + 5(2.19) - 3(0.19) - 9(-1.31) - 8(-1.06) = 44.75$$

Hence

$$\begin{aligned} SS_{\text{block}} &= SS_{\text{regression}}(m, b_i, t_j) - SS_{\text{regression}}(m, t_j) \\ &= 83.40 - 44.75 = 38.65 \end{aligned}$$

agreeing with 38.6 of Table 4.6.

4.6 MISSING VALUES

Occasionally in a randomized block design an observation is lost. A vial may break, an animal may die, or a tire may disintegrate, so that there occurs one or more missing observations in the data. For a single-factor completely randomized design this presents no problem, since the analysis of variance can be run with unequal n_j's. But for a two-way analysis this means a loss of orthogonality, since for some blocks the $\sum_j t_j$ no longer equals zero, and for some treatment the $\sum_i b_i$ no longer equals zero. When blocks and treatments are orthogonal, the block totals are added over all treatments and vice versa. If one or more observations are missing, the usual procedure is to replace the value with one which makes the sum of the squares of the errors a minimum.

In the tire brand test example, suppose that the brand C tire on car III blew out and was ruined before completing the 20,000 miles. The resulting data appear in Table 4.8 where y is put in place of this missing value.

Table 4.8 Missing Value Example

	Brand				
Car	A	B	C	D	$T_{i.}$
I	4	1	-1	0	4
II	1	1	-1	-2	-1
III	0	0	y	-2	$y - 2$
IV	0	-5	-4	-4	-13
$T_{.j}$	5	-3	$y - 6$	-8	$y - 12 = T_{..}$

Now,

$$\text{SS}_{\text{error}} = \text{SS}_{\text{total}} - \text{SS}_{\text{treatment}} - \text{SS}_{\text{block}}$$

$$= \sum_i \sum_j Y_{ij}^2 - \sum_j \frac{T_{.j}^2}{n} - \sum_i \frac{T_{i.}^2}{k} + \frac{T_{..}^2}{nk}$$

For this example,

$$\text{SS}_{\text{error}} = 4^2 + 1^2 + \cdots + y^2 + \cdots + (-4)^2 - \frac{(5)^2 + (-3)^2 + (y - 6)^2 + (-8)^2}{4} - \frac{(4)^2 + (-1)^2 + (y - 2)^2 + (-13)^2}{4} + \frac{(y - 12)^2}{16}$$

In order to find the y value which will minimize this expression, it is differentiated with respect to y and set equal to zero. As all constant terms have their derivatives zero,

$$\frac{d(\text{SS}_{\text{error}})}{dy} = 2y - \frac{2(y - 6)}{4} - \frac{2(y - 2)}{4} + \frac{2(y - 12)}{16} = 0$$

Solving gives

$$16y - 4y + 24 - 4y + 8 + y - 12 = 0$$

$$9y = -20$$

$$y = -\frac{20}{9} = -2.2$$

If this value is now used in the y position, the resulting ANOVA table is as shown in Table 4.9. This is an approximate ANOVA as the sums of squares are slightly biased. The resulting ANOVA is not too different from before, but the degrees of freedom for the error term are reduced by one, since there are only 15 actual observations, and y is determined from these 15 readings.

Table 4.9 ANOVA for Tire Brand Test Example Adjusted for a Missing Value (approximate)

Source	df	SS	MS
Brands τ_j	3	28.7	9.5
Cars β_i	3	38.3	12.7
Error ε_{ij}	8	11.2	1.4
Totals	14	78.2	

This procedure can be used on any reasonable number of missing values by differentiating the error sum of squares partially with respect to each such missing value and setting it equal to zero, giving as many equations as unknown values to be solved for these missing values.

In general, for one missing value y_{ij} it can be shown that

$$y_{ij} = \frac{nT'_{i.} + kT'_{.j} - T'_{..}}{(n-1)(k-1)} \tag{4.6}$$

where the primed totals $T'_{i.}$, $T'_{.j}$, and $T'_{..}$ are the totals indicated without the missing value y. In our example,

$$y_{ij} = y_{33} = \frac{4(-2) + 4(-6) - (-12)}{(3)(3)}$$

$$y_{33} = \frac{-20}{9} = -2.2 \text{ (as before)}$$

Another approach to the missing-value problem is to set up the least squares normal equations without the missing value and solve them as usual.

They will, of course, be more difficult to solve now since the sums of t's and b's in all equations are not zero. This procedure will give an exact solution as shown below. See Kempthorne [11] p. 97.

The least squares normal equations are

$$\begin{aligned}
-12 &= 15m + 4t_1 + 4t_2 + 3t_3 + 4t_4 + 4b_1 + 4b_2 + 3b_3 + 4b_4 \\
5 &= 4m + 4t_1 + b_1 + b_2 + b_3 + b_4 \\
-3 &= 4m + 4t_2 + b_1 + b_2 + b_3 + b_4 \\
-6 &= 3m + 3t_3 + b_1 + b_2 + b_4 \\
-8 &= 4m + 4t_4 + b_1 + b_2 + b_3 + b_4 \\
4 &= 4m + t_1 + t_2 + t_3 + t_4 + 4b_1 \\
-1 &= 4m + t_1 + t_2 + t_3 + t_4 + 4b_2 \\
-2 &= 3m + t_1 + t_2 + t_4 + 3b_3 \\
-13 &= 4m + t_1 + t_2 + t_3 + t_4 + 4b_4
\end{aligned} \tag{4.7}$$

Since $\sum_i b_i = \sum_j t_j = 0$, the above equations may be rewritten as

$$\begin{aligned}
-12 &= 15m + t_1 + t_2 + t_4 + b_1 + b_2 + b_4 \\
5 &= 4m + 4t_1 \\
-3 &= 4m + 4t_2 \\
-6 &= 3m + 3t_3 + b_1 + b_2 + b_4 \\
-8 &= 4m + 4t_4 \\
4 &= 4m + 4b_1 \\
-1 &= 4m + 4b_2 \\
-2 &= 3m + t_1 + t_2 + t_4 + 3b_3 \\
-13 &= 4m + 4b_4
\end{aligned}$$

Adding the fourth and eighth equations,

$$-8 = 6m + 2t_3 + 2b_3$$

And, dividing by 2,

$$-4 = 3m + t_3 + b_3$$

Adding this to the first equation in the above set,

$$-16 = 18m \quad \text{or} \quad m = -8/9 \quad \text{or} \quad -32/36$$

Substituting this value in the set of equations above,

$t_1 = 77/36$, $t_2 = 5/36$, $t_3 = -42/36$, $t_4 = -40/36$ which add to zero
$b_1 = 68/36$, $b_2 = 23/36$, $b_3 = -6/36$, $b_4 = -85/36$ which also add to zero

The sum of squares of regression on m, b, and t is

$$\begin{aligned} SS_{\text{regression}}(m, b_i, t_j) &= mT_{..} + \sum_i b_i T_{i.} + \sum_j t_j T_{.j} \\ &= (-32/36)(-12) + (77/36)(5) + (5/36)(-3) \\ &\quad + (-42/36)(-6) + (-40/36)(-8) + (68/36)(4) \\ &\quad + (23/36)(-1) + (-6/36)(-2) \\ &\quad + (-85/36)(-13) = 74.778 \end{aligned}$$

$$\begin{aligned} SS_{\text{error}} &= \sum_i \sum_j Y_{ij}^2 - SS_{\text{regression}}(m, b, t) \\ &= 86 - 74.778 = 11.222 \text{ (as before)} \end{aligned}$$

To test $H_0: \tau_j = 0$ for all j, the reduced model is $Y_{ij} = m' + b'_i + e'_{ij}$ and the normal equations are

$$\begin{aligned} -12 &= 15m' + b'_1 + b'_2 + b'_4 \\ 4 &= 4m' + 4b'_1 \\ -1 &= 4m' + 4b'_2 \\ -2 &= 3m' + 3b'_3 \\ -13 &= 4m' + 4b'_4 \end{aligned}$$

Solving this set of equations,

$$m' = -19/24, \quad b'_1 = 43/24, \quad b'_2 = 13/24, \quad b'_3 = 3/24, \quad b'_4 = -59/24$$

and the sum of squares of regression on m' and b' is

$$\begin{aligned} SS_{\text{regression}}(m', b'_i) &= (-19/24)(-12) + (43/24)(4) + (13/24)(-1) \\ &\quad + (3/24)(-2) + (-59/24)(-13) = 47.833 \end{aligned}$$

Hence,

$$\begin{aligned} SS_{\text{treatment}} &= SS_{\text{regression}}(m, b, t) - SS_{\text{regression}}(m', b') \\ &= 74.778 - 47.833 = 26.945 \end{aligned}$$

To test $H_1: \beta_i = 0$ for all i, the reduced model is $Y_{ij} = m'' + t''_j + e''_{ij}$ and the normal equations are

$$\begin{aligned} -12 &= 15m'' + t''_1 + t''_2 + t''_4 \\ 5 &= 4m'' + 4t''_1 \\ -3 &= 4m'' + 4t''_2 \\ -6 &= 3m'' + 3t''_3 \\ -8 &= 4m'' + 4t''_4 \end{aligned}$$

Solving this set of equations,

$$m'' = -7/8, \quad t''_1 = 17/8, \quad t''_2 = 1/8, \quad t''_3 = -9/8, \quad t''_4 = -9/8$$

and the sum of squares of regression on m'' and t'' is

$$SS_{\text{regression}}(m'', t''_j) = (-7/8)(-12) + (17/8)(5) + (1/8)(-3) + (-9/8)(-6) + (-9/8)(-8) = 36.500$$

Hence,

$$SS_{\text{block}} = SS_{\text{regression}}(m, b, t) - SS_{\text{regression}}(m'', t'') = 74.778 - 36.500 = 38.278$$

Summarizing these exact results along with the approximate results of Table 4.9 carried to three decimal places for comparisons give the values of Table 4.10.

Table 4.10 Comparison of Approximate and Exact Results for Table 4.8

Source	df	SS (Approximate)	SS (Exact)
Brands τ_j	3	28.708	26.945
Cars β_i	3	38.308	38.278
Error ε_{ij}	8	11.222	11.222
Totals	14	78.238	76.445

The results are very similar and the approximate method is usually satisfactory. With the use of high-speed computers, it is a simple matter to get the inverse of a matrix such as given by Equations (4.7) and solve for the unknown parameters. Thus an exact solution may be easily programmed.

4.7 RANDOMIZED INCOMPLETE BLOCKS—RESTRICTION ON EXPERIMENTATION

Method for Balanced Blocks

In some randomized block designs it may not be possible to apply all treatments in every block. If there were, for example, six brands of tires to test in the preceding example, only four could be tried on a given car (not using the spare), and such a block would be incomplete, having only four out of the six treatments in it.

Take the problem of determining the effect on current flow of four treatments applied to the coils of TV tube filaments. As each treatment application requires some time, it is not possible to run several observations of these treatments in one day. If days were taken as blocks, all four treatments must be run in random order on each of several days in order to have a randomized block design. After checking, it is found that even four treatments cannot be completed in a day; three are the most that can be run. The question then is: Which treatments are to be run on the first day, which on the second, and so forth, if information is desired on all four treatments?

The solution to this problem is to use a balanced incomplete block design. An *incomplete block design* is simply one in which there are more treatments than can be put in a single block. A *balanced incomplete block design* is an incomplete block design in which every pair of treatments occurs the same number of times in the experiment. Tables of such designs may be found in Fisher and Yates [9]. The number of blocks necessary for balancing will depend upon the number of treatments that can be run in a single block.

For the example mentioned there are four treatments and only three treatments can be run in a block. The balanced design for this problem requires four blocks (days) as shown in Table 4.11.

Table 4.11 Balanced Incomplete Block Design for TV Filament Example

Block	Treatment				
(days)	A	B	C	D	$T_{i.}$
1	2	—	20	7	29
2	—	32	14	3	49
3	4	13	31	—	48
4	0	23	—	11	34
$T_{.j}$	6	68	65	21	$160 = T_{..}$

In this design only treatments A, C, and D are run on the first day; B, C, and D on the second day, and so forth. Note that each pair of treatments, such as AB, occurs together twice in the experiment. A and B occur together on days three and four; C and D occur together on days one and two; and so on. As in randomized complete block designs, the order in which the three treatments are run on a given day is completely randomized.

The analysis of such a design is easier if some new notation is introduced. Let

b = number of blocks in the experiment ($b = 4$)

t = number of treatments in the experiment ($t = 4$)

k = number of treatments per block ($k = 3$)

r = number of replications of a given treatment throughout the experiment ($r = 3$)

N = total number of observations
$= bk = tr$ ($N = 12$)

λ = number of times each pair of treatments appears together throughout the experiment
$= r(k - 1)/(t - 1)$ ($\lambda = 2$)

Table 4.11 has the current readings after they have been coded by subtracting 513 milliamperes and the block and treatment totals. The analysis for a balanced incomplete block design proceeds as follows:

1. Calculate the total sum of squares as usual:

$$SS_{\text{total}} = \sum_i \sum_j Y_{ij}^2 - \frac{T_{..}^2}{N}$$

$$= 3478 - \frac{(160)^2}{12} = 1344.67$$

2. Calculate the block sum of squares, ignoring treatments:

$$SS_{\text{block}} = \sum_{i=1}^{b} \frac{T_{i.}^2}{k} - \frac{T_{..}^2}{N}$$

$$= \frac{(29)^2 + (49)^2 + (48)^2 + (34)^2}{3} - \frac{(160)^2}{12} = 100.67$$

3. Calculate treatment effects, adjusting for blocks:

$$SS_{\text{treatment}} = \frac{\sum_{j=1}^{t} Q_j^2}{k\lambda t} \tag{4.8}$$

where

$$Q_j = kT_{.j} - \sum_i n_{ij} T_{i.} \tag{4.9}$$

where $n_{ij} = 1$ if treatment j appears in block i, and $n_{ij} = 0$ if treatment j does *not* appear in block i. Note that $\sum_i n_{ij} T_{i.}$ is merely the sum of all block totals which contain treatment j.

For the data given,

$$Q_1 = 3(6) - (29 + 48 + 34) = 18 - 111 = -93$$
$$Q_2 = 3(68) - 131 = 73$$
$$Q_3 = 3(65) - 126 = 69$$
$$Q_4 = 3(21) - 112 = -49$$
$$\overline{0}$$

Note that

$$\sum_{j=1}^{t} Q_j = 0$$

which is always true. Then

$$SS_{\text{treatment}} = \frac{(-93)^2 + (73)^2 + (69)^2 + (-49)^2}{3(2)4} = 880.83$$

4. Calculate the error sum of squares by subtraction

$$\begin{aligned}SS_{\text{error}} &= SS_{\text{total}} - SS_{\text{block}} - SS_{\text{treatment}}\\ &= 1344.67 - 100.67 - 880.83 = 363.17\end{aligned}$$

Table 4.12 summarizes in an ANOVA table.

Table 4.12 ANOVA for Incomplete Block Design Example

Source	df	SS	MS
Blocks (days)	3	100.67	—
Treatments (adjusted)	3	880.83	293.61
Error	5	363.17	72.63
Totals	11	1344.67	

An F test gives $F_{3,5} = 293.61/72.63 = 4.04$, which is not significant at the 5-percent level (Appendix Table D).

In Table 4.12 the error degrees of freedom are determined by subtraction rather than as the product of the block and treatment degrees of freedom. However, this error degrees of freedom is seen to be the product of treatment and block degrees of freedom (9) if 4 is subtracted for the four missing values in the design.

In some incomplete block designs it may be desirable to test for a block effect. The mean square for blocks was not computed in Table 4.12 since it had not been adjusted for treatments. In the case of a *symmetrical balanced incomplete randomized block design*, where $b = t$, the block sum of squares may be adjusted in the same manner as the treatment sums of squares

$$Q'_i = rT_{i.} - \sum_j n_{ij}T_{.j}$$

$$Q'_1 = 3(29) - 92 = -5$$

$$Q'_2 = 3(49) - 154 = -7$$

$$Q'_3 = 3(48) - 139 = +5$$

$$Q'_4 = 3(34) - 95 = +7$$

$$\sum_i Q'_i = 0$$

$$SS_{\text{block}} = \sum_{i=1}^{b} (Q'_i)^2/r\lambda b = \frac{(-5)^2 + (-7)^2 + (5)^2 + (7)^2}{3(2)4} = 6.17$$

and

$$SS_{\text{treatment(unadjusted)}} = \frac{6^2 + 68^2 + 65^2 + 21^2}{3} - \frac{(160)^2}{12} = 975.34$$

The results of this adjustment and the one in treatments may now be summarized for this symmetrical case as shown in Table 4.13.

Table 4.13 ANOVA for Incomplete Block Design Example for Both Treatments and Blocks

Source	df	SS	MS
Blocks (adjusted)	3	6.17	2.06
Blocks	(3)	(100.67)	—
Treatments (adjusted)	3	880.83	293.61
Treatments	(3)	(975.34)	—
Error	5	363.17	72.63
Totals	11	1344.67	

The terms in parentheses are inserted only to show how the error term was computed for one adjusted effect and one unadjusted effect

$$\begin{aligned} SS_{\text{error}} &= SS_{\text{total}} - SS_{\text{treatment (adjusted)}} - SS_{\text{block}} \\ &= 1344.67 - 880.83 - 100.67 = 363.17 \end{aligned}$$

or

$$\begin{aligned} SS_{\text{error}} &= SS_{\text{total}} - SS_{\text{treatment}} - SS_{\text{block (adjusted)}} \\ &= 1344.67 - 975.34 - 6.17 = 363.17 \end{aligned}$$

It should be noted also that the final sum of squares values used in Table 4.13 to get the mean square values do not add up to the total sum of squares. This is characteristic of a nonorthogonal design. The F test for blocks was not run because its value is obviously extremely small, which indicates no day-to-day effect on current flow.

For nonsymmetrical or unbalanced designs, the general regression method may be a useful alternative. If contrasts are to be computed for an incomplete block design, it can be shown that the sum of squares for a contrast is given by

$$SS_{C_m} = \frac{(C_m)^2}{(\sum_{j=1}^{t} c_{jm}^2)k\lambda t} \tag{4.10}$$

where contrasts C_m are made on the Q_j's rather than $T_{.j}$'s.

As an example consider the following orthogonal contrasts on the data of Table 4.11:

$$\begin{aligned}
C_1 &= Q_1 - Q_2 &&= -166 \\
C_2 &= Q_1 + Q_2 - 2Q_3 &&= -158 \\
C_3 &= Q_1 + Q_2 + Q_3 - 3Q_4 &&= 196
\end{aligned}$$

The corresponding sums of squares are

$$\begin{aligned}
SS_{C_1} &= \frac{(-166)^2}{(2)(3)(2)(4)} = 574.08 \\
SS_{C_2} &= \frac{(-158)^2}{(6)(3)(2)(4)} = 173.36 \\
SS_{C_3} &= \frac{(196)^2}{(12)(24)} = 133.39 \\
& \qquad\qquad\qquad\quad \overline{880.83}
\end{aligned}$$

which checks with the adjusted sums of squares for treatments. Comparing each of the above sums of squares with their 1 df against the error mean square in Table 4.12 indicates that contrast C_1 is the only one of these three that is significant at the 5-percent level of significance. This indicates a real difference in current flow between treatments A and B, even though treatments in general showed no significant difference in current flow.

General Regression Approach to Randomized Incomplete Block Design

Applying the general regression significance test to the data in Table 4.11, the normal equations are

$$\begin{aligned}
160 &= 12m + 3t_1 + 3t_2 + 3t_3 + 3t_4 + 3b_1 + 3b_2 + 3b_3 + 3b_4 \\
29 &= 3m + t_1 + t_3 + t_4 + 3b_1 \\
49 &= 3m + t_2 + t_3 + t_4 + 3b_2 \\
48 &= 3m + t_1 + t_2 + t_3 + 3b_3 \\
34 &= 3m + t_1 + t_2 + t_4 + 3b_4 \qquad (4.11) \\
6 &= 3m + 3t_1 + b_1 + b_3 + b_4 \\
68 &= 3m + 3t_2 + b_2 + b_3 + b_4 \\
65 &= 3m + 3t_3 + b_1 + b_2 + b_3 \\
21 &= 3m + 3t_4 + b_1 + b_2 + b_4
\end{aligned}$$

Since

$$\sum_i b_i = \sum_j t_j = 0$$

from the first equation in Equation (4.11),

$$m = 160/12$$

The method for solving for the t's is as follows. Multiply the sixth equation in Equation (4.11) by 3 and add it to the third equation:

$$
\begin{aligned}
\text{6th} \times 3\text{: } 18 &= 9m + 9t_1 \qquad\qquad\qquad + 3b_1 \qquad + 3b_3 + 3b_4 \\
\text{3rd: } 49 &= 3m \qquad + t_2 + t_3 + t_4 \qquad + 3b_2 \\
\hline
\text{adding: } 67 &= 12m + 9t_1 + t_2 + t_3 + t_4 + 3\sum_i^4 b_i
\end{aligned}
$$

or

$$67 = 12m + 8t_1 + \sum_{j=1}^{4} t_j + 3\sum_{i=1}^{4} b_i$$

or

$$t_1 = \frac{67 - 12m}{5}$$

Since $\sum t_j = \sum b_i = 0$,

$$t_1 = \frac{67 - 12(160/12)}{8} = \frac{67 - 160}{8} = \frac{-93}{8}$$

In the same manner, using the seventh and second equations gives

$$t_2 = \frac{233 - 160}{8} = \frac{73}{8}$$

Using the eighth and fifth equations gives

$$t_3 = \frac{229 - 160}{8} = \frac{69}{8}$$

Using the ninth and fourth equations gives

$$t_4 = \frac{111 - 160}{8} = \frac{-49}{8}$$

and

$$\sum_{j=1}^{4} t_j = 0$$

Note that these t_j's are proportional to the Q_j's, as $t_j = Q_j/8$.

By a similar elimination of t's, the b's can be obtained. The second and seventh equations give

$$
\begin{aligned}
\text{2nd} \times 3\text{: } 87 &= 9m + 3t_1 \qquad + 3t_3 + 3t_4 + 9b_1 \\
\text{7th: } 68 &= 3m \qquad + 3t_2 \qquad\qquad + b_2 + b_3 + b_4 \\
\hline
\text{adding: } 155 &= 12m + 3\sum_j t_j + 8b_1 + \sum_i b_i
\end{aligned}
$$

$$b_1 = \frac{155 - 12m}{8} = \frac{155 - 160}{8} = -\frac{5}{8}$$

Likewise,

$$b_2 = \frac{153 - 160}{8} = -\frac{7}{8}$$

$$b_3 = \frac{165 - 160}{8} = \frac{5}{8}$$

$$b_4 = \frac{167 - 160}{8} = \frac{7}{8}$$

and

$$\sum_{i=1}^{4} b_i = 0$$

Again these b_i's are proportional to the Q_i''s.

$$\begin{aligned}
\text{SS}_{\text{regression}}(m, b_i, t_j) &= mT_{..} + \sum_i b_i T_{i.} + \sum_j t_j T_{.j} \\
&= (160/12)(160) + (-5/8)(29) + (-7/8)(49) \\
&\quad + (5/8)(48) + (7/8)(34) + (-93/8)(6) + (73/8)(68) \\
&\quad + (69/8)(65) + (-49/8)(21) = 3114.83
\end{aligned}$$

$$\begin{aligned}
\text{SS}_{\text{error}} &= \sum_i \sum_j Y_{ij}^2 - \text{SS}_{\text{regression}}(m, b, t) \\
&= 3478 - 3114.83 = 363.17 \text{ (as before)}
\end{aligned}$$

To test $H_0: \tau_j = 0$ for all j, the reduced model is $Y_{ij} = m' + b_i' + e_{ij}'$, and the normal equations are

$$\begin{aligned}
160 &= 12m' + 3b_1' + 3b_2' + 3b_3' + 3b_4' \\
29 &= 3m' + 3b_1' \\
49 &= 3m' + 3b_2' \\
48 &= 3m' + 3b_3' \\
34 &= 3m' + 3b_4'
\end{aligned}$$

Solving $m' = 160/12$ (as before) but

$$b_1' = \frac{29 - 3m}{3} = -\frac{11}{3}$$

$$b_2' = \frac{49 - 3m}{3} = \frac{9}{3}$$

$$b_3' = \frac{48 - 3m}{3} = \frac{8}{3}$$

$$b_4' = \frac{34 - 3m}{3} = -\frac{6}{3}$$

$$\sum b_i' = 0$$

Note that the b_i''s are different from the b_i's so that

$$\begin{aligned} SS_{\text{regression}}(m, b_i') &= (160/12)(160) + (29)(-11/3) + (49)(9/3) \\ &\quad + 48(8/3) + 34(-6/3) = 2234.00 \\ SS_{\text{treatment (adjusted)}} &= SS_{\text{regression}}(m, b, t) - SS_{\text{regression}}(m, b') \\ &= 3114.83 - 2234.00 = 880.83 \text{ (as before)} \end{aligned}$$

A similar procedure would give $SS_{\text{block (adjusted)}}$ if desired.

When complete balance is not possible in a design, one may often achieve partial balance. In such partially balanced incomplete block designs (PBIBs), the λ value is not constant since some pairs of treatments may occur a specific number of times in the experiment and other pairs occur a different number of times giving λ_1 and λ_2. These designs are discussed in Kempthorne [11, ch. 27].

4.8 SUMMARY

Experiment	*Design*	*Analysis*
I. Single factor		
	1. Completely randomized	1. One-way ANOVA
	$Y_{ij} = \mu + \tau_j + \varepsilon_{ij}$	
	2. Randomized block	2.
	$Y_{ij} = \mu + \tau_j + \beta_i + \varepsilon_{ij}$	
	a. Complete	*a*. Two-way ANOVA
	b. Incomplete, balanced	*b*. Special formulas
	c. Incomplete, general	*c*. Regression method

PROBLEMS

4.1 The effects of four types of graphite coaters on light box readings are to be studied. As these readings might differ from day to day, observations are to be taken on each of the four types every day for three days. The order of testing of the four types on any given day can be randomized. The results are

	Graphite Coater Type			
Day	*M*	*A*	*K*	*L*
1	4.0	4.8	5.0	4.6
2	4.8	5.0	5.2	4.6
3	4.0	4.8	5.6	5.0

Analyze these data as a randomized block design and state your conclusions.

4.2 Set up orthogonal contrasts among coater types and analyze for Problem 4.1.

4.3 Use the Newman–Keuls range test to compare four coater type means for Problem 4.1.

4.4 Solve Problem 4.1 by the general regression method.

4.5 If the reading on type K for the second day was missing, what missing value should be inserted and what is the analysis now?

4.6 Data on screen color difference on a television tube measured in degrees Kelvin are to be compared for four operators. On a given day only three operators can be used in the experiment. A balanced incomplete block design gave results as follows:

	Operator			
Day	A	B	C	D
Monday	780	820	800	—
Tuesday	950	—	920	940
Wednesday	—	880	880	820
Thursday	840	780	—	820

Do a complete analysis of these data and discuss your findings with regard to differences between operators.

4.7 Run orthogonal contrasts on the operators in Problem 4.6.

4.8 Determine the missing values in the table for Problem 4.6.

4.9 Use these missing values to obtain the error sum of squares for Problem 4.6.

4.10 Complete an analysis of variance of Problem 4.6 using the error sum of squares as determined in Problem 4.9. Compare your results with those in Problem 4.6.

4.11 Set up at least four contrasts in Problem 4.1 and use the Scheffé method to test the significance of these.

4.12 Perform an exact test for treatments (graphite coater types) in Problem 4.5 and compare with the approximate method used by inserting a missing value.

4.13 At the Bureau of Standards an experiment was to be run on the wear resistance of a new experimental shoe leather as compared to the standard leather in use by the Army. It was decided to equip several men with one shoe of the experimental type leather and the other shoe with standard leather and after many weeks in the field, compare wear on the two types. Considering each man as a block, suggest a reasonable number of men to be used in this experiment and outline its possible analysis.

4.14 If three experimental types of leather were to be tested along with the standard in Problem 4.13, it is obvious that only two of the four types can be tested on one man. Set up a balanced incomplete block design for this situation. Insert some arbitrary numerical values and complete an ANOVA table for your data.

4.15 Given the following data:

Block	Treatment			
	A	*B*	*C*	*D*
1	238	213	—	279
2	218	—	226	207
3	—	208	232	210

Assume randomization of treatments onto the blocks and comment on the following suggested methods of analysis.

1. "Get your Q values on treatments and the corrected SS for treatments and proceed with the ANOVA table."
2. "Fill in the missing values by the method of least squares and then do a regular (complete block) analysis with these values inserted."
3. "Set up a general regression significance test and determine the necessary sums of squares."

4.16 A study was to be made of a new process for refining oil. The process was to consist of treatment X and treatment Y. In addition to trying X alone and Y alone, they wanted to try X followed by Y, Y followed by X, X followed by Y one day later, Y followed by X one day later, and the present process as a control. This meant seven treatments. Their equipment only allowed for three treatments with each charge stock. Set up a balanced design for this experiment using as few charge stocks as possible.

4.17 Considering the seven treatments described in Problem 4.16, set up contrasts which should be of interest to the experimenter. Can a meaningful orthogonal set be proposed?

5
Single-Factor Experiments—Latin and Other Squares

5.1 INTRODUCTION

The reader may have wondered about a possible position effect in the problem on testing tire brands in Chapter 4. Experience shows that rear tires get different wear than front tires and even different sides of the same car may show different amounts of tread wear. In the randomized block design of Chapter 4, the four brands were randomized onto the four wheels of each car with no regard for position. The effect of position on wear could be balanced out by rotating the tires every 5000 miles giving each brand 5000 miles on each wheel. However, if this is not feasible, the positions can impose another restriction on the randomization in such a way that each brand is not only used once on each car but also only once in each of the four possible positions: left front, left rear, right front, and right rear.

5.2 LATIN SQUARES

A design in which each treatment appears once and only once in each row (position) and once and only once in each column (cars) is called a *Latin square design*. Interest is still centered on one factor, treatments, but two restrictions are placed on the randomization. An example of one such 4×4 Latin square is shown in Table 5.1.

Table 5.1 4×4 Latin Square Design

	Car			
Position	I	II	III	IV
1	*C*	*D*	*A*	*B*
2	*B*	*C*	*D*	*A*
3	*A*	*B*	*C*	*D*
4	*D*	*A*	*B*	*C*

Such a design is only possible when the number of levels of both restrictions equals the number of treatment levels. In other words, it must be a square. It is not true that all randomization is lost in this design, as the particular Latin square to be used on a given problem may be chosen at random from several possible Latin squares of the required size. Tables of such squares are found in Fisher and Yates [9].

The analysis of the data in a Latin square design is a simple extension of previous analyses where the data are now added in a third direction—positions. If the data of Table 4.5 in Chapter 4 are imposed on the Latin square of Table 5.1, the result could be that shown in Table 5.2, letters still representing tire brand.

Table 5.2 Latin Square Design Data on Tire Wear

	Car				
Position	I	II	III	IV	$T_{..k}$
1	$C = -1$	$D = -2$	$A = 0$	$B = -5$	-8
2	$B = 1$	$C = -1$	$D = -2$	$A = 0$	-2
3	$A = 4$	$B = 1$	$C = -3$	$D = -4$	-2
4	$D = 0$	$A = 1$	$B = 0$	$C = -4$	-3
$T_{i..}$	4	-1	-5	-13	$T_{...} = -15$

Treatments totals of A, B, C, D are

$$T_{.j.}: 5, \quad -3, \quad -9, \quad -8$$

where the model is now

$$Y_{ijk} = \mu + \beta_i + \tau_j + \gamma_k + \varepsilon_{ijk}$$

and γ_k represents the positions effect. Since the only new totals are for positions, a position sum of squares can be computed as

$$SS_{\text{position}} = \frac{(-8)^2 + (-2)^2 + (-2)^2 + (-3)^2}{4} - \frac{(-15)^2}{16} = 6.2$$

and

$$\begin{aligned} SS_{\text{error}} &= SS_{\text{total}} - SS_{\text{treatment}} - SS_{\text{car}} - SS_{\text{position}} \\ &= 80.9 - 30.6 - 38.6 - 6.2 = 5.5 \end{aligned}$$

Table 5.3 is the ANOVA table for these data.

Once again another restriction placed on the randomization has further reduced the experimental error, although the position effect is not significant

Table 5.3 Latin Square ANOVA

Source	df	SS	MS	EMS
Brands τ_j	3	30.6	10.2	$\sigma_\varepsilon^2 + 4\phi_\tau$
Cars β_i	3	38.6	12.9	$\sigma_\varepsilon^2 + 4\sigma_\beta^2$
Positions γ_k	3	6.2	2.1	$\sigma_\varepsilon^2 + 4\phi_\gamma$
Error ε_{ijk}	6	5.5	0.9	σ_ε^2
Totals	15	80.9		

at the 5-percent level. This further reduction of error variance is attained at the expense of degrees of freedom, since now the estimate of σ_ε^2 is based on only 6 df instead of 9 df as in the randomized block design. This means less precision in estimating this error variance. But the added restrictions should be made if the environmental conditions suggest them. After discovering that position had no significant effect, some investigators might "pool" the position sum of squares with the error sum of squares and obtain a more precise estimate of σ_ε^2, namely, 1.3, as given in Table 4.6 of Chapter 4. However, there is a danger in "pooling," as it means "accepting" a hypothesis of no position effect, and the investigator has no idea about the possible error involved in "accepting" a hypothesis. Naturally if the degrees of freedom on the error term are reduced much below that of Table 5.3, there will have to be some pooling to get a reasonable yardstick for assessing other effects.

5.3 GRAECO-LATIN SQUARES

In some experiments still another restriction may be imposed on the randomization and the design may be a Graeco-Latin square such as Table 5.4 exhibits.

Table 5.4 Graeco-Latin Square Design

	Car			
Position	I	II	III	IV
1	$A\alpha$	$B\beta$	$C\gamma$	$D\delta$
2	$B\gamma$	$A\delta$	$D\alpha$	$C\beta$
3	$C\delta$	$D\gamma$	$A\beta$	$B\alpha$
4	$D\beta$	$C\alpha$	$B\delta$	$A\gamma$

In this design the third restriction is at levels α, β, γ, δ, and not only do these each appear once and only once in each row and each column, but they appear once and only once with each level of treatment *A*, *B*, *C*, or *D*. The model for this would be

$$Y_{ijkm} = \mu + \beta_i + \tau_j + \gamma_k + \omega_m + \varepsilon_{ijkm}$$

where ω_m is the effect of the latest restriction with levels α, β, γ, and δ. An outline of the analysis appears in Table 5.5.

Table 5.5 Graeco Latin ANOVA Outline

Source	df
β_i	3
τ_j	3
γ_k	3
ω_m	3
ε_{ijkm}	3
Total	15

Such a design may not be very practical, as only 3 df are left for the error variance.

5.4 YOUDEN SQUARES

When the conditions for a Latin square are met except for the fact that only three treatments are possible (for example, because in one block only three positions are available) and where there are four blocks altogether, the design is an incomplete Latin square. This design is called a *Youden square*. One such Youden square is illustrated in Table 5.6.

Table 5.6 Youden Square Design

	Position		
Block	1	2	3
I	*A*	*B*	*C*
II	*D*	*A*	*B*
III	*B*	*C*	*D*
IV	*C*	*D*	*A*

Note that the addition of a column (*D*, *C*, *A*, *B*) would make this a Latin square if another position were available. A situation calling for a Youden square might occur if four materials are to be tested on four machines but there were only three heads on each machine whose orientation might affect the results.

The analysis of a Youden square proceeds like the incomplete block analysis. Assuming hypothetical values for some measured variable Y_{ijk} where

$$Y_{ijk} = \mu + \beta_i + \tau_j + \gamma_k + \varepsilon_{ijk}$$

with

$$i = 1, \cdots, 4 \qquad j = 1, 2, \cdots, 4 \qquad k = 1, 2, 3$$

we might have the data of Table 5.7.

Table 5.7 Youden Square Design Data

	Position			
Block	1	2	3	$T_{i..}$
I	$A = 2$	$B = 1$	$C = 0$	3
II	$D = -2$	$A = 2$	$B = 2$	2
III	$B = -1$	$C = -1$	$D = -3$	-5
IV	$C = 0$	$D = -4$	$A = 2$	-2
$T_{..k}$	-1	-2	$+1$	$-2 = T_{...}$

In this table

$$t = b = 4$$

$$r = k = 3$$

$$\lambda = 2$$

Treatment totals of A, B, C, D are

$$T_{.j.}: 6, \quad 2, \quad -1, \quad -9$$

From this,

$$\text{SS}_{\text{total}} = \sum_i \sum_j \sum_k Y_{ijk}^2 - \frac{T_{...}^2}{N} = 48 - \frac{(-2)^2}{12} = 47.67$$

Position effect may first be ignored, since every position occurs once and only once in each block and once with each treatment, so that positions are orthogonal to blocks and treatments:

$$\text{SS}_{\text{block (ignoring treatments)}} = \frac{(3)^2 + (2)^2 + (-5)^2 + (-2)^2}{3} - \frac{(-2)^2}{12}$$

$$= \frac{42}{3} - \frac{1}{3} = \frac{41}{3} = 13.67$$

For treatment sum of squares adjusted for blocks we get

$$Q_1 = 3(+6) - 3 = 15$$

$$Q_2 = 3(\ \ 2) - 0 = 6$$

$$Q_3 = 3(-1) - (-4) = 1$$

$$Q_4 = 3(-9) - (-5) = -22$$

and

$$\sum_i Q_i = 0$$

$$SS_{\text{treatment}} = \frac{(15)^2 + (6)^2 + (1)^2 + (-22)^2}{(4)(6)} = 31.08$$

$$SS_{\text{position}} = \frac{(-1)^2 + (-2)^2 + (1)^2}{4} - \frac{(-2)^2}{12} = 1.17$$

$$\begin{aligned} SS_{\text{error}} &= SS_{\text{total}} - SS_{\text{block}} - SS_{\text{treatment (adjusted)}} - SS_{\text{position}} \\ &= 47.67 - 13.67 - 31.08 - 1.17 \\ &= 1.75 \end{aligned}$$

The analysis appears in Table 5.8.

Table 5.8 Youden Square ANOVA

Source	df	SS	MS
Treatments τ_j (adjusted)	3	31.08	10.36
Blocks β_i	3	13.67	—
Positions γ_k	2	1.17	0.58
Error ε_{ijk}	3	1.75	0.58
Totals	11	47.67	

The position effect here is not significant and it might be desirable to pool it with the error, getting

$$s_\varepsilon^2 = \frac{1.17 + 1.75}{2 + 3} = \frac{2.92}{5} = 0.58$$

as an estimate of error variance, with 5 df. Then the treatment effect is highly significant. The block mean square not given as blocks must be adjusted by treatments if block effects are to be assessed. The procedure is the same as shown in Chapter 4 on incomplete block designs.

5.5 SUMMARY

Experiment	*Design*	*Analysis*
I. Single factor	1. Completely randomized	1. One-way ANOVA
	$Y_{ij} = \mu + \tau_j + \varepsilon_{ij}$	
	2. Randomized block	2.
	$Y_{ij} = \mu + \tau_j + \beta_i + \varepsilon_{ij}$	
	a. Complete	*a*. Two-way ANOVA
	b. Incomplete, balanced	*b*. Special ANOVA
	c. Incomplete, general	*c*. Regression method
	3. Latin square	3.
	a. $Y_{ijk} = \mu + \beta_i + \tau_j + \gamma_k + \varepsilon_{ijk}$	
	a. Complete	*a*. Three-way ANOVA
	b. Incomplete, Youden square	*b*. Special ANOVA (like 2*b*)
	4. Graeco-Latin square	4. Four-way ANOVA
	$Y_{ijkm} = \mu + \beta_i + \tau_j + \gamma_k + \omega_m + \varepsilon_{ijkm}$	

It might be emphasized once again that, so far, interest has been centered on a single factor (treatments) for the discussion in Chapters 3, 4, and 5. The special designs simply represent restrictions on the randomization. In the next chapter, two or more factors will be considered.

PROBLEMS

5.1 In a research study at Purdue University on metal-removal rate, five electrode shapes *A*, *B*, *C*, *D*, and *E* were studied. The removal was accomplished by an electric discharge between the electrode and the material being cut. For this experiment five holes were cut in five workpieces, and the order of electrodes was arranged so that only one electrode shape was used in the same position on each of the five workpieces. Thus, the design was a Latin square design, with workpieces (strips) and positions on the strip as restrictions on the randomization. Several variables were studied, one of which was the Rockwell hardness of metal where each hole was to be cut. The results were as follows:

	Position				
Strip	1	2	3	4	5
I	*A*(64)	*B*(61)	*C*(62)	*D*(62)	*E*(62)
II	*B*(62)	*C*(62)	*D*(63)	*E*(62)	*A*(63)
III	*C*(61)	*D*(62)	*E*(63)	*A*(63)	*B*(63)
IV	*D*(63)	*E*(64)	*A*(63)	*B*(63)	*C*(63)
V	*E*(62)	*A*(61)	*B*(63)	*C*(63)	*D*(62)

Analyze these data and test for an electrode effect, position effect, and strip effect on Rockwell hardness.

5.2 The times in hours necessary to cut the holes in Problem 5.1 were recorded as follows:

	Position				
Strip	1	2	3	4	5
I	*A*(3.5)	*B*(2.1)	*C*(2.5)	*D*(3.5)	*E*(2.4)
II	*E*(2.6)	*A*(3.3)	*B*(2.1)	*C*(2.5)	*D*(2.7)
III	*D*(2.9)	*E*(2.6)	*A*(3.5)	*B*(2.7)	*C*(2.9)
IV	*C*(2.5)	*D*(2.9)	*E*(3.0)	*A*(3.3)	*B*(2.3)
V	*B*(2.1)	*C*(2.3)	*D*(3.7)	*E*(3.2)	*A*(3.5)

Analyze these data for the effect of electrodes, strips, and positions on time.

5.3 Analyze the electrode effect further in Problem 5.2 and make some statement as to which electrodes are best if the shortest cutting time is the most desirable factor.

5.4 A composite measure of screen quality was made on screens using four lacquer concentrations, four standing times, four acryloid concentrations (A, B, C, D), and four acetone concentrations (α, β, Γ, Δ). A Graeco-Latin square design was used with data recorded as follows:

	Lacquer Concentration			
Standing Time	$\frac{1}{2}$	1	$1\frac{1}{2}$	2
30	$C\beta(16)$	$B\Gamma(12)$	$D\Delta(17)$	$A\alpha(11)$
20	$B\alpha(15)$	$C\Delta(14)$	$A\Gamma(15)$	$D\beta(14)$
10	$A\Delta(12)$	$D\alpha(6)$	$B\beta(14)$	$C\Gamma(13)$
5	$D\Gamma(9)$	$A\beta(9)$	$C\alpha(8)$	$B\Delta(9)$

Do a complete analysis of these data.

5.5 If in Problem 5.2 it were found that there is no room on the strips for a cut in the fifth position, consider the data for four positions only and analyze as a Youden square.

5.6 Explain why a three-level Graeco-Latin square is not a feasible design.

5.7 In a chemical plant, five experimental treatments are to be used on a basic raw material in an effort to increase the chemical yield. Since batches of raw material may differ, five batches are chosen at random. Also the order of the experiments may effect yield as well as the operator who performs the experiment. Considering order and operators as further restrictions on randomization, set up a suitable design for this experiment which will be least expensive to run. Outline its analysis.

5.8 If another restriction is placed on the data of Problem 5.7, again at five levels (say the tests are run in five different kettles), design an augmented experiment. This might be called hyper-Graeco-Latin square. Outline its analysis and comment.

5.9 In Problem 4.16, consider that the experiments are to be run on three days with each charge stock so that there are three treatments on each charge stock and three treatments per day as well. Set up the layout for this design and outline its analysis.

5.10 If a Youden square is completed to form a Latin square, the additional rows (or columns) form what is called the complementary Youden square. Determine the complementary Youden squares for Problems 5.5 and 5.9.

5.11 For an $m \times m$ Latin square, prove that a missing value may be estimated by

$$Y_{ijk} = \frac{m(T'_{i..} + T'_{.j.} + T'_{..k}) - 2T'_{...}}{(m - 1)(m - 2)}$$

where the primes indicate totals for row, treatment, and column containing the missing value and $T'_{...}$ is the grand total of all actual observations.

6

Factorial Experiments

6.1 INTRODUCTION

In the preceding three chapters, all of the experiments involved only one factor and its effect on a measured variable. Several different designs were considered, but all of these represented restrictions on the randomization where interest was still centered on the effect of a single factor.

Suppose there are now two factors of interest to the experimenter, for example, the effect of both temperature and altitude on the current flow in a small computer. One traditional method is to hold altitude constant and vary the temperature and then hold temperature constant and change the altitude, or, in general, hold all factors constant except one and take current flow readings for several levels of this one factor, then choose another factor to vary, holding all others constant, and so forth. In order to examine this type of experimentation, consider a very simple example in which temperature is to be set at 25°C and 55°C only and altitudes of 0-K (K = 10,000 ft) and 3-K (30,000 ft) are to be used. If one factor is to be varied at a time, the altitude may be set for sea level or 0-K and the temperature varied from 25°C to 55°C. Suppose the current flow changed from 210 mA to 240 mA with this temperature increase. Now there is no way to assess whether or not this 30-mA increase is real or due to chance. Unless there is available some previous estimate of the error variability, the experiment must be repeated in order to obtain an estimate of the error or chance variability within the experiment. If the experiment is now repeated and the readings are 205 mA and 230 mA as the temperature is varied from 25°C to 55°C, it seems obvious, without any formal statistical analysis, that there is a real increase in current flow, since for each repetition the increase is large compared to the variation in current flow within a given temperature. Graphically these results appear as in Figure 6.1.

Four experiments have now been run to determine the effect of temperature at 0-K altitude only. To check the effect of altitude, the temperature can

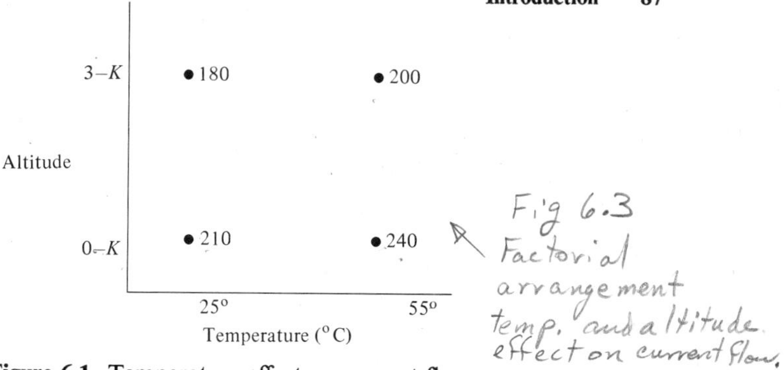

Figure 6.1 Temperature effect on current flow.

be held at 25°C and the altitude varied to 3-K by adjustment of pressure in the laboratory. Using the results already obtained at 0-K (assuming they are representative), two observations of current flow are now taken at 3-K with the results shown in Figure 6.2.

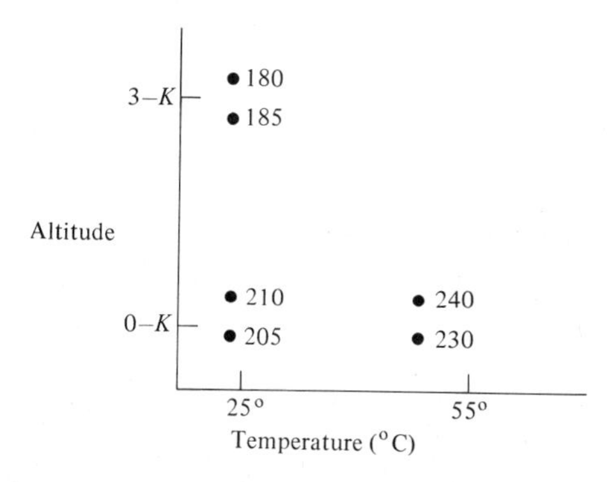

Figure 6.2 Temperature and altitude effect on current flow.

From these experiments the temperature increase is seen to increase the current flow an average of

$$\frac{30 \text{ mA} + 25 \text{ mA}}{2} = 27.5 \text{ mA}$$

and the increase in altitude decreases the current flow on the average of

$$\frac{30 \text{ mA} + 20 \text{ mA}}{2} = 25 \text{ mA}$$

This information is gained after six experiments have been performed and no information is available on what would happen at a temperature of 55°C and an altitude of 3-K.

An alternative experimental arrangement would be a factorial arrangement where each temperature level is combined with each altitude and only four experiments are run. Results of four such experiments might appear as in Figure 6.3.

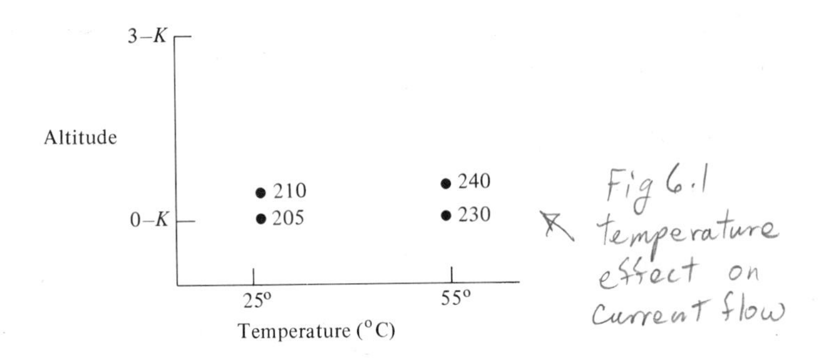

Figure 6.3 Factorial arrangement of temperature and altitude effect on current flow.

With this experiment, one estimate of temperature effect on current flow is 240 − 210 mA = 30 mA at 0-K, and another estimate is 200 − 180 mA = 20 mA at 3-K. Hence, two estimates can be made of temperature effect [average (30 + 20)/2 = 25 mA] using all four observations without necessarily repeating any observation at the same point. Using the same four observations, two estimates of altitude effect can be determined: 180 − 210 = −30 mA at 25°C, and 200 − 240 = −40 mA at 55°C, an average decrease of 35 mA for a 3-K increase in altitude. Here, with just four observations instead of six, valid comparisons have been made on both temperature and altitude, and in addition some information has been obtained as to what happens at 55°C and altitude 3-K.

From this simple example, some of the advantages of a factorial experiment can be seen:

1. More efficiency than one-factor-at-a-time experiments (here four-sixths or two-thirds the amount of experimentation).
2. All data are used in computing both effects. (Note that all four observations are used in determining the average effect of temperature and the average effect of altitude.)

3. Some information is gleaned on possible interaction between the two factors. (In the example, the increase in current flow of 20 mA at 3-K was about the same order of magnitude as the 30 mA increase at 0-K. If these increases had differed considerably, interaction might be said to be present.)

These advantages are even more pronounced as the number of levels of the two factors is increased. A *factorial experiment* is one in which all levels of a given factor are combined with all levels of every other factor in the experiment. Thus, if four temperatures are considered at three altitudes, a 4×3 factorial experiment would be run requiring 12 different experimental conditions. In the example above, if the current flow were 160 mA for 55°C and 3-K, the results could be shown as in Figure 6.4.

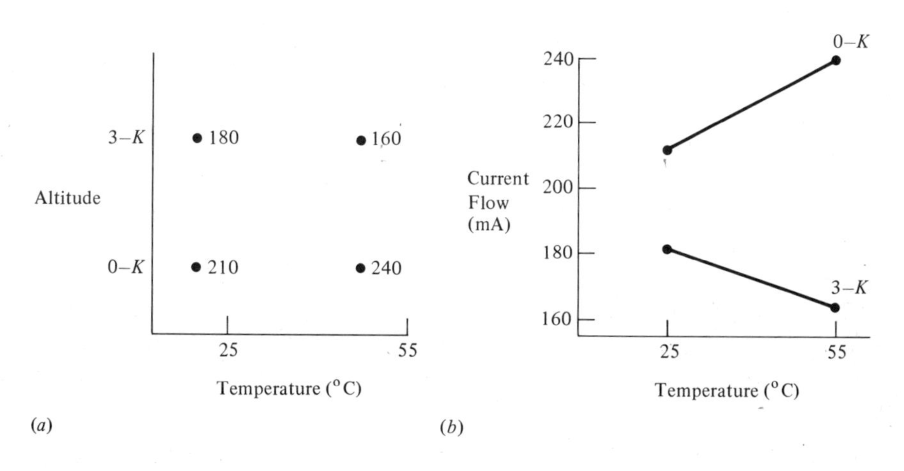

Figure 6.4 Interaction in a factorial experiment.

From Figure 6.4 (*b*) note that as the temperature is increased from 25 to 55°C at 0-K, the current flow increases by 30 mA, but at 3-K for the same temperature increase the current flow *decreases* by 20 mA. When a change in one factor produces a different change in the response variable at one level of another factor than at other levels of this factor, there is an *interaction* between the two factors. This is also observable in Figure 6.4 (*b*) as the two altitude lines are not parallel. If the data of Figure 6.3 are plotted, we get the relations pictured in Figure 6.5.

It is seen that the lines are much more nearly parallel and no interaction is said to be present. An increase in temperature at 0-K produces about the same increase in current flow (30 mA) as at 3-K (20 mA). Now a word of

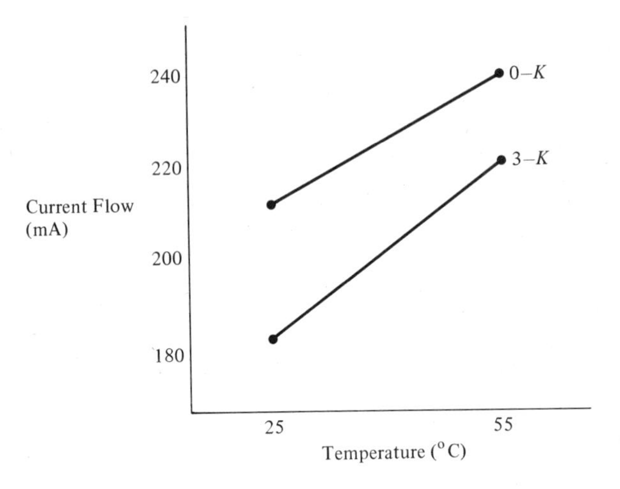

Figure 6.5 No-interaction temperature diagram in temperature–altitude study.

warning is necessary. Lines can be made to look nearly parallel or quite diverse depending on the scale chosen; therefore, it is necessary to run statistical tests in order to determine whether or not the interaction is statistically significant. It is only possible to test the significance of such interaction if more than one observation is taken for each experimental condition. The graphic procedure shown above is given merely to give some insight into what interaction is and how it might be displayed for explaining the factorial experiment. The following example will illustrate the computation of interaction.

Example 6.1 An experiment is to be conducted to determine the effects of three different types of phosphor and two types of face-plate glass on the light output in a TV tube. Light output is the measured variable; it is measured in microamperes (μA). The value of cathode current required to give 30 foot-lamberts of light output is what is actually to be recorded. As only three phosphor types are available and two types of face-plate glass, the experiment would be a 3×2 factorial experiment. Both factors are at fixed levels, since these are the only types available, and both factors are also qualitative.

For the design of the experiment, it was agreed to take three observations under each of the six (3×2) experimental conditions, as previous data showed that results on cathode current were quite repeatable with respect to detectable differences between phosphor types and glass types. The experimenter agreed that it would be no problem to completely randomize the order for running the 18 experiments—three at each of six experimental conditions. In fact the six "cells" could be numbered, 1, 2, $\cdots$, 6, and dice rolled to determine which experimental conditions would be run first, which second and so on, until all 18 were completed. To see this procedure in more detail, consider the sample layout in Table 6.1.

Table 6.1 Data Layout for Phosphor, Glass-Type Experiment

	Phosphor Type					
Glass Type	*A*		*B*		*C*	
1	1	9	3	14	5	1
		12		15		6
		13		16		7
2	2	4	4	3	6	2
		8		5		10
		18		17		11

Dice are rolled and the results are as follows: 5, 6, 4, 2, 4, 5, 5, 2, 1, 6, 6, 1, 1, 3, X, X, 3, X, 3, X, 4, where a repeated value (X) is ignored after three such values have been obtained, and the last observation is not needed since it must be a 2 in order to have just three observations per cell. In Table 6.1 the order of experimentation dictated by the dice toss is given (condition 5 is first, condition 6 is second, and so on) by numbers, 1, 2, $\cdots$, 18. This illustrates only one possible method for complete randomization. A table of random numbers could be used, or simply, six numbers on chips could be drawn out at random (replacing each time) to decide which of the six conditions to run first, second, third, and so on. This may seem like so much busywork, but it is very important to employ some objective method of randomization because lack of randomization may seriously influence the results of the experiment. If, for example, there were a change in the house voltage during the last third of the test and all of the type C phosphor were run after this change occurred, it would be impossible to determine whether or not an increase in cathode current was due to type C phosphor or to this change in voltage. By randomization, Table 6.1 shows that for the last third of the experiment the numbers 13, 14, 15, 16, 17, and 18 are quite well spread throughout the experiment. The above scheme shows one way to completely randomize a 3×2 factorial with three observations per cell, the only restriction being that three observations will be made under the same experimental conditions. It is a great advantage in the analysis if the number of observations per cell can be kept equal.

The experiment is now a 3×2 factorial experiment with three observations per cell executed as a completely randomized design. The mathematical model can be evolved from the completely randomized model of Chapter 3 as

$$Y_{ij} = \mu + \tau_j + \varepsilon_{ij} \tag{6.1}$$

In this experiment there are $3 \times 2 = 6$ experimental conditions, often called *treatment combinations*. Hence, there are 5 df between these six treatment combinations and $2 \times 6 = 12$ df within treatments. A one-way ANOVA layout appears in Table 6.2.

Table 6.2 ANOVA Layout

Source	df	SS
Between treatments	5	
Within treatments	12	
Total	17	

However, since the two factors are arranged in a factorial manner, or are *crossed*, as some say, 2 df may be assigned to phosphor type (columns of Table 6.1) and 1 df may be assigned to glass type (rows of Table 6.1). This leaves $5 - 2 - 1 = 2$ df between treatment combinations, or between cells not accounted for. These 2 df are associated with the interaction between phosphor and glass types. The model is then

$$Y_{ijk} = \mu + P_i + G_j + PG_{ij} + \varepsilon_{k(ij)} \tag{6.2}$$

where

P_i = phosphor type $\quad i = 1, 2, 3$
G_j = glass type, $\quad j = 1, 2$
PG_{ij} = interaction between P and G
$\varepsilon_{k(ij)}$ = random error within cell $i, j \quad$ where $k = 1, 2, 3$

Parentheses are used to indicate that the three observations are within each of the six cells. Sometimes these observations are said to be "nested" within the cell. This notation will be most advantageous in later chapters. English letters are used here in the population model instead of Greek letters so that one may designate the factor by the letter that that factor represents (P for phosphor, G for glass). It should be realized that in the remainder of the book these letters are really population effects and hypotheses are to be tested on them. An ANOVA layout now appears as in Table 6.3.

From the models and the ANOVA layout it should be noted that the two main effects (P and G) and their interaction ($P \times G$) come from a breakdown of the treatment effect and do not come from the random error. In Chapters 4 and 5, the block and position effects were taken from the random error in an attempt to reduce this error because of design limitations. In the factorial

Table 6.3 Second ANOVA Layout

Source	df		SS
Between treatments	5		
Phosphor types		2	
Glass types		1	
$P \times G$ interaction		2	
Within treatments	12		
Total	17		

experiment the design is still completely randomized and that model fits the situation with no adjustment of the error term.

For the analysis of this experiment, data were collected according to the randomization scheme of Table 6.1. After coding each value by subtracting 260 μA and dividing each reading by 5, the coded results are as shown in Table 6.4.

Table 6.4 Phosphor–Glass-Type Data (Coded)

	Phosphor Type		
Glass Type	*A*	*B*	*C*
1	4	8	2
	6	10	5
	5	7	6
2	−6	0	−8
	−5	−4	−7
	−4	−5	−6

To simplify the analysis, the six treatment combinations could be arranged as a single classification ANOVA as in Table 6.5.

Table 6.5 Phosphor–Glass-Type Data in One-Way Arrangement

	Treatment						
	P_AG_1	P_AG_2	P_BG_1	P_BG_2	P_CG_1	P_CG_2	
	4	−6	8	0	2	−8	
	6	−5	10	−4	5	−7	
	5	−4	7	−5	6	−6	
$T_{.j}$	15	−15	25	−9	13	−21	$T_{..} = 8$
$\sum_{i=1}^{3} Y_{ij}^2$	77	77	213	41	65	149	$\sum_i \sum_j Y_{ij}^2 = 622$

In Table 6.5 the subscripts on the P and G indicate the levels of these factors in each treatment combination. Using the methods given in Section 3.3, a one-way ANOVA gives the values shown in Table 6.6.

Table 6.6 One-Way ANOVA on Phosphor–Glass-Type Data

Source	df	SS
Between treatments	5	585.11
Within treatments	12	33.33
Totals	17	618.44

As the 5 df between treatments can be broken down into phosphor type, glass type, and interaction degrees of freedom, so can the sums of squares. Returning to Table 6.4, the phosphor-type totals (columns) are given by

$$T_{i..}: 0, \quad 16, \quad -8 \qquad T_{...} = 8$$

and the sum of squares for this main effect is

$$SS_P = \frac{0^2 + (16)^2 + (-8)^2}{6} - \frac{(8)^2}{18}$$

$$= 53.33 - 3.56 = 49.77$$

The glass-type totals (rows) are given by

$$T_{.j.}: 53, \quad -45 \qquad T_{...} = 8$$

and the sum of squares for this main effect is

$$SS_G = \frac{(53)^2 + (-45)^2}{9} - \frac{(8)^2}{18}$$

$$= 537.11 - 3.56 = 533.55$$

using the methods of Section 4.2 of a two-way ANOVA. Totaling these two sums of squares, $49.77 + 533.55 = 583.32$, leaves $585.11 - 583.32 = 1.79$ for the $P \times G$ interaction sum of squares. Summarizing this in an ANOVA table, we get Table 6.7.

In interpreting the above results, three different hypotheses may be tested:

$$H_1: P_i = 0 \text{ for all } i \text{ (no phosphor-type effect)}$$

$$F_{2,12} = \frac{24.88}{2.78} = 8.95$$

Table 6.7 ANOVA for Phosphor–Glass-Type Problem: 3 × 2 with Three Observations per Cell

Source	df	SS	MS
Phosphor type	2	49.77	24.88
Glass type	1	533.55	533.55
$P \times G$ interaction	2	1.79	0.89
Within cells or error	12	33.33	2.78
Totals	17	618.44	

which is significant at the 1-percent level (see Appendix Table D).

$$H_2\colon G_j = 0 \text{ for all } j \text{ (no glass-type effect)}$$

$$F_{1,12} = \frac{533.55}{2.78} = 191.92$$

which is *highly* significant.

$$H_3\colon PG_{ij} = 0 \text{ for all } i \text{ and } j \text{ (no } P \times G \text{ interaction effect)}$$

$$F_{2,12} = \frac{0.89}{2.78}$$

which is less than 1 and hence not significant.

Since both factors are fixed, all tests are made by using the error mean square in the denominator of the F test. (Variations of this procedure are discussed in Chapter 10). The results of these three tests show that glass type has a very decided effect on cathode current and hence on light output, and that phosphor type also affects light output, but there is no significant interaction. This latter may be interpreted to mean that as light output changes for the three phosphor types, these changes are about the same for each glass type. Graphically (using coded cell totals from Table 6.4) we get Figure 6.6.

The large gap between the two glass-type curves illustrates the sizeable glass-type effect. The trend in each curve over the three phosphor types indicates a phosphor type effect, and the fact that the two curves are nearly parallel demonstrates the presence of little or no interaction.

Since only two glass types were used, it is obvious that the second glass type, G_2, is the better as the current flow is considerably less, which means that the light output of these tubes is higher than for glass type 1. To compare the three phosphor types, a Newman–Keuls test may be used on the coded phosphor type averages. These are for phosphor types A, B, and C

$$\overline{Y}_{i..}\colon 0, \quad 2.67, \quad -1.33$$

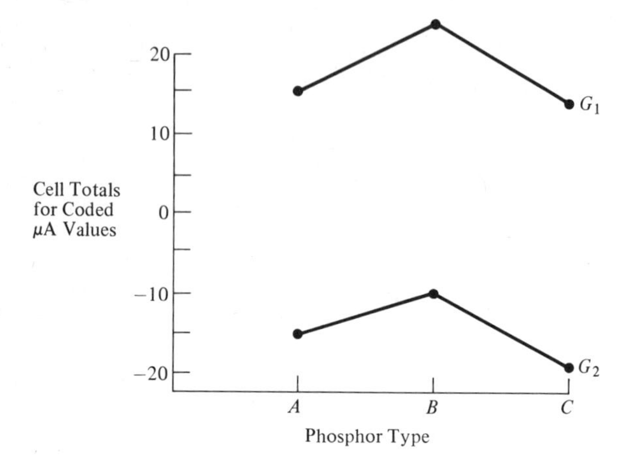

Figure 6.6 Phosphor–glass-type plot.

Using the method given in Section 3.3, we have for phosphor types C, A, and B,

$$\text{means:} \ -1.33, \quad 0, \quad 2.67$$

and

$$\text{error mean square} = 2.78 \text{ (with 12 df)}$$

$$\text{standard error of } \overline{Y}_{i..} = \sqrt{\frac{2.78}{6}} = 0.68$$

Tabled ranges are Z_p, 12 df (0.05) (Appendix Table E)

$$Z_{2,12}(0.05) = 3.08 \qquad R_{2,12} = (3.08)(0.68) = 2.09$$

and

$$Z_{3,12}(0.05) = 3.77 \qquad R_{3,12} = (3.23)(0.68) = 2.56$$

Testing phosphor-type differences gives

$$B - C \doteq 2.67 - (-1.33) = 4.00 > 2.56^*$$

$$B - A = 2.67 - \quad 0 \quad = 2.67 > 2.09^*$$

$$A - C = \quad 0 \quad - (-1.33) = 1.33 < 2.09$$

where one asterisk indicates significance at the 5-percent level. At this level of significance B differs from both A and C, whereas there is no significant difference between A and C. It is therefore concluded that phosphor type B

is inferior since it requires a significantly higher cathode current and hence lower light output than A or C.

Combining these results on the two factors (since there is no significant interaction), we find that the best combination of glass type and phosphor type to give maximum light output (minimum cathode current) would be glass type 2 and either phosphor type A or C, whichever is more economical to use.

It should be pointed out that in calculating interaction it is not necessary to rewrite Table 6.4 as in Table 6.5, but only to note that the interaction sum of squares is found by subtracting the sum of squares for each main effect from the cell sum of squares. These cells are formed by all levels of these two main effects.

It might also be noted that in the model for this problem, it is assumed that the errors $\varepsilon_{k(ij)}$ are NID $(0, \sigma_\varepsilon^2)$. This means that the variances within each of the six cells in this problem are assumed to have come from normal populations with equal variances. Since the ranges of the coded data in the six cells are 2, 2, 3, 5, 4, and 2, this seems like a safe assumption.

6.2 ANOVA RATIONALE

For a two-factor factorial experiment with n observations per cell, run as a completely randomized design, a general model would be

$$Y_{ijk} = \mu + A_i + B_j + AB_{ij} + \varepsilon_{k(ij)} \tag{6.3}$$

where A and B represent the two factors, $i = 1, 2, \cdots, a$ levels of factor A, $j = 1, 2, \cdots, b$ levels of factor B, and $k = 1, 2, \cdots, n$ observations per cell. In terms of population means this becomes

$$\begin{aligned} Y_{ijk} - \mu_{...} \equiv{} & (\mu_{i..} - \mu_{...}) + (\mu_{.j.} - \mu_{...}) \\ & + (\mu_{ij.} - \mu_{i..} - \mu_{.j.} + \mu_{...}) + (Y_{ijk} - \mu_{ij.}) \end{aligned} \tag{6.4}$$

where $\mu_{ij.}$ represents the true mean of the i, j cell or treatment combination. Justification for the interaction term in the model comes from subtracting A and B main effects from the cell effect as follows:

$$(\mu_{ij.} - \mu_{...}) - (\mu_{i..} - \mu_{...}) - (\mu_{.j.} - \mu_{...}) = \mu_{ij.} - \mu_{i..} - \mu_{.j.} + \mu_{...}$$

If each mean is now replaced by its sample estimate, the resulting sample model is

$$\begin{aligned} Y_{ijk} - \bar{Y}_{...} ={} & (\bar{Y}_{i..} - \bar{Y}_{...}) + (\bar{Y}_{.j.} - \bar{Y}_{...}) \\ & + (\bar{Y}_{ij.} - \bar{Y}_{i..} - \bar{Y}_{.j.} + \bar{Y}_{...}) + (Y_{ijk} - \bar{Y}_{ij.}) \end{aligned}$$

If this expression is now squared and summed over i, j, and k, all cross products vanish, and the results give

$$\sum_i^a \sum_j^b \sum_k^n (Y_{ijk} - \bar{Y}_{...})^2 = \sum_i^a \sum_j^b \sum_k^n (\bar{Y}_{i..} - \bar{Y}_{...})^2$$
$$+ \sum_i^a \sum_j^b \sum_k^n (\bar{Y}_{.j.} - \bar{Y}_{...})^2$$
$$+ \sum_i^a \sum_j^b \sum_k^n (\bar{Y}_{ij.} - \bar{Y}_{i..} - \bar{Y}_{.j.} + \bar{Y}_{...})^2$$
$$+ \sum_i^a \sum_j^b \sum_k^n (Y_{ijk} - \bar{Y}_{ij.})^2$$

which again expresses the idea that the total sum of squares can be broken down into the sum of squares between means of factor A, plus the sum of squares between means of factor B, plus the sum of squares of $A \times B$ interaction, plus the error sum of squares (or within cell sum of squares). Each sum of squares is seen to be independent of the others. Hence, if any such sum of squares is divided by its associated degrees of freedom, the results are independently chi-square distributed, and F tests may be run.

The degree-of-freedom breakdown would be

$$(abn - 1) \equiv (a - 1) + (b - 1) + (a - 1)(b - 1) + ab(n - 1)$$

the interaction being cell df $= (ab - 1)$ minus the main effect df, $(a - 1)$ and $(b - 1)$ or $(ab - 1) - (a - 1) - (b - 1) = ab - a - b + 1 = (a - 1)(b - 1)$, and within each cell the degrees of freedom are $n - 1$ and there are ab such cells giving $ab(n - 1)$ df for error. An ANOVA table can now be set up expanding and simplifying the sum of squares expressions using totals (Table 6.8).

The formulas for sum of squares in Table 6.8 provide good computational formulas for a two-way ANOVA with replication. The error sum of squares might be rewritten as

$$SS_{error} = \sum_i^a \sum_j^b \left[\sum_k^n Y_{ijk}^2 - \frac{T_{ij.}^2}{n} \right]$$

which points up the fact that the sum of squares within each of the $a \times b$ cells is being pooled or added for all such cells. This depends on the assumption that the variance within all cells came from populations with equal variance. The interaction sum of squares can also be rewritten as

$$\left(\sum_i^a \sum_j^b \frac{T_{ij.}^2}{n} - \frac{T_{...}^2}{nab} \right) - \left(\sum_i^a \frac{T_{i..}^2}{nb} - \frac{T_{...}^2}{nab} \right) - \left(\sum_j^b \frac{T_{.j.}^2}{na} - \frac{T_{...}^2}{nab} \right)$$

which shows again that interaction is calculated by subtracting the main effect sum of squares from the cell sum of squares.

Table 6.8 General ANOVA for Two-Factor Factorial with n Replications per Cell

Source	df	SS	MS
Factor A_i	$a - 1$	$\sum_i^a \frac{T_{i..}^2}{nb} - \frac{T_{...}^2}{nab}$	Each SS divided by its df
Factor B_j	$b - 1$	$\sum_j^b \frac{T_{.j.}^2}{na} - \frac{T_{...}^2}{nab}$	
$A \times B$ interaction	$(a - 1)(b - 1)$	$\sum_i^a \sum_j^b \frac{T_{ij.}^2}{n} - \sum_i^a \frac{T_{i..}^2}{nb} - \sum_j^b \frac{T_{.j.}^2}{na} + \frac{T_{...}^2}{nab}$	
Error $\varepsilon_{k(ij)}$	$ab(n - 1)$	$\sum_i^a \sum_j^b \sum_k^n Y_{ijk}^2 - \sum_i^a \sum_j^b \frac{T_{ij.}^2}{n}$	
Totals	$abn - 1$	$\sum_i^a \sum_j^b \sum_k^n Y_{ijk}^2 - \frac{T_{...}^2}{nab}$	

Example 6.2 To extend the factorial idea a bit further, consider a problem with three factors. Such a problem was presented in Chapter 1 on the effect of tool type, angle of bevel, and type of cut on power consumption for ceramic tool cutting. Reference to this problem will point out the phases of experiment, design, and analysis as followed in Example 6.1.

It is a 2 × 2 × 2 factorial experiment with four observations per cell run in a completely randomized manner. The mathematical model is

$$Y_{ijkm} = \mu + T_i + B_j + TB_{ij} + C_k + TC_{ik} + BC_{jk} + TBC_{ijk} + \varepsilon_{m(ijk)}$$

where TBC_{ijk} represents a three-way interaction.

The data for this example are given in Table 1.1 and the ANOVA table in Table 1.2. That this analysis is a simple extension of the methods used on Example 6.1 will be shown with the coded data from Table 1.1 (see Table 6.9).

Table 6.9 shows the total for each small cell and the sum of the squares of the cell observations. These results will be useful in doing the ANOVA. By this time we should be able to set up the steps in the analysis without recourse to formulas in dot notation.

First the total sum of squares: Add the squares of all readings (the circled numbers) and subtract a correction term. The grand total is −13 squared, divided by the number of observations (32):

$$SS_{\text{total}} = 441 - \frac{(-13)^2}{32} = 435.72$$

Table 6.9 Coded Ceramic Tool Data of Table 1.2, Code: 2(X–28.0)

	Tool Type 1		Tool Type 2		
	Bevel Angle		Bevel Angle		
Type of Cut	15°	30°	15°	30°	
Continuous	2 −3 5 (42) −2 — 2	1 1 4 (99) 9 — 15	0 1 0 (37) −6 — −5	3 8 2 (77) 0 — 13	25
Interrupted	0 −6 −3 (54) −3 — −12	−2 2 −1 (10) −1 — −2	−7 −6 0 (101) −4 — −17	−1 0 −2 (21) −4 — −7	−38
Totals	−10	13	−22	+6	−13

Tool type sum of squares: Add for each tool type. The totals are +3 and −16; square these and divide by the number of observations per type (16), add these results for both types, and subtract the correction term. Thus,

$$SS_{\text{tool type}} = \frac{3^2 + (-16)^2}{16} - \frac{(-13)^2}{32} = 11.28$$

Bevel angle sum of squares: Same procedure on the totals for each bevel angle, −32 and 19:

$$SS_{\text{bevel angle}} = \frac{(-32)^2 + (19)^2}{16} - \frac{(-13)^2}{32} = 81.28$$

Type of cut sum of squares: Same procedure, with cut totals of 25 and −38:

$$SS_{\text{type of cut}} = \frac{(25)^2 + (-38)^2}{16} - \frac{(-13)^2}{32} = 124.03$$

For the $T \times B$ interaction, ignore type of cut and use cell totals for the $T \times B$ cells. These are −10, 13, −22, +6:

$$SS_{T \times B \text{ interaction}} = \frac{(-10)^2 + (13)^2 + (-22)^2 + (6)^2}{8} - \frac{(-13)^2}{32} - 11.28 - 81.28 = 0.78$$

For $T \times C$ interaction, ignore bevel angle and the cell totals become 17, −14, 8, −24:

$$SS_{T \times C \text{ interaction}} = \frac{(17)^2 + (-14)^2 + (8)^2 + (-24)^2}{8} - \frac{(-13)^2}{32}$$

$$- 11.28 - 124.03 = 0.03$$

For $B \times C$ interaction, ignore tool type and the cell totals become −3, 28, −29, −9:

$$SS_{B \times C \text{ interaction}} = \frac{(-3)^2 + (28)^2 + (-29)^2 + (-9)^2}{8} - \frac{(-13)^2}{32}$$

$$- 81.28 - 124.03 = 3.78$$

For the three-way interaction $T \times B \times C$, consider the totals of the smallest cells, 2, 15, −5, 13, −12, −2, −17, and −7. From this cell sum of squares subtract *not only* the main effect sum of squares *but also* the three two-way interaction sums of squares. Thus,

$$SS_{T \times B \times C \text{ interaction}}$$

$$= \frac{(2)^2 + (15)^2 + (-5)^2 + (13)^2 + (-12)^2 + (-2)^2 + (-17)^2 + (-7)^2}{4}$$

$$- \frac{(-13)^2}{32} - 11.28 - 81.28 - 124.03 - 0.78 - 0.03 - 3.78 = 0.79$$

By subtraction

$$SS_{\text{error}} = 213.75$$

These results are displayed in Table 6.10.

Table 6.10 ANOVA for Ceramic Tool Problem

Source	df	SS	MS
Tool type T_i	1	11.28	11.28
Bevel angle B_j	1	81.28	81.28
$T \times B$ interaction TB_{ij}	1	0.78	0.78
Type of cut C_k	1	124.03	124.03
$T \times C$ interaction TC_{ik}	1	0.03	0.03
$B \times C$ interaction BC_{jk}	1	3.78	3.78
$T \times B \times C$ interaction TBC_{ijk}	1	0.79	0.79
Error $\varepsilon_{m(ijk)}$	24	213.75	8.91
Totals	31	435.72	

If the results displayed in Table 6.10 are compared with those in Table 1.2, they appear to differ considerably. Actually they give the same F test results, but in Table 6.10 the data were coded involving multiplication by 2; the data of Table 1.2 are uncoded. Multiplication by 2 will multiply the variance or mean square by 4, so if all mean square values in Table 6.10 are divided by 4, the results are the same as in Table 1.2. For example, on tool types $11.28/4 = 2.82$, and error $213.75/4 = 53.44$. It is worth noting that any decoding is unnecessary for determining the F ratios. However, if one wishes confidence limits on the original data or components of variance on the original data, it may be necessary to decode the results.

The interpretation of the results of this example are given in Chapter 1. The purpose of presenting it again in this chapter is to show that factorial experiments with three or more factors can easily be analyzed by simple extension of the methods of this chapter.

6.3 REMARKS

Since the examples in this chapter have contained several replications within a cell, it would be well to examine a situation involving only one observation per cell. In this case $k = 1$, and the model is written as

$$Y_{ij} = \mu + A_i + B_j + AB_{ij} + \varepsilon_{ij}$$

A glance at the last two terms indicates that we cannot distinguish between the interaction and the error—they are hopelessly confounded. Then the only reasonable situation for running one observation per cell is one in which past experience generally assures us that there is no interaction. In such a case, the model is written as

$$Y_{ij} = \mu + A_i + B_j + \varepsilon_{ij}$$

It may also be noted that this model looks very much like the model for a randomized block design for a single-factor experiment (Chapter 4). In Chapter 4 that model was written as

$$Y_{ij} = \mu + \tau_j + \beta_i + \varepsilon_{ij}$$

Even though the models do look alike and an analysis would be run in the same way, this latter is a single-factor experiment—treatments are the factor—and β_i represents a restriction on the randomization. In the factorial model, there are two factors of interest, A_i and B_j, and the design is completely randomized. It is, however, assumed in the randomized-block situation that there is no interaction between treatments and blocks. This is often a more reasonable assumption for blocks and treatments since blocks are often chosen at random. For a two-factor experiment an interaction between A and B may very well be present, and some external information must be available in order to assume that no such interaction exists. If the experimenter is not sure about interaction, he must take more than one observation per cell and test the hypotheses of no interaction.

As the number of factors increases, however, the presence of higher order interactions is much more unlikely, so it is fairly safe to assume no four-way, five-way, $\cdots$, interactions. Even if these were present, they would be difficult to explain in practical terms.

6.4 SUMMARY

Experiment	*Design*	*Analysis*
I. Single factor		
	1. Completely randomized $Y_{ij} = \mu + \tau_j + \varepsilon_{ij}$	1. One-way ANOVA
	2. Randomized block $Y_{ij} = \mu + \tau_j + \beta_i + \varepsilon_{ij}$	2.
	a. Complete	*a*. Two-way ANOVA
	b. Incomplete, balanced	*b*. Special ANOVA
	c. Incomplete, general	*c*. Regression method
	3. Latin square $Y_{ijk} = \mu + \beta_i + \tau_j + \gamma_k + \varepsilon_{ijk}$	3.
	a. Complete	*a*. Three-way ANOVA
	b. Incomplete, Youden square	*b*. Special ANOVA (like 2*b*)
	4. Graeco-Latin square $Y_{ijkm} = \mu + \beta_i + \tau_j + \gamma_k + \omega_m + \varepsilon_{ijkm}$	4. Four-way ANOVA
II. Two or more factors		
A. Factorial (crossed)		
	1. Completely randomized $Y_{ijk} = \mu + A_i + B_j + AB_{ij} + \varepsilon_{k(ij)} \cdots$ for more factors	1.
	a. General case	*a*. ANOVA with interactions

PROBLEMS

6.1 To determine the effect of exhaust index (in seconds) and pump heater voltage (in volts) on the pressure inside a vacuum tube (in microns of mercury), three exhaust indexes and two voltages are chosen at fixed values. For each combination of exhaust index and voltage, two tests are made. The order of experimentation is completely randomized. The results are as follows:

Pump Heater	Exhaust Index (seconds)		
Voltage	60	90	150
127	0.048	0.028	0.007
	0.058	0.033	0.015
220	0.062	0.014	0.006
	0.054	0.010	0.009

Do an analysis of variance on these data and test the effect of exhaust index, heater voltage, and interaction on the pressure.

6.2 Plot the results of Problem 6.1 to show that your conclusions are reasonable.

6.3 For any significant effects in Problem 6.1, test further between the levels of the significant factors.

6.4 For Problem 6.1, what combination of exhaust index and heater voltage would you recommend if a minimum pressure is desired? Explain your choice.

6.5 Adhesive force on gummed material was determined under three fixed humidity and three fixed temperature conditions. Four readings were made under each set of conditions. The experiment was completely randomized and the results set out in an ANOVA table as follows:

Source	df	SS	MS
Humidity		9.07	
Temperature		8.66	
$H \times T$ interaction		6.07	
Error			
Total		52.30	

Complete this table.

6.6 For the data in Problem 6.5, test all indicated hypotheses and state your conclusions.

6.7 Set up a mathematical model for the experiment in Problem 6.5 and indicate the hypotheses to be tested in terms of your model.

6.8 The object of an experiment is to determine thrust forces in drilling at different speeds, feeds, and in different materials. Five speeds are used, three feeds, and two materials with two samples tested under each set of conditions. The order of the experiment is completely randomized and the levels of all factors are fixed. The following data are recorded on thrust forces after subtracting 200 from all readings:

Material	Feed	Speed 100	220	475	715	870
B_{10}	0.004	122	108	108	66	80
		110	85	60	50	60
	0.008	332	276	248	248	276
		330	310	295	275	310
	0.014	640	612	543	612	696
		500	500	450	610	610
V_{10}	0.004	192	136	122	108	136
		170	130	85	75	75
	0.008	386	333	318	472	499
		365	330	330	350	390
	0.014	810	779	810	893	1820
		725	670	750	890	890

Do a complete analysis of this experiment and state your conclusions.

6.9 Plot any results in Problem 6.8 which are significant.

6.10 Set up tests on means where suitable and draw conclusions from Problem 6.8.

6.11 In an experiment for testing rubber materials, interest centered on the effect of the mix (*A*, *B*, or *C*), the laboratory involved (1, 2, 3, or 4), and the temperature (145, 155, 165°C) on the time in minutes to a 2-in.-lb rise above the minimum time. Assuming a completely randomized design, do an ANOVA on the data below:

	Temperature (°C) 145			155			165		
	Mix			Mix			Mix		
Laboratory	*A*	*B*	*C*	*A*	*B*	*C*	*A*	*B*	*C*
1	11.2	11.2	11.5	6.7	6.8	7.0	4.8	4.8	5.0
	11.1	11.5	11.4	6.8	6.7	7.0	4.8	4.9	4.9
2	11.8	12.3	12.3	7.3	7.5	7.5	5.3	5.4	5.3
	11.8	12.3	11.9	7.2	7.7	7.3	5.3	5.2	5.3
3	11.5	12.3	12.7	6.6	7.1	7.8	5.0	5.3	5.2
	11.6	12.0	12.5	6.9	7.2	7.3	5.0	5.0	5.0
4	11.5	11.8	12.7	7.2	6.7	7.1	4.5	4.7	4.5
	11.3	11.7	12.7	6.9	7.0	7.0	4.6	4.5	4.5

6.12 Plot any significant interactions in Problem 6.11 and discuss the conditions which should produce the minimum time to a 2-in.-lb rise.

6.13 Explain why complete randomization might be difficult in Problem 6.11.

6.14 An industrial engineer presented two types of stimuli (two-dimensional and three-dimensional films) of two different jobs (1 and 2) to each of five analysts. Each analyst was presented each job-stimulus film twice and the order of the whole experiment was considered completely randomized. The engineer was interested in the consistency of analyst ratings of four sequences within each job stimulus presentation. The variable recorded is the log variance of the four sequences since log variance is more likely to be normally distributed than the variance. Data showed:

	Stimulus			
	Two-Dimensional		Three-Dimensional	
Analyst	Job 1	Job 2	Job 1	Job 2
1	1.42	1.40	1.00	0.92
	1.25	1.44	0.90	0.93
2	1.59	1.83	1.46	1.43
	2.09	2.02	1.68	1.02
3	1.26	1.80	0.85	1.33
	1.48	1.75	1.43	1.48
4	1.76	0.98	1.21	0.73
	1.47	1.23	1.25	1.22
5	1.43	1.17	1.60	1.31
	1.72	1.18	2.03	1.12

Do a complete analysis and summarize your results.

6.15 Make further tests on Problem 6.14 as suggested by the ANOVA results.

6.16 The following results were reported on a study of "factors which affect the salary of high-school teachers of commercial subjects:"

Source	df	SS
Sex	1	239,763
Size of school	2	423,056
Years in position	5	564,689
Sex by size interaction	2	18,459
Sex × years interaction	5	85,901
Size × years interaction	10	240,115
Sex × size × years interaction	10	151,394
Within classes	153	1,501,642

Indicate the mathematical model for the above study. Show a possible data layout. Complete the ANOVA table and comment on any significant results.

6.17 Four factors are studied for their effect on the luster of plastic film. These factors are (1) film thickness (1 or 2 mils), (2) drying conditions (regular or special), (3) length of wash (20, 30, 40, or 60 minutes), and (4) temperature of wash (92°C or 100°C). Two observations of film luster are taken under each set of conditions. Assuming complete randomization, analyze the data below:

	Regular Dry				Special Dry			
Minutes	92°C		100°C		92°C		100°C	
1-mil Thickness								
20	3.4	3.4	19.6	14.5	2.1	3.8	17.2	13.4
30	4.1	4.1	17.5	17.0	4.0	4.6	13.5	14.3
40	4.9	4.2	17.6	15.2	5.1	3.3	16.0	17.8
60	5.0	4.9	20.9	17.1	8.1	4.3	17.5	13.9
2-mil Thickness								
20	5.5	3.7	26.6	29.5	4.5	4.5	25.6	22.5
30	5.7	6.1	31.6	30.2	5.9	5.9	29.2	29.8
40	5.6	5.6	30.5	30.2	5.5	5.8	32.6	27.4
60	7.2	6.0	31.4	29.6	8.0	9.9	33.5	29.5

6.18 Plot and discuss any significant interactions found in Problem 6.17. What conditions would you recommend for maximum luster and why?

7

2^n Factorial Experiments

7.1 INTRODUCTION

In the last chapter, factorial experiments were considered and a general method for their analysis was given. There are a few special cases which are of considerable interest in future designs. One of these is the case of n factors where each factor is at just two levels. These levels might be two extremes of temperature, two extremes of pressure, two time values, two machines, and so on. Although this may seem like a rather trivial case since only two levels are involved, it is, nevertheless, very useful for at least two reasons: to introduce notation and concepts useful when more involved designs are discussed and to illustrate what main effects and interactions there really are in this simple case. It is also true that in practice many experiments are run at just two levels of each factor. The ceramic tool cutting example of Chapters 1 and 6 is a $2 \times 2 \times 2$, or 2^3 factorial, with four observations per cell. Throughout this chapter, the two levels will be considered as fixed levels. This is quite reasonable because the two levels are chosen at points near the extremes, rather than at random.

7.2 2^2 FACTORIAL

The simplest case to consider is one in which two factors are of interest and each factor is set at just two levels. This is a $2 \times 2 = 2^2$ factorial, and the design will be considered as completely randomized. The example in Section 6.1 is of this type. The factors are temperature and altitude, and each is set at two levels: temperature at 25°C and 55°C and altitude at 0-K and 3-K. This gives four treatment combinations, displayed in Figure 6.3, where the response variable is the current flow in milliamperes. In order to generalize a bit, consider temperature as factor A and altitude as factor B. The model for this completely randomized design would be

$$Y_{ij} = \mu + A_i + B_j + AB_{ij} + \varepsilon_{ij} \tag{7.1}$$

where $i = 1, 2$ and $j = 1, 2$ in this case. Unless there is some replication, of course, no assessment of interaction can be made independent of error. From the data of Figure 6.3, when both temperature A and altitude B are at their low levels, the response is 210 mA. This may be designated by subscripts on AB as follows: $A_0B_0 = 210$ mA. Following this notation, A_1B_0 means A at its high level and B at its low level, or temperature 55°C and altitude 0-K. The response is $A_1B_0 = 240$ mA. Likewise, $A_0B_1 = 180$ mA and $A_1B_1 = 200$ mA. Since a 2^n experiment is encountered so often in the literature, most authors have adopted another notation for these treatment combinations. For this new notation, just the subscripts on AB are used as exponents on the small letters ab. If both factors are at their low levels, $a^0b^0 = (1)$, and (1) represents the response of both factors at their low level. $a^0b^1 = b$ represents B at its high level and A at its low level, $a^1b^0 = a$ represents the high level of A and low level of B, and $a^1b^1 = ab$ represents the response when both factors are at their high levels. This notation can easily be extended to more factors, provided only two levels of each factor are involved.

For the example of Figure 6.3, the treatment combinations can be represented by the vertices of a square as in Figure 7.1.

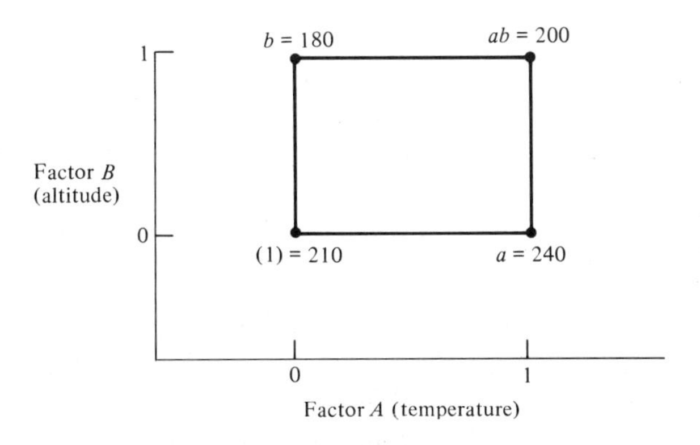

Figure 7.1 2^2 factorial experiment.

In Figure 7.1, the low and high levels of factors A and B are represented by 0 and 1, respectively, on the A and B axes. The intersection of these levels in the plane of the figure shows the four treatment combinations. For example: $00 = (1)$ represents both factors at their low levels, $10 = a$ represents A high, B low, and so on, $01 = b$, $11 = ab$. These expressions are only symbolic and are to be considered as merely a mnemonic device to simplify the design and its analysis. The normal order for writing these treatment combinations is (1), a, b, ab. Note that (1) is written first, then the high level of each factor with the low level of the other (a, b), and the fourth term is the algebraic product of the second and third (ab). When a third factor is introduced, it is placed at the end of this sequence and then multiplied by all of its predecessors. For example, if factor C is also present at two levels, the treatment combinations are

$$(1),\ a,\ b,\ ab,\ c,\ ac,\ bc,\ abc$$

which can be represented as the vertices of a cube.

Returning to Figure 7.1, we find that the *effect of a factor* is defined as the change of response produced by a change in the level of that factor. At the low level of B, the effect of A is then 240 − 210 or $a - (1)$, whereas the effect of A at the high level of B is 200 − 180 or $ab - b$. The average effect of A is then

$$A = \tfrac{1}{2}[a - (1) + ab - b]$$

or

$$A = \tfrac{1}{2}[-(1) + a - b + ab]$$

Note that the coefficients on this A effect are all $+1$ when A is at its high level in the treatment combinations ($+a$, $+ab$) and the coefficients are all -1 when A is at its low level as in (1) and b. Note also that

$$2A = -(1) + a - b + ab$$

is a contrast as defined in Section 3.3 (the sum of its coefficients are $-1 + 1 - 1 + 1 = 0$). This concept will be useful, as the sum of squares due to this contrast or effect can easily be determined.

The average effect of B, based on low level of A $[180 - 210 = b - (1)]$ and high level of A $(200 - 240 = ab - a)$, is

$$B = \tfrac{1}{2}[b - (1) + ab - a]$$

or

$$B = \tfrac{1}{2}[-(1) - a + b + ab]$$

and again it is seen that the same four responses are used, but $+1$ coefficients are on the treatment combinations for the high level of B and -1 coefficients for the low level of B. Also $2B$ is a contrast.

To determine the effect of the interaction between A and B, note that at the high level of B the A effect is $ab - b$, and at the low level of B the A effect is $a - (1)$. If these two effects differ, there is an interaction between A and B. Thus the interaction is the average *difference* between these two differences

$$\begin{aligned} AB &= \tfrac{1}{2}\{(ab - b) - [a - (1)]\} \\ &= \tfrac{1}{2}[ab - b - a + (1)] \end{aligned}$$

or

$$= \tfrac{1}{2}[(1) - a - b + ab]$$

Here again the same four treatment combinations are used with a different combination of coefficients. Note that $2AB$ is also a contrast, and all contrasts $2A$, $2B$, $2AB$ are orthogonal to each other. Summarizing in normal order gives

$$2A = -(1) + a - b + ab$$

$$2B = -(1) - a + b + ab$$

$$2AB = +(1) - a - b + ab$$

Note that the interaction effect takes the responses on one diagonal of the square with $+1$ coefficients and the responses on the other diagonal with -1 coefficients. Note also that the coefficients for the interaction effect can be found by multiplying the corresponding coefficients of the two main effects. As the only coefficients used are $+1$'s and -1's, the proper coefficients on the treatment combinations for each main effect and interaction can be determined from Table 7.1.

Table 7.1 Coefficients for Effects in a 2^2 Factorial Experiment

Treatment	Effect		
Combination	A	B	AB
(1)	−	−	+
a	+	−	−
b	−	+	−
ab	+	+	+

From Table 7.1 the orthogonality of the effects is easily seen, as well as the generation of the interaction coefficients from the main effect coefficients.

Another approach to the interaction between A and B would be to consider that at the high level of A the effect of B is $ab - a$, and at the low level of A the effect of B is $b - (1)$, so that the average difference is

$$\begin{aligned} AB &= \tfrac{1}{2}\{(ab - a) - [b - (1)]\} \\ &= \tfrac{1}{2}[ab - a - b + (1)] \\ &= \tfrac{1}{2}[+(1) - a - b + ab] \end{aligned}$$

which is the same expression as given before.

For the response data of Figure 7.1, the effects are

$$A = \tfrac{1}{2}(-210 + 240 - 180 + 200) = 25 \text{ mA}$$

$$B = \tfrac{1}{2}(-210 - 240 + 180 + 200) = -35 \text{ mA}$$

$$AB = \tfrac{1}{2}(+210 - 240 - 180 + 200) = -5 \text{ mA}$$

Since $2A$, $2B$, and $2AB$ are contrasts, the sum of squares due to a contrast is

$$SS_{C_m} = \frac{(\text{contrast})^2}{n \sum c_{jm}^2}$$

where n is the number of observations in each total (here $n = 1$). Also $\sum c_{jm}^2 = 1 + 1 + 1 + 1 = 4$ (or 2^2). From this definition

$$SS_A = \frac{[2(25)]^2}{4} = 625$$

$$SS_B = \frac{[2(-35)]^2}{4} = 1225$$

$$SS_{AB} = \frac{[2(-5)]^2}{4} = 25$$

$$SS_{\text{total}} = 1875$$

Since each effect and the interaction has but 1 df and there is no measure of error as only one observation was taken in each cell, this ANOVA is quite trivial, but it does show another approach to analysis based on effects. The simple ANOVA table would be that shown by Table 7.2.

Table 7.2 ANOVA for a 2^2 Factorial with No Replication

Source	df	SS	MS
A_i	1	625	625
B_j	1	1225	1225
Error or AB_{ij}	1	25	25
Totals	3	1875	

If the general methods of Chapter 6 are used on the data in Figure 7.1 (coded by subtracting 200), the results are those in Table 7.3.

Table 7.3 Coded Data of Figure 7.1

	Factor A		
Factor B	0	1	Totals
0	+10	+40	+50
1	−20	0	−20
Totals	−10	+40	+30

$$SS_{\text{total}} = (10)^2 + (-20)^2 + (40)^2 + (0)^2 - \frac{(30)^2}{4} = 1875$$

$$SS_A = \frac{(-10)^2 + (40)^2}{2} - \frac{(30)^2}{4} = 850 - 225 = 625$$

$$SS_B = \frac{(50)^2 + (-20)^2}{2} - \frac{(30)^2}{4} = 1450 - 225 = 1225$$

$$SS_{AB} \text{ (by subtraction)} = 1875 - 625 - 1225 = 25$$

which are the same sums of squares as given in Table 7.2. Even though no separate measure of error is available, a glance at the mean squares of Table 7.2 shows that both main effects are large compared to the interaction effect.

The results of this section can easily be extended to cases where there are r replications in the cells, by using cell totals for the responses at (1), a, b, and ab and adjusting the sum of squares for the effects accordingly (see Example 7.1).

Example 7.1 An example of a 2^2-factorial experiment with two replications per cell is considered here, using hypothetical responses to illustrate the principles of the previous section. Consider the data of Table 7.4.

Table 7.4 2^2 Factorial with Two Replications

	Factor *A*				
Factor *B*	0		1		Totals
0	4		2		
	6		−2		
		10		0	+10
1	3		−4		
	7		−6		
		10		−10	0
Totals		+20		−10	+10

Using the general methods of Chapter 6, the sums of squares are

$$SS_{\text{total}} = 4^2 + 6^2 + 3^2 + 7^2 + 2^2 + (-2)^2 + (-4)^2 + (-6)^2$$

$$- \frac{(10)^2}{8} = 170 - 12.5 = 157.5$$

$$SS_A = \frac{(20)^2 + (-10)^2}{4} - 12.5 = 112.5$$

$$SS_B = \frac{(10)^2 + (0)^2}{4} - 12.5 = 12.5$$

$$SS_{A \times B \text{ interaction}} = \frac{(10)^2 + (10)^2 + (0)^2 + (-10)^2}{2} - 12.5 - 112.5$$

$$- 12.5 = 12.5$$

$$SS_{\text{error}} = 157.5 - 112.5 - 12.5 - 12.5 = 20.0$$

These results could be displayed in an ANOVA table. Using the methods of Section 7.2, we get, with the treatment combinations, the total response

$$(1) = 10 \qquad a = 0 \qquad b = 10 \qquad ab = -10$$

and the contrasts

$$4A = -10 + 0 - 10 + (-10) = -30$$

$$4B = -10 - 0 + 10 + (-10) = -10$$

$$4AB = +10 - 0 - 10 + (-10) = -10$$

The coefficient 4 used with each response represents the two individual responses at each level. In general the coefficient of these effects is $r \cdot 2^{n-1}$, where r is the number of replications and n the number of factors. Here $r = 2$ and $n = 2$. The sum of squares for these three contrasts are

$$SS_A = \frac{(4A)^2}{r \cdot 2^2} = \frac{(-30)^2}{2 \cdot 4} = \frac{900}{8} = 112.5$$

$$SS_B = \frac{(4B)^2}{r \cdot 2^2} = \frac{(-10)^2}{2 \cdot 4} = \frac{100}{8} = 12.5$$

$$SS_{AB} = \frac{(4AB)^2}{r \cdot 2^2} = \frac{(-10)^2}{2 \cdot 4} = \frac{100}{8} = 12.5$$

since

$$2^2 = \sum_{i=1}^{4} c_{jm}^2 = 1 + 1 + 1 + 1 = 4$$

These results are the same as given by the general method. The total sum of squares must be calculated as usual in order to get the error sum of squares by subtraction.

For this special case of a 2^n-factorial experiment, Yates [20] developed a rather simple scheme for computing these contrasts. The method can best be illustrated on Example 7.1, using Table 7.5.

Table 7.5 Yates Method on a 2^2 Factorial

Treatment Combination	Response	(1)	(2)	SS
(1)	10	10	$10 =$ total	12.5
a	0	0	$-30 = 4A$	112.5
b	10	-10	$-10 = 4B$	12.5
ab	-10	-20	$-10 = 4AB$	12.5

In Table 7.5, list all treatment combinations in the first column. Place the total response to each of these treatment combinations in the second column. For the third column, labeled (1), add the responses in pairs, for example, $10 + 0 = 10$ for the first two and $10 - 10 = 0$ for the next two; this completes half of column (1). For the second half, subtract the responses in pairs, always subtracting the first from the second, for example, $0 - 10 = -10$; $-10 - (10) = -20$. This completes column (1). Column (2) is determined in the same manner as column (1) using the column (1) results: $10 + 0 = 10$; $-10 - 20 = -30$; $0 - 10 = -10$; $-20 - (-10) = -10$. Proceed in this same manner until the nth column is reached: (1), (2), (3), $\cdots$, (n). In this case $n = 2$, so there are just two columns. The values in column n are the contrasts, where the first entry is the grand total of all readings; the one corresponding to a is $r \cdot 2^{n-1} \cdot A$, b is $r \cdot 2^{n-1} \cdot B$, and ab is $r \cdot 2^{n-1} \cdot AB$. When the results of the last column [column (n)] are squared and divided by $r \cdot 2^n$, the results are the sums of squares as shown above. The first sum of squares $(\text{total})^2/8$ is the correction term for the grand mean given in Chapter 6. The Yates method reduces the analysis to simply adding and subtracting numbers. It is very useful provided the experiment is a 2^n factorial with r replications per cell.

As proof for the Yates method on a 2^2 factorial, go through the steps above using the treatment combination symbols for the responses as in Table 7.6.

Table 7.6 Yates Method on a 2^2 in General; One Observation per Cell

Treatment Combination	(1)	(2)
(1)	$(1) + a$	$(1) + a + b + ab =$ total
a	$b + ab$	$-(1) + a - b + ab = 2A$
b	$a - (1)$	$-(1) - a + b + ab = 2B$
ab	$ab - b$	$(1) - a - b + ab = 2AB$

It is obvious that the last column does give the proper treatment contrasts.

7.3 2^3 FACTORIAL

Considering a third factor C, also at two levels, the experiment will be a $2 \times 2 \times 2$ or 2^3 factorial, again run in a completely randomized manner. The treatment combinations are now (1), a, b, ab, c, ac, bc, abc, and they may be represented as vertices of a cube as in Figure 7.2.

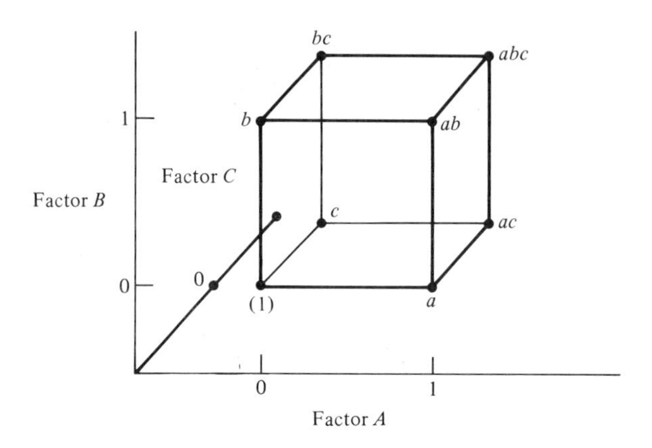

Figure 7.2 2^3 factorial arrangements.

With these $8 = 2^3$ observations, or $8r$ if there are replications, the main effects and each interaction may be expressed by using the proper coefficients (-1 or $+1$) on these eight responses. For the effect of factor A, consider all responses in the right-hand plane (a, ab, ac, abc) with plus signs and all in the left-hand plane [(1), b, bc, c] with minus signs as these will show the effect of increasing the level of A. Or,

$$4A = -(1) + a - b + ab - c + ac - bc + abc$$

For factor B, consider responses in the lower and higher planes of the cube

$$4B = -(1) - a + b + ab - c - ac + bc + abc$$

The AB interaction is determined by the difference in the A effect from level 0 of B to level 1 of B, regardless of C

$$\text{at } B_0\text{: } 2A \text{ effect } a + ac - c - (1)$$

$$\text{at } B_1\text{: } 2A \text{ effect } abc + ab - b - bc$$

The difference in these is the interaction

$$4AB = [abc + ab - b - bc] - [a + ac - c - (1)]$$

$$4AB = abc + ab - b - bc - a - ac + c + (1)$$

or

$$4AB = +(1) - a - b + ac + c - ac - bc + abc$$

which gives the same signs as the products of corresponding signs on $4A$ and $4B$.

For factor C, compare the responses in the back plane of the cube to those in the front plane of the cube

$$4C = c + bc + ac + abc - (1) - b - a - ab$$

or

$$4C = -(1) - a - b - ab + c + ac + bc + abc$$

The AC and BC interactions can be determined as AB was, and it can be seen that the resulting interaction effects are

$$4AC = +(1) - a + b - ab - c + ac - bc + abc$$

$$4BC = +(1) + a - b - ab - c - ac + bc + abc$$

To determine the ABC interaction, consider the BC interaction at level 0 of A versus the BC interaction at level 1 of A. Any difference in these is an ABC interaction

$$2BC \text{ interaction at } A_0\text{: } +(1) - b - c + bc$$

$$2BC \text{ interaction at } A_1\text{: } +a - ab - ac + abc$$

Their difference is

$$4ABC = (a - ab - ac + abc) - [(1) - b - c + bc)]$$

or

$$4ABC = -(1) + a + b - ab + c - ac - bc + abc$$

which can also be obtained from multiplying the coefficients of A and BC, or B and AC, or C and AB. These others can also be used to get the ABC interaction, but the results are the same. Summarizing gives us Table 7.7.

Table 7.7 Coefficients for Effects in a 2^3 Factorial Experiment

Treatment Combination	Effect							
	Total	A	B	AB	C	AC	BC	ABC
(1)	+	−	−	+	−	+	+	−
a	+	+	−	−	−	−	+	+
b	+	−	+	−	−	+	−	+
ab	+	+	+	+	−	−	−	−
c	+	−	−	+	+	−	−	+
a	+	+	−	−	+	+	−	−
bc	+	−	+	−	+	−	+	−
abc	+	+	+	+	+	+	+	+

Table 7.7 illustrates once again the orthogonality of the effects and can easily be extended to 4, 5, and more factors if each factor is at two levels only. In a 2^3-factorial experiment, the sum of squares is given by

$$SS_{\text{effect}} = \frac{(\text{contrast})^2}{r \cdot 2^3} = \frac{(\text{contrast})^2}{8r}$$

and again the Yates method leads to an easy computation of the contrasts.

Example 7.2 In Chapters 1 and 6, the problem on power requirements for cutting with ceramic tools was analyzed in detail. It is readily seen that this is a $2 \times 2 \times 2 = 2^3$ factorial with four replications per cell. This problem could be analyzed by the special methods of this chapter. From Table 6.9 on coded data the treatment combinations might be summarized as in Table 7.8.

Table 7.8 Ceramic Tool Data in 2^3 Form

	Tool Type			
	1		2	
	Bevel Angle		Bevel Angle	
Type of Cut	15°	30°	15°	30°
Continuous	(1)	*b*	*a*	*ab*
	2	15	−5	13
Interrupted	*c*	*bc*	*ac*	*abc*
	−12	−2	−17	−7

Tool type = factor *A*, bevel = factor *B*, cut = factor *C*.

In Table 7.8, the totals of the four replications for each treatment combination have been entered in their corresponding cells. By the Yates method we get the entries in Table 7.9.

Table 7.9 Yates Method on Ceramic Tool Data

Treatment Combination	Response	(1)	(2)	(3)	SS
(1)	2	−3	25	−13 = total	5.28
a	−5	28	−38	−19 = 16*A*	11.28
b	15	−29	−9	51 = 16*B*	81.28
ab	13	−9	−10	5 = 16*AB*	0.78
c	−12	−7	31	−63 = 16*C*	124.03
ac	−17	−2	20	−1 = 16*AC*	0.03
bc	−2	−5	5	−11 = 16*BC*	3.78
abc	−7	−5	0	−5 = 16*ABC*	0.78

These sums of squares are seen to be in substantial agreement with those of Table 6.10. We must resort to the individual readings, however, to determine the total sum of squares and then the error sum of squares. This was done in Table 6.10, and the interpretation of this problem is given in Chapter 6. The purpose of repeating the problem here was merely to show the use of the Yates method on a 2^3-factorial experiment.

7.4 2^n—REMARKS

The methods shown for 2^2 and 2^3 factorials may easily be extended to 2^n factorials where n factors are each considered at two levels. The contrasts are determined from the responses to the treatment combinations by associating plus signs with high levels of a factor and minus signs with low levels; the contrasts for interactions are found by multiplication of corresponding coefficients.

The general relationships for 2^n factorials with r replications per cell are

$$\text{contrast} = r \cdot 2^{n-1} \text{ (effect)}$$

or

$$\text{effect} = \frac{1}{r \cdot 2^{n-1}} \text{ (contrast)}$$

and

$$\text{SS}_{\text{contrast}} = \frac{(\text{contrast})^2}{r \cdot 2^n}$$

A general ANOVA would be as in Table 7.10, but not in "normal" order.

Table 7.10 ANOVA for a 2^n Factorial with r Replications

Source		df	
Main effects	A	1	n
	B	1	
	C	1	
	⋮	⋮	
2-factor interactions	AB	1	$C(n, 2) = \frac{n(n-1)}{2}$
	AC	1	
	BC	1	
	⋮	⋮	
3-factor interactions	ABC	1	$C(n, 3) = \frac{n(n-1)(n-2)}{6}$
	ABD	1	
	BCD	1	
	⋮	⋮	
4-factor interactions, and so on			
Sum of all treatment combinations			$2^n - 1$
Residual or error			$2^n(r - 1)$
Total			$r \cdot 2^n - 1$

7.5 SUMMARY

Experiment	*Design*	*Analysis*
I. Single factor		
	1. Completely randomized $Y_{ij} = \mu + \tau_j + \varepsilon_{ij}$	1. One-way ANOVA
	2. Randomized block $Y_{ij} = \mu + \beta_i + \tau_j + \varepsilon_{ij}$	2.
	a. Complete	*a*. Two-way ANOVA
	b. Incomplete, balanced	*b*. Special ANOVA
	c. Incomplete, general	*c*. Regression method
	3. Latin square $Y_{ijk} = \mu + \beta_i + \tau_j + \gamma_k + \varepsilon_{ijk}$	3.
	a. Complete	*a*. Three-way ANOVA
	b. Incomplete, Youden square	*b*. Special ANOVA (like 2*b*)
	4. Graeco-Latin square $Y_{ijkm} = \mu + \beta_i + \tau_j + \gamma_k + \omega_m + \varepsilon_{ijkm}$	4. Four-way ANOVA
II. Two or more factors		
A. Factorial (crossed)		
	1. Completely randomized $Y_{ijk} = \mu + A_i + B_j + AB_{ij} + \varepsilon_{k(ij)} \cdots$ for more factors	1.
	a. General case	*a*. ANOVA with interactions
	b. 2^n case	*b*. Yates method or general ANOVA; use (1), *a*, *b*, *ab*, $\cdots$

PROBLEMS

7.1 For the 2^2 factorial with three observations per cell given below (hypothetical data), do an analysis by the method of Chapter 6.

	Factor *A*	
Factor *B*	A_1	A_2
B_1	0	4
	2	6
	1	2
B_2	−1	−1
	−3	−3
	1	−7

7.2 Redo Problem 7.1 by the Yates method and compare results.

7.3 In an experiment on chemical yield, three factors were studied, each at two levels. The experiment was completely randomized and the factors were known only as A, B, and C. The results were

A_1				A_2			
B_1		B_2		B_1		B_2	
C_1	C_2	C_1	C_2	C_1	C_2	C_1	C_2
1595	1745	1835	1838	1573	2184	1700	1717
1578	1689	1823	1614	1592	1538	1815	1806

Analyze by the methods of Chapter 6.

7.4 Analyze Problem 7.3 by the Yates method and compare your results.

7.5 Plot any results from Problem 7.3 which might be meaningful from a management point of view.

7.6 The results of Problem 7.3 lead to another experiment with four factors each at two levels, with the following data:

A_1							
B_1				B_2			
C_1		C_2		C_1		C_2	
D_1	D_2	D_1	D_2	D_1	D_2	D_1	D_2
1985	2156	1694	2184	1765	1923	1806	1957
1592	2032	1712	1921	1700	2007	1758	1717
A_2							
B_1				B_2			
C_1		C_2		C_1		C_2	
D_1	D_2	D_1	D_2	D_1	D_2	D_1	D_2
1595	1578	2243	1745	1835	1863	1614	1917
2067	1733	1745	1818	1823	1910	1838	1922

Analyze these data by the general methods of Chapter 6.

7.7 Analyze Problem 7.6 using the Yates method.

7.8 Plot from Problem 7.6 any results you think are meaningful.

7.9 Any factor whose levels are powers of 2 such as 4, 8, 16, $\cdots$, may be considered as pseudo factors in the form $2^2, 2^3, 2^4, \cdots$. Yates' method can then be used if all factor levels are 2 or powers of 2 and the resulting sums of squares for the pseudo factors can be added to give the sums of squares for the actual factors and appropriate interactions. Try out this scheme on the data of Problem 6.17.

7.10 Sketch rough graphs for the following situations where each factor is at 2 levels only:

1. Main effects A and B both significant, AB interaction not significant.
2. Main effect A significant, B not significant, and AB significant.
3. Both main effects not significant but interaction significant.

7.11 A systematic test was made to determine the effects on the coil breakdown voltage of the following six variables, each at 2 levels as indicated:

1. Firing furnace: Number 1 or 3
2. Firing temperature: 1650°C or 1700°C
3. Gas humidification: Yes or no
4. Coil outside diameter: Large or small
5. Artificial chipping: Yes or no
6. Sleeve: Number 1 or 2

Assuming complete randomization, devise a data sheet for this experiment, outline its ANOVA and explain your error term if only one coil is to be used for each of the treatment combinations.

7.12 For a 2^3 factorial, can you sketch a graph (or graphs) to show no significant 2-factor interactions, but a significant 3-factor interaction?

8

Qualitative and Quantitative Factors

8.1 LINEAR REGRESSION

Before discussing quantitative and qualitative levels of one factor and several factors, consider a brief review of linear regression. In this problem the life (in hours) of a small electrical part is believed to be affected by the temperature at which these parts are activated. The temperature range of concern is from 0 to 100°C. The following fixed temperature levels were chosen: 0, 25, 50, 75, and 100°C. Data were collected in a completely randomized manner on 15 parts, with three being activated at each of the five temperature levels. Life test results are shown in Table 8.1.

Table 8.1 Life Test Data

	Activation Temperature (°C)					
	0	25	50	75	100	
Life (hours)	53	60	67	65	58	
	50	62	70	68	62	
	47	58	73	62	60	
$T_{.j}$	150	180	210	195	180	$T_{..} = 915$
$\sum_{i=1}^{3} Y_{ij}^2$	7518	10,808	14,718	12,693	10,808	$\sum_{i}^{3} \sum_{j}^{5} Y_{ij}^2 = 56,545$

The ANOVA model for this design is

$$Y_{ij} = \mu + \tau_j + \varepsilon_{ij}$$

as in Chapter 3. The ANOVA results are given in Table 8.2.

The F ratio of 165/7 is obviously highly significant and one concludes that temperature indeed has a strong effect on the life of these parts.

Table 8.2 Life Test ANOVA

Source	df	SS	MS
Between temperatures	4	660	165
Error	10	70	7
Total	14	730	

One could now make certain contrasts among the temperature means or apply a Newman–Keuls test. However, since the temperature factor is at quantitative levels and these levels are equispaced, one might suspect some functional relationship between life and temperature. If such a relationship can be found, life might be predicted from temperature at least in the range from 0 to 100°C.

In seeking a model for the relationship between Y_{ij} and the temperature (call it X_j), plot the results from Table 8.1 and sketch in a straight line as a first approximation (Figure 8.1). In sketching in this straight line, one tries to come as near as possible to the mean Y's for each X (shown by $\times$). From Figure 8.1, it is obvious that the straight line is not too good a "fit" to these average points. However, as a first approximation, a linear regression model would give

$$Y_{ij} \equiv \mu + (\mu_{y/x} - \mu) + (\mu_{.j} - \mu_{y/x}) + (Y_{ij} - \mu_{.j})$$

where μ is the effect of the overall mean, $\mu_{y/x}$ represents the predicted mean of Y for each X, based on the model considered, $\mu_{.j} - \mu_{y/x}$ represents the departure of the treatment means from the regression model, and $Y_{ij} - \mu_{.j}$ is the usual error within each treatment. If there is no regression model involved, the second and third expressions on the right combine to give $\mu_{.j} - \mu$ or τ_j which is the ANOVA model.

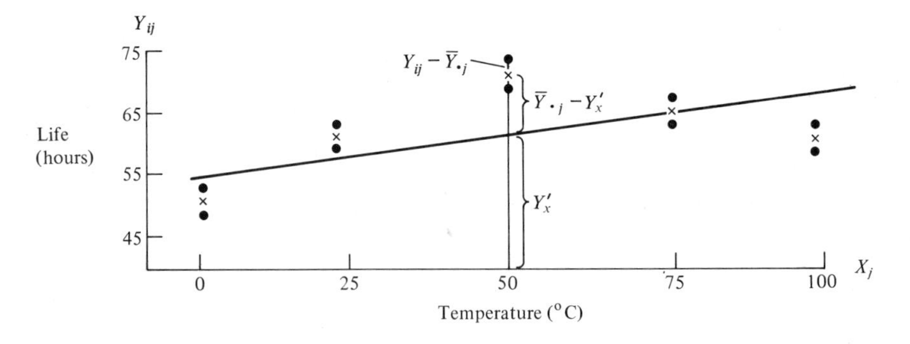

Figure 8.1 Life versus temperature.

When the sample estimates are substituted in this model,

$$Y_{ij} = \bar{Y}_{..} + (Y'_x - \bar{Y}_{..}) + (\bar{Y}_{.j} - Y'_x) + (Y_{ij} - \bar{Y}_{.j}) \qquad (8.1)$$

and these components can be labeled as in Figure 8.1. For a straight-line model,

$$Y'_x = b_0 + b_1 X_j \qquad (8.2)$$

where b_0 is the intercept and b_1 the slope. To determine b_0 and b_1, the *method of least squares* is used as in the general regression models of previous chapters. Here the sum of squares of the deviation of each Y_{ij} from its predicted $Y(Y'_x)$ is to be minimized. (The results are the same if the sum of squares of departures from regression, $\bar{Y}_{.j} - Y'_x$, are minimized.)

$$Y_{ij} - Y'_x = Y_{ij} - b_0 - b_1 X_j$$

$$\sum_i \sum_j (Y_{ij} - Y'_x)^2 = \sum_i \sum_j (Y_{ij} - b_0 - b_1 X_j)^2$$

Differentiating with respect to b_0 and b_1,

$$-2 \sum_i \sum_j (Y_{ij} - b_0 - b_1 X_j) = 0$$

$$-2 \sum_i \sum_j (Y_{ij} - b_0 - b_1 X_j) X_j = 0$$

Simplifying, the least squares normal equations are

$$\begin{aligned} \sum_i \sum_j Y_{ij} &= b_0 N + b_1 n \sum_j X_j \\ \sum_i \sum_j X_j Y_{ij} &= b_0 n \sum_j X_j + b_1 n \sum_j X_j^2 \end{aligned} \qquad (8.3)$$

From Table 8.1,

$$\sum_i \sum_j Y_{ij} = T_{..} = 915, \qquad N = 15, \qquad n = 3$$

$$\sum_j X_j = 0 + 25 + \cdots + 100 = 250$$

$$\sum_j X_j^2 = 0^2 + 25^2 + \cdots + 100^2 = 18{,}750$$

$$\sum_i \sum_j X_j Y_{ij} = 0(53 + 50 + 47) + 25(60 + 62 + 58) + \cdots + 100(58 + 62 + 60) = 47{,}625$$

Substituting

$$915 = 15b_0 + 750b_1$$

$$47{,}625 = 750b_0 + 56{,}250b_1$$

Solving simultaneously for b_0 and b_1 gives

$$b_0 = 56 \quad \text{and} \quad b_1 = 0.1$$

and the linear regression model is

$$Y'_x = 56 + 0.1X_j$$

Substituting the observed X's in this model yields the predicted values and departures from the mean values as shown in Table 8.3.

Table 8.3 Departures from Linear Regression

X_j	Y'_x	$\bar{Y}_{.j}$	$\bar{Y}_{.j} - Y'_x$
0	56.0	50	−6.0
25	58.5	60	1.5
50	61.0	70	9.0
75	63.5	65	1.5
100	66.0	60	−6.0
Total			0

It is noted that these departures from linear regression add to zero but are of fairly high numerical value. In order to determine how much of the variability among temperature means can be accounted for by this linear regression, the sum of squares due to linear regression is calculated

$$SS_{\text{regression}}(b_0, b_1) = \sum_i \sum_j (Y'_x - \bar{Y}_{..})^2 = \sum_i \sum_j (b_0 + b_1 X_j - \bar{Y}_{..})^2$$

From the first of the least squares normal equations (8.3),

$$b_0 = \frac{\sum_i \sum_j Y_{ij} - b_1 n \sum_j X_j}{N}$$

$$= \bar{Y}_{..} - b_1 \sum_j X_j/k = \bar{Y}_{..} - b_1 \bar{X}$$

Hence,

$$SS_{\text{regression}}(b_0, b_1) = \sum_i \sum_j (\bar{Y}_{..} - b_1\bar{X} + b_1 X_j - \bar{Y}_{..})^2$$

$$= b_1^2 n \sum_j (X_j - \bar{X})^2$$

$$= b_1^2 n \left[\sum_j X_j^2 - \left(\sum X_j \right)^2 \Big/ k \right]$$

From the data of Table 8.1,

$$SS_{\text{regression}}(b_0, b_1) = 3(0.1)^2[18{,}750 - (250)^2/5] = 187.50$$

This sum of squares due to linear regression carries 1 df and the analysis becomes as shown in Table 8.4.

Table 8.4 ANOVA with Linear Regression

Source	df		SS	MS
Between temperatures	4		660	
Due to linear regression		1	187.50	187.50
Departure from linear regression		3	472.50	157.50
Error	10		70	7.00
Totals	14		730	

Applying F tests to the data of Table 8.4 shows a highly sighificant linear effect and also a highly significant departure from linear regression. This indicates that the regression model should be revised to allow for curvilinear regression.

Before examining a curvilinear fit, a few well-known statistics can be obtained from Table 8.4. The proportion of the total sum of squares (730) accounted for by linear regression (187.50) is r^2, the square of the *Pearson product–moment correlation* coefficient. Here $r^2 = 187.50/730 = 0.2568$ or $r = 0.51$, hardly a high correlation even though statistically significant. This result can easily be checked by one of the common formulas for Pearson r. If the sum of squares for departure from linear regression is pooled with the sum of squares for error, the resulting term $(472.50 + 70 = 542.50)$ is called the sum of squares of errors of estimate from a straight line. Its mean square $(542.50/13 = 41.73)$ is called the variance error of estimate and its square root (6.46), the *standard error of estimate*. When linear regression is sufficient to explain the variance of the data, this standard error of estimate $s_{y.x}$ is used to set confidence limits on the slope of the line β_1, on the true means $\mu_{y/x}$ for each X, and on the true observation Y_{ij} for each X.

If the sum of squares due to treatments (temperatures in this problem) is divided by the total sum of squares, the resulting statistic is usually called *eta squared* (η^2) which indicates how much of the total variability could be accounted for by a curve through all the means. Here $\eta^2 = 660/730 = 0.9041$ which indicates that a fourth-degree equation (since there are five means) could account for 90.4 percent of the variability, whereas linear regression $(Y'_x = 56 + 0.1X_j)$ only accounts for 25.68 percent. This is another way of showing the need for some higher degree model.

The methods giving the results above are general and can be used on any linear curve-fitting problem. However, some of the arithmetic is a bit cumbersome. When the X variable is set at equispaced levels, the results can be determined in a much simpler fashion.

Code the X values using $u_j = (X_j - \bar{X})/c$ where $\bar{X}$ is the mean and c the interval width on the X variable. Here $u_j = (X_j - 50)/25$. The u values are now $u_j = -2, -1, 0, +1, +2$, and the regression model is

$$Y'_x = b'_0 + b'_1 u_j \tag{8.4}$$

For this model, the least squares normal equations are

$$\begin{aligned} \sum_i \sum_j Y_{ij} &= b'_0 N + b'_1 n \sum_j u_j \\ \sum_i \sum_j u_j Y_{ij} &= b'_0 n \sum_j u_j + b'_1 n \sum_j u_j^2 \end{aligned} \tag{8.5}$$

But the u_j's have been chosen so that $\sum_j u_j = 0$. This makes the solutions of Equations (8.5)

$$\begin{aligned} b'_0 &= \frac{\sum_i \sum_j Y_{ij}}{N} = \bar{Y}_{..} \\ b'_1 &= \frac{\sum_i \sum_j u_j Y_{ij}}{n \sum_j u_j^2} \end{aligned} \tag{8.6}$$

It is readily seen that the u_j's simply serve as coefficients on the Y_{ij}'s (or on their sums for a given j) which makes $\sum_i \sum_j u_j Y_{ij}$ a contrast on the treatment totals $\sum_i Y_{ij} = T_{.j}$. Substituting

$$b'_0 = \bar{Y}_{..} = 915/15 = 61$$

$$b'_1 = \frac{(-2)(150) + (-1)(180) + (0)(210) + (1)(195) + (2)(180)}{(3)(10)}$$

$$b'_1 = 75/30 = 2.5$$

and the model is

$$Y'_x = 61 + 2.5u_j$$

That this is the same model can be seen by substituting for u_j,

$$\begin{aligned} Y'_x &= 61 + 2.5\left(\frac{X_j - 50}{25}\right) = 61 + 0.1X_j - 5 \\ &= 56 + 0.1X_j \end{aligned}$$

Since $\sum_i \sum_j u_j Y_{ij}$ is a contrast, the sum of squares due to a contrast is

$$SS_C = \frac{C^2}{n \sum_j c_{jm}^2} = \frac{(\sum_i \sum_j u_j Y_{ij})^2}{n \sum_j u_j^2} = \frac{(75)^2}{30} = 187.50$$

which is the sum of squares due to linear regression.

8.2 CURVILINEAR REGRESSION

When a linear regression model is not sufficient to explain all the significant variation in the means, the next logical step is to consider a second-degree (or quadratic) model:

$$Y'_x = b_0 + b_1 X_j + b_2 X_j^2 \tag{8.7}$$

By the methods of least squares, the normal equations are

$$\begin{aligned}
\sum_i \sum_j Y_{ij} &= b_0 N + b_1 n \sum_j X_j + b_2 n \sum_j X_j^2 \\
\sum_i \sum_j X_j Y_{ij} &= b_0 n \sum_j X_j + b_1 n \sum_j X_j^2 + b_2 n \sum_j X_j^3 \\
\sum_i \sum_j X_j^2 Y_{ij} &= b_0 n \sum_j X_j^2 + b_1 n \sum_j X_j^3 + b_2 n \sum_j X_j^4
\end{aligned} \tag{8.8}$$

These equations can be solved for b_0, b_1, and b_2, but it is difficult. Again coded by u_j's,

$$Y'_x = b'_0 + b'_1 u_j + b'_2 u_j^2 \tag{8.9}$$

$$\begin{aligned}
\sum_i \sum_j Y_{ij} &= b'_0 N + b'_1 n \sum_j u_j + b'_2 n \sum_j u_j^2 \\
\sum_i \sum_j u_j Y_{ij} &= b'_0 n \sum_j u_j + b'_1 n \sum_j u_j^2 + b'_2 n \sum_j u_j^3 \\
\sum_i \sum_j u_j^2 Y_{ij} &= b'_0 n \sum_j u_j^2 + b'_1 n \sum_j u_j^3 + b'_2 n \sum_j u_j^4
\end{aligned} \tag{8.10}$$

Because of the choice of the u_j's, the sums of all odd powers of the u_j's are zero ($\sum u_j = \sum u_j^3 = 0$). The equations become

$$\begin{aligned}
\sum_i \sum_j Y_{ij} &= b'_0 N + b'_2 n \sum_j u_j^2 \\
\sum_i \sum_j u_j Y_{ij} &= b'_1 n \sum_j u_j^2 \\
\sum_i \sum_j u_j^2 Y_{ij} &= b'_0 n \sum_j u_j^2 + b'_2 n \sum_j u_j^4
\end{aligned}$$

Substituting,

$$\begin{aligned}
915 &= 15b'_0 + 30b'_2 \\
75 &= 30b'_1 \\
1695 &= 30b'_0 + 102b'_2
\end{aligned}$$

Solving,

$$b'_0 = 67.42, \qquad b'_1 = 2.5, \qquad b'_2 = -3.21$$

and

$$Y' = 67.42 + 2.5u_j - 3.21u_j^2$$

which can be rewritten in terms of X_j if desired. Substituting the u_j values, the predicted points and departures from quadratic regression are as given in Table 8.5.

Table 8.5 Departures from Second-Degree Regression

X_j	u_j	Y'_x	$\bar{Y}_{.j}$	$\bar{Y}_{.j} - Y'_x$
0	−2	49.58	50	0.42
25	−1	61.71	60	−1.71
50	0	67.42	70	2.58
75	1	66.71	65	−1.71
100	2	59.58	60	0.42
Total				0

In graphical form they are shown in Figure 8.2, which indicates a much better fit than the straight-line model. For the sum of squares due to this second-degree equation,

$$\begin{aligned} SS_{\text{regression}}(b'_0, b'_1, b'_2) &= \sum_i \sum_j (Y'_x - \bar{Y}_{..})^2 \\ &= \sum_i \sum_j (b'_0 + b'_1 u_j + b'_2 u_j^2 - \bar{Y}_{..})^2 \end{aligned}$$

From the first of the normal equations,

$$b'_0 = \bar{Y}_{..} - \frac{b'_2 \sum u_j^2}{k}$$

Substituting,

$$\begin{aligned} SS_{\text{regression}}(b'_0, b'_1, b'_2) &= \sum_i \sum_j \left(-b'_2 \sum_j u_j^2/k + b'_1 u_j + b'_2 u_j^2 \right)^2 \\ &= \sum_i \sum_j \left(b'_2 \left[u_j^2 - \sum_j u_j^2/k \right] + b'_1 u_j \right)^2 \\ &= b_2'^2 n \sum_j \left(u_j^2 - \sum_j u_j^2/k \right)^2 + b_1'^2 n \sum_j u_j^2 \end{aligned}$$

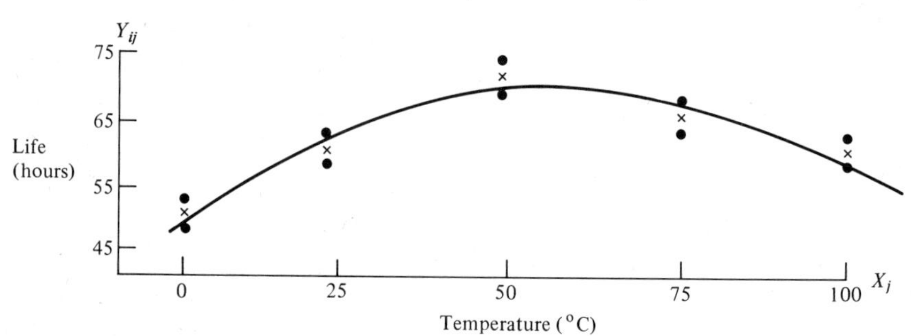

Figure 8.2 Life versus temperature.

since the cross-product terms are zero as $\sum u_j = 0$. Substituting,

$$\begin{aligned} SS_{\text{regression}}(b'_0, b'_1, b'_2) &= (-3.21)^2(3)(14) + (2.5)^2(3)(10) \\ &= 433.93 + 187.50 = 621.43 \end{aligned}$$

Note that the first term, 433.93, is the amount of variability accounted for by the second-degree term since 187.50 was the amount accounted for by a straight line. Summarizing we obtain Table 8.6.

Table 8.6 ANOVA with Second-Degree Regression

Source	df		SS		MS
Between temperatures	4		660		
Due to linear regression b_1		1		187.50	187.50**
Due to quadratic over linear regression b_2/b_1		1		433.93	433.93**
Departure from quadratic model		2		38.57	19.28
Error	10		70		7.00
Total	14		730		

** Two asterisks indicate significance at the 1-percent level.

Here the F ratio shows highly significant linear and quadratic effects and the departure from the quadratic model is not significant at the 5-percent significance level. One might very well stop here and conclude that the quadratic model accounts for all the significant effects of temperature on life of this part. The life can also be predicted quite accurately by the model

$$Y'_x = 67.42 + 2.5u_j - 3.21u_j^2$$

where

$$u_j = \frac{X_j - 50}{25}$$

Another statistic often reported is the proportion of total variation accounted for by the highest degree equation which is significant. In this case the quadratic model accounts for 621.43/730 or 85.13 percent of the variability. This quantity is called the *correlation index* R^2. When linear regression accounts for all the significant variability, $R^2 = r^2$. Note, too, that $\eta^2 - r^2$ measures the departure from linear regression and $\eta^2 - R^2$, the departure from curvilinear regression.

8.3 ORTHOGONAL POLYNOMIALS

In the above discussions on linear and quadratic regression, it was noted that the problem could be solved much easier if the X_j values were equispaced and coded such that $\sum_j u_j = \sum_j u_j^3 = 0$. The advantage of this type of curve fitting are even greater as the degree of the polynomial increases. If

$$Y'_x = f(X, X^2, X^3, \cdots, X^k)$$

is a polynomial in X, it has been shown that it may be rewritten in the form

$$Y'_x = A_0\xi' + A_1\xi'_1 + A_2\xi'_2 + \cdots + A_k\xi'_k \tag{8.11}$$

where each ξ'_i is a polynomial of degree i and all polynomials, such as ξ'_m and ξ'_q, are orthogonal to each other. The advantage of writing the model in this form is that additional polynomials of higher degree may be added, which are independent of the ones already considered. The values of these polynomials are given for the first five powers of u as

$$\begin{aligned}
\xi'_0 &= 1 \\
\xi'_1 &= \lambda_1 u \\
\xi'_2 &= \lambda_2\left[u^2 - \frac{k^2-1}{12}\right] \\
\xi'_3 &= \lambda_3\left[u^3 - u\left(\frac{3k^2-7}{20}\right)\right] \\
\xi'_4 &= \lambda_4\left[u^4 - \frac{u^2}{14}(3k^2-13) + \frac{3}{560}(k^2-1)(k^2-9)\right] \\
\xi'_5 &= \lambda_5\left[u^5 - \frac{5u^3}{18}(k^2-7) + \frac{u}{1008}(15k^4-230k^2+407)\right]
\end{aligned} \tag{8.12}$$

where k is the number of levels of the factor X, and the λ's are chosen so that the ξ'_i's are integers for all u. Because of the complications of these formulas and the practical need to test for significant high-order regression models, the values of the ξ'_i's and λ's have been tabled by Fisher and Yates [9], and a section is reproduced as Appendix Table F.

The A_i's are given by

$$A_i = \frac{\sum_{i,j} Y_{ij}\,\xi'_i}{\sum_{i,j} (\xi'_i)^2} \tag{8.13}$$

In order to see how these formulas and tables apply, consider a quadratic model with just three points on the X scale. The u_j's would be $-1, 0, +1$. As these are already integers, $\lambda_1 = 1$ and ξ'_1 takes on values $-1, 0, +1$ for each j. For ξ'_2, $(k^2 - 1)/12 = (9 - 1)/12 = 2/3$ and the u^2 values are $+1, 0, +1$ which makes $[u^2 - (k^2 - 1)/12] = 1/3, -2/3, 1/3$ so λ_2 is taken as equal to 3 in order to make the ξ'_2 integers. Then the ξ'_2 are $+1$, -2, $+1$. Appendix Table F, gives these values for $k = 3$. Also given in the table are $\sum (\xi'_i)^2$ for each set of ξ'_i and the λ_i. In the expression for A_i, each Y_{ij} is multiplied by the corresponding ξ'_i, and since $\sum_i \xi'_i = 0$, this expression is a contrast. Since the coefficients $-1, 0, +1$ are orthogonal to

1, −2, 1, these provide two orthogonal contrasts on the Y_{ij}'s (or their totals, $T_{.j}$) and the sum of squares for such contrasts is given by

$$SS_C = \frac{(\sum_i \sum_j Y_{ij} \xi_i')^2}{n \sum_i (\xi_i')^2}$$

One needs only to apply the coefficients given in Appendix Table F, to the treatment totals of a single-factor problem in order to determine the sum of squares for linear, quadratic, cubic, quartic, and so on, effects up to the $(k - 1)$th-degree equation. Equations (8.11), (8.12), and (8.13) are only needed if the prediction model $Y_x' = f(u, u^2, u^3, \cdots, u^{k-1})$ is desired.

To illustrate this, consider the example of Sections 8.1 and 8.2. Since $k = 5$ levels of temperature, Table F indicates the following coefficients for the linear, quadratic, cubic, and quartic effects:

						$\sum_i (\xi_i')^2$	λ
Linear	−2	−1	0	1	2	10	1
Quadratic	2	−1	−2	−1	2	14	1
Cubic	−1	2	0	−2	1	10	5/6
Quartic	1	−4	6	−4	1	70	35/12
$T_{.j}$	150	180	210	195	180		

Applying each set of coefficients to the treatment totals gives

Linear $(-2)(150) + (-1)(180) + (0)(210) + (1)(195) + (2)(180) = 75$

Quadratic $(+2)(150) + (-1)(180) + (-2)(210) + (-1)(195) + (2)(180) = -135$

Cubic $(-1)(150) + (2)(180) + (0)(210) + (-2)(195) + (1)(180) = 0$

Quartic $(+1)(150) + (-4)(180) + (6)(210) + (-4)(195) + (1)(180) = 90$

The corresponding sums of squares are

$$SS_{\text{linear}} = \frac{(75)^2}{3(10)} = 187.50$$

$$SS_{\text{quadratic}} = \frac{(-135)^2}{3(14)} = 433.93$$

$$SS_{\text{cubic}} = \frac{(0)^2}{3(10)} = 0.00$$

$$SS_{\text{quartic}} = \frac{(90)^2}{3(70)} = 38.57$$

and the ANOVA can be written as in Table 8.7.

Table 8.7 ANOVA for Polynomial Model

Source	df		SS		MS
Between temperatures	4		660		—
T_{linear}		1		187.50	187.50**
$T_{\text{quadratic}}$		1		433.93	433.93**
T_{cubic}		1		0.00	0.00
T_{quartic}		1		38.57	38.57*
Error	10		70		7.00
Total	14		730		

* One asterisk indicates significance at the 5-percent level; two, the 1-percent level.

This table confirms the results in Sections 8.1 and 8.2 but whereas the departure from the quadratic model was not significant in Table 8.6, this table shows a slightly significant quartic effect at the 5-percent level of significance. The analysis might be terminated here, but if a prediction equation is desired, Equations (8.11), (8.12), and (8.13) must be used. The A_i's are

$$A_0 = \frac{\sum_{i,j} Y_{ij}(1)}{\sum_{i,j}(1)^2} = \frac{\sum_{i,j} Y_{ij}}{N} = \bar{Y}_{..} = 61$$

$$A_1 = \frac{75}{3(10)} = 2.5$$

$$A_2 = \frac{(-135)}{3(14)} = -3.21$$

$$A_3 = \frac{0}{3(10)} = 0$$

$$A_4 = \frac{90}{3(70)} = 0.43$$

and the ξ_i''s are

$$\xi_0' = 1$$

$$\xi_1' = 1u$$

$$\xi_2' = 1\left[u^2 - \frac{(5)^2 - 1}{12}\right] = u^2 - 2$$

$$\xi_3' = \frac{5}{6}\left[u^3 - u\left(\frac{75 - 7}{20}\right)\right] = \frac{5}{6}u^3 - \frac{17}{6}u = \frac{1}{6}(5u^3 - 17u)$$

$$\xi_4' = \frac{35}{12}\left[u^4 - \frac{u^2}{14}(62) + \frac{3}{560}(24)(16)\right] = \frac{1}{12}(35u^4 - 155u^2 + 72)$$

Substituting in Equation (8.11),

$$Y'_x = 61 + 2.5u + (-3.21)(u^2 - 2) + 0.43(1/12)(35u^4 - 155u^2 + 72)$$
$$= 69.99 + 2.5u - 8.74u^2 + 1.25u^4$$

as the quartic model which should pass through all five temperature means.

Table 8.8 Departures from Quartic Regression

X_j	u_j	Y'_x	$\bar{Y}_{.j}$	$\bar{Y}_{.j} - Y'_x$
0	−2	50.03	50	−0.03
25	−1	60.00	60	0.00
50	0	69.99	70	0.01
75	1	65.00	65	0.00
100	2	60.03	60	−0.03

Table 8.8 checks within the accuracy of the data.

If a quadratic model seems satisfactory, compute only the first three terms in Equation (8.11):

$$Y'_x = 61 + 2.5u_j + (-3.21)(u_j^2 - 2)$$
$$= 67.42 + 2.5u_j - 3.21u_j^2$$

which checks with the model computed in Section 8.2.

8.4 QUANTITATIVE AND QUALITATIVE FACTORS

In Sections 8.1 to 8.3, a single factor was considered at quantitative levels only. Qualitative levels of a single factor were discussed in Chapter 3 and for two or more factors in Chapter 6. When two or more factors are involved in an experiment, some may be at qualitative levels and some at quantitative levels. Two factors are considered in the sections which follow.

8.5 TWO FACTORS—ONE QUALITATIVE, ONE QUANTITATIVE

A problem will now be considered where one factor is set at qualitative levels and the other at quantitative (and equispaced) levels.

We wish to determine the effect of both depth and position in a tank (Figure 8.3) on the concentration of a cleaning solution in ounces per gallon. Concentrations are measured at three depths from the surface of the tank, 0 in., 15 in., and 30 in. At each depth measurements are taken at five different lateral positions in the tank. These are considered as five qualitative positions, although probably some orientation measure might be made on them. At

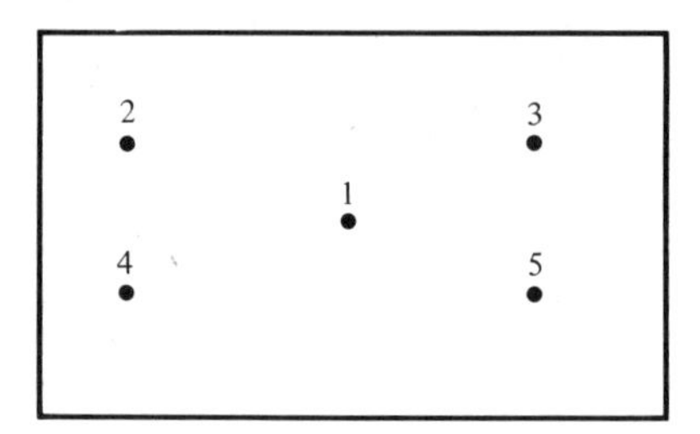

Figure 8.3

each depth and position, two observations are taken. This is then a 5 × 3 factorial with two replications per cell (total of 30 observations). The design is a completely randomized design and the model is

$$Y_{ijk} = \mu + D_i + P_j + DP_{ij} + \varepsilon_{k(ij)}$$

with

$$i = 1, 2, 3 \qquad j = 1, 2, \cdots, 5 \qquad k = 1, 2$$

where D_i represents the depth and P_j the position. The data collected are shown in Table 8.9.

Table 8.9 Cleaning-Solution Concentration Data

	Depth from Top of Tank D_i		
Position P_j	0 in.	15 in.	30 in.
1	5.90	5.90	5.94
	5.91	5.89	5.80
2	5.90	5.89	5.75
	5.91	5.89	5.83
3	5.94	5.91	5.86
	5.90	5.91	5.83
4	5.93	5.94	5.83
	5.91	5.90	5.89
5	5.90	5.94	5.83
	5.87	5.90	5.86

In this case where one factor is set at quantitative levels (D_i) and the other at qualitative levels (P_j), the first step is to run a two-factor analysis of variance just as if both were qualitative factors. Coding the data of Table 8.9 by subtracting 5.90 and multiplying by 100 gives Table 8.10.

Table 8.10 Coded Cleaning-Solution Concentration Data with Totals

Position P_j	Depth from Top of Tank D_i						$T_{.j.}$
	0 in.		15 in.		30 in.		
1	0		0		4		
	1		−1		−10		
		1		−1		−6	−6
2	0		−1		−15		
	1		−1		−7		
		1		−2		−22	−23
3	4		1		−4		
	0		1		−7		
		4		2		−11	−5
4	3		4		−7		
	1		0		−1		
		4		4		−8	0
5	0		4		−7		
	−3		0		−4		
		−3		4		−11	−10
$T_{i..}$		7		7		−58	$T_{...} = -44$
$\sum_{k=1}^{2} \sum_{j=1}^{5} Y_{ijk}^2$		37		37		570	$\sum_i^3 \sum_j^5 \sum_k^2 Y_{ijk}^2 = 644$

From these coded data,

$$SS_{\text{total}} = 644 - \frac{(-44)^2}{30} = 579.47$$

$$SS_D = \frac{(7)^2 + (7)^2 + (-58)^2}{10} - \frac{(-44)^2}{30} = 281.67$$

$$SS_P = \frac{(-6)^2 + (-23)^2 + (-5)^2 + (0)^2 + (-10)^2}{6} - \frac{(-44)^2}{30} = 50.47$$

$$SS_{D \times P \text{ interaction}} = \frac{(1)^2 + (1)^2 + (4)^2 + \cdots + (-11)^2}{2} - \frac{(-44)^2}{30} - 281.67 - 50.47 = 58.33$$

$$SS_{\text{error}} = 579.47 - 281.67 - 50.47 - 58.33 = 189.00$$

giving the ANOVA table of Table 8.11.

Table 8.11 ANOVA for Cleaning-Solution Concentration Problem

Source	df	SS	MS
Depths D_i	2	281.67	140.83
Positions P_j	4	50.47	12.62
$D \times P$ interaction DP_{ij}	8	58.33	7.29
Error $\varepsilon_{k(ij)}$	15	189.00	12.60
Totals	29	579.47	

From Table 8.11 it is seen that only depth produced a significant effect on concentration of solution as

$$F_{2,15} = \frac{140.83}{12.60} = 11.24$$

which is significant at the 1-percent level of significance. Since the depth effect is significant and since the three depths are equispaced, it may be worthwhile to extract a linear and quadratic depth effect to learn how concentration varies with depth. The coefficients for $k = 3$ levels are shown in Table 8.12.

Table 8.12 Orthogonal Coefficients

				$\sum (\xi_i')^2$	λ
Linear	-1	0	$+1$	2	1
Quadratic	$+1$	-2	$+1$	6	3
$T_{i..}$	7	7	-58		

Applying these coefficients to the depth totals $T_{i..}$ gives

$$\text{linear effect of depth} = -1(7) + 0(7) + 1(-58) = -65$$

$$\text{quadratic effect of depth} = +1(7) - 2(7) + 1(-58) = -65$$

Since these are orthogonal contrasts, their sums of squares are

$$SS_{\text{linear}} = \frac{(-65)^2}{10(2)} = 211.25$$

$$SS_{\text{quadratic}} = \frac{(-65)^2}{10(6)} = \underline{70.42}$$

$$281.67$$

and their total is the depth sum of squares.

Even though the $D \times P$ interaction is not significant, there may be an interaction between the linear effect of depth and positions or between the quadratic effect of depth and positions. To compute these interactions, the linear effect of depth is determined at each position and these effects are then compared to see whether or not they differ. The same procedure is followed for the quadratic effect of depth at each position.

Applying the linear coefficients of -1, 0, $+1$ at each position gives

$$\begin{aligned}
P_1&: -1(1) + 0(-1) + 1(-6) &= -7\\
P_2&: -1(1) + 0(-2) + 1(-22) &= -23\\
P_3&: -1(4) + 0(2) + 1(-11) &= -15\\
P_4&: -1(4) + 0(4) + 1(-8) &= -12\\
P_5&: -1(-3) + 0(4) + 1(-11) &= \underline{-8}\\
&& -65
\end{aligned}$$

To compare these five effects, determine the sum of squares between them

$$\frac{(-7)^2 + (-23)^2 + (-15)^2 + (-12)^2 + (-8)^2}{2(2)} - \frac{(-65)^2}{10(2)} = 41.50$$

Similarly, for the quadratic effect of depth at the five positions, we have

$$\begin{aligned}
P_1&: +1(1) - 2(-1) + 1(-6) &= -3\\
P_2&: +1(1) - 2(-2) + 1(-22) &= -17\\
P_3&: +1(4) - 2(2) + 1(-11) &= -11\\
P_4&: +1(4) - 2(4) + 1(-8) &= -12\\
P_5&: +1(-3) - 2(4) + 1(-11) &= \underline{-22}\\
&& -65
\end{aligned}$$

Comparing these five quadratic effects gives

$$\frac{(-3)^2 + (-17)^2 + (-11)^2 + (-12)^2 + (-22)^2}{2(6)} - \frac{(-65)^2}{10(6)} = 16.83$$

Note that the sum of these two sums of squares (41.50 + 16.83) equals the $D \times P$ interaction sum of squares (58.33) as it should. Summarizing for this problem with one quantitative and one qualitative factor, the complete ANOVA breakdown is as shown in Table 8.13.

The results in Table 8.13 indicate a strong depth effect, with the linear depth effect significant at the 1-percent level (**) and the quadratic depth effect significant at the 5-percent level (*). There is no position effect nor

Table 8.13 Complete ANOVA for Cleaning-Solution Concentration Problem

Source	df		SS		MS
Depths	2		281.67		
Linear		1		211.25	211.25**
Quadratic		1		70.42	70.42*
Positions	4		50.47		12.62
$D \times P$ interaction	8		58.33		
$D_{\text{linear}} \times P$		4		41.50	10.38
$D_{\text{quadratic}} \times P$		4		16.83	4.21
Error	15		189.00		12.60
Totals	29		579.47		

interaction between depth and position. These results seem reasonable from a graph of cell totals versus depth and positions (Figure 8.4).

This graph shows little interaction, as the curves are quite parallel (statistically speaking) and there is little difference between the five position curves. The depth effect is quite obvious, and concentration is seen to drop off with increasing depth but on more of a curve than a straight line. Further investigation is suggested at depths between those already studied. The lack of any position effect or interaction should mean that this new experiment could be run at only one position; the results should then be the same at all five positions.

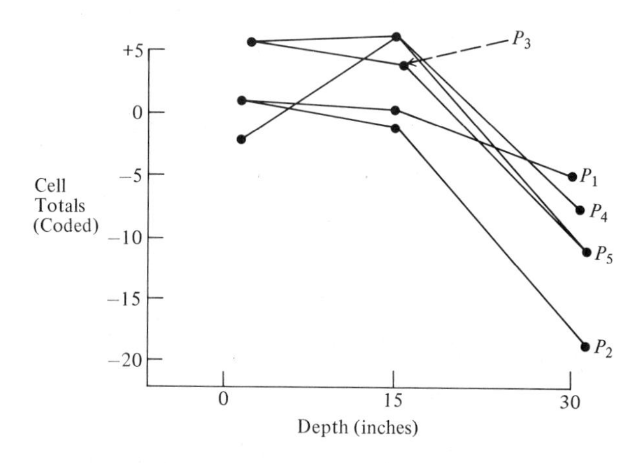

Figure 8.4 Graph of cleaning solution concentration problem.

If one wishes the equation for predicting concentration (coded) from depth,

$$u = \frac{X - 15}{15}$$

and

$$A_0 = \bar{Y}_{..} = -44/30 = -1.47$$

$$A_1 = -65/10(2) = -3.25$$

$$A_2 = -65/10(6) = -1.08$$

and

$$\xi'_0 = 1$$

$$\xi'_1 = 1u$$

$$\xi'_2 = 3(u^2 - 2/3) = 3u^2 - 2$$

and

$$Y'_x = -1.47 - 3.25u - 1.08(3u^2 - 2)$$

$$= 0.69 - 3.25u - 3.24u^2$$

giving predictions as follows:

X_j	u_j	Y'_x	$\bar{Y}_.$
0	−1	0.70	0.70
15	0	0.69	0.70
30	1	−5.80	−5.80

which match exactly as they should.

8.6 TWO FACTORS—BOTH QUANTITATIVE

If both factors in a two-factor factorial are at quantitative levels, each factor can be broken down into its linear, quadratic, or cubic effects, and all combinations of interaction can be determined, such as linear by linear, linear by quadratic, and quadratic by quadratic. To illustrate the procedure for analyzing a factorial experiment where both factors are at quantitative (and equispaced) levels, consider a problem in which we wish to study the effect of knife-edge radius R in inches and feedroll force F in pounds per inch on the energy necessary to cut 1-inch lengths of alfalfa. The measured variable (energy) was measured in 100 inch-pounds per pound of dry matter. The resulting data are given in Table 8.14.

Table 8.14 Data for Alfalfa Cutting Problem

Feedroll Force F_j (lb/in)	Knife-Edge Radius R_i (inches) 0.000		0.005		0.010		0.015		$T_{.j.}$
5	29		98		44		84		
	30		128		81		100		
	20		67		77		63		
		79		293		202		247	821
10	22		35		53		103		
	26		80		93		90		
	16		29		59		98		
		64		144		205		291	704
15	18		49		59		80		
	17		68		103		91		
	11		61		128		77		
		46		178		289		248	761
20	38		68		87		86		
	31		74		116		113		
	21		47		90		81		
		90		189		293		280	852
$T_{i..}$		279		804		989		1066	$T_{...} = 3138$

Here the experiment was run in a completely random order with three replications per cell. The model was

$$Y_{ijk} = \mu + R_i + F_j + RF_{ij} + \varepsilon_{k(ij)}$$

with

$$i = 1, 2, 3, 4 \qquad j = 1, 2, 3, 4 \qquad k = 1, 2, 3$$

Both the knife-edge radius and the feedroll force were set at four equispaced quantitative levels. In analyzing these data, an analysis is first made as if both factors were qualitative. Using the data of Table 8.14, the following statistics were computed:

$$SS_{\text{total}} = 254{,}656 - \frac{(3138)^2}{48} = 49{,}509.25$$

$$SS_{\text{radius}} = \frac{(279)^2 + (804)^2 + (989)^2 + (1066)^2}{12} - \frac{(3138)^2}{48}$$

$$= 31{,}414.42$$

$$\text{SS}_{\text{force}} = \frac{(821)^2 + (704)^2 + (761)^2 + (852)^2}{12} - \frac{(3138)^2}{48}$$

$$= 1076.75$$

$$\text{SS}_{R \times F \text{ interaction}} = \frac{(79)^2 + (64)^2 + \cdots + (280)^2}{3} - \frac{(3138)^2}{48} - 31{,}414.42$$

$$- 1076.75 = 6474.08$$

$$\text{SS}_{\text{error}} = 49{,}509.25 - 31{,}414.42 - 1076.75 - 6474.08$$

$$= 10{,}544.00$$

The ANOVA is summarized in Table 8.15.

Table 8.15 First ANOVA for Alfalfa Problem

Source	df	SS	MS	F
R_i	3	31,414.42	10,471.47	31.78**
F_j	3	1076.75	358.92	1.09
$(R \times F)_{ij}$	9	6474.08	719.34	2.18
$\varepsilon_{k(ij)}$	32	10,544.00	329.50	
Totals	47	49,509.25		

** Two asterisks indicate significance at the 1-percent level.

The results of this first analysis show that the knife-edge radius R_i has a highly significant effect on the energy requirements. There is no apparent force effect and the interaction is not quite significant at the 5-percent level of significance.

Since both effects are quantitative, they will be broken down further in an attempt to see how the energy requirement might be related to each factor. Even though feedroll force F_j is not significant here, it will nevertheless be broken into its component parts to illustrate the method of analysis. As both factors are at four equispaced quantitative levels, the proper coefficients for each factor total are given by the orthogonal polynomials of Appendix Table F.

					$\sum (\xi_i')^2$	λ
Linear	−3	−1	+1	+3	20	2
Quadratic	+1	−1	−1	+1	4	1
Cubic	−1	+3	−3	+1	20	10/3

Applying these coefficients to both the radius totals and force totals, linear, quadratic, and cubic effects and sums of squares can be determined as follows:

Sums of Squares

$$R_L = -3(279) - 1(804) + 1(989) + 3(1066) = 2546$$

$$SS_{R_L} = \frac{(2546)^2}{12(20)} = 27{,}008.82$$

$$R_Q = +1(279) - 1(804) - 1(989) + 1(1066) = -448$$

$$SS_{R_Q} = \frac{(-448)^2}{12(4)} = 4{,}181.33$$

$$R_C = -1(279) + 3(804) - 3(989) + 1(1066) = 232$$

$$SS_{R_C} = \frac{(232)^2}{12(20)} = 224.27$$

$$SS_R = 31{,}414.42$$

$$F_L = -3(821) - 1(704) + 1(761) + 3(852) = 150$$

$$SS_{F_L} = \frac{(150)^2}{12(20)} = 93.75$$

$$F_Q = +1(821) - 1(704) - 1(761) + 1(852) = 208$$

$$SS_{F_Q} = \frac{(208)^2}{12(4)} = 901.33$$

$$F_C = -1(821) + 3(704) - 3(761) + 1(852) = -140$$

$$SS_{F_C} = \frac{(-140)^2}{12(20)} = 81.67$$

$$SS_F = 1{,}076.75$$

Each of these effects has 1 df, and each may be tested for significance. The $R \times F$ interaction with its 9 df can be broken down into nine single df components as follows:

$$R_L \times F_L \qquad R_Q \times F_L \qquad R_C \times F_L$$

$$R_L \times F_Q \qquad R_Q \times F_Q \qquad R_C \times F_Q$$

$$R_L \times F_C \qquad R_Q \times F_C \qquad R_C \times F_C$$

In practice it is often difficult to interpret some of these higher polynomial interactions, and sometimes only the linear by linear, linear by quadratic, and quadratic by linear interactions are computed and the remaining ones lumped into a residual interaction. Since the objective of this problem is to illustrate the method of analysis, all nine components will be computed. A simple computing scheme for these single degree-of-freedom interactions involves only the cell totals and proper coefficients, since cell totals form the basis for interaction effects. The proper coefficients for these cell totals can be found by multiplying the corresponding coefficients of the main effects. To illustrate this, consider the $R_L \times F_L$ interaction arrayed as in Table 8.16.

Table 8.16 Linear by Linear Interaction

	R_L			
F_L	−3	−1	+1	+3
−3	9 79	3 293	−3 202	−9 247
−1	3 64	1 144	−1 205	−3 291
+1	−3 46	−1 178	+1 289	+3 248
+3	−9 90	−3 189	+3 293	+9 280

Multiplying the coefficients of R_L by those of F_L gives the coefficients in the upper left-hand corners of the cells in Table 8.16. These coefficients are then applied to the corresponding cell totals. The results give a contrast:

$$\begin{aligned} R_L \times F_L = {} & 9(79) + 3(64) - 3(46) - 9(90) + 3(293) + 1(144) \\ & - 1(178) - 3(189) - 3(202) - 1(205) + 1(289) \\ & + 3(293) - 9(247) - 3(291) + 3(248) + 9(280) = 758 \end{aligned}$$

The sum of squares due to $R_L \times F_L$ is then

$$SS_{R_L \times F_L} = \frac{(758)^2}{3[9^2 + 3^2 + (-3)^2 + \cdots + 9^2]} = \frac{(758)^2}{3(400)} = 478.80$$

For $R_L \times F_Q$, use the linear coefficients on R multiplied by the quadratic coefficients on F as in Table 8.17.

Table 8.17 Linear by Quadratic Interaction

	R_L				
F_Q	−3	−1	+1	+3	
+1	−3	−1	+1	+3	
−1	+3	+1	−1	−3	$\sum_{i,j} c_{ij}^2 = 80$
−1	+3	+1	−1	−3	
+1	−3	−1	+1	+3	

These cell coefficients are then applied to the cell totals

$$R_L \times F_Q = -3(79) + 3(64) + \cdots + 3(280) = -372$$

$$SS_{R_L \times F_Q} = \frac{(-372)^2}{3(80)} = 576.60$$

For $R_L \times F_C$, the cell coefficients are as shown in Table 8.18.

Table 8.18 Linear by Cubic Interaction

	R_L				
F_C	−3	−1	+1	+3	
−1	3	1	−1	−3	
+3	−9	−3	3	9	$\sum_{i,j} c_{ij}^2 = 400$
−3	9	3	−3	−9	
+1	−3	−1	1	3	

$$R_L \times F_C = 3(79) - 9(64) + \cdots + 3(280) = 336$$

$$SS_{R_L \times F_C} = \frac{(336)^2}{3(400)} = 94.08$$

For $R_Q \times F_L$, the cell coefficients are as shown in Table 8.19.

Table 8.19 Quadratic by Linear Interaction

	R_Q				
F_L	+1	−1	−1	+1	
−3	−3	3	3	−3	
−1	−1	1	1	−1	$\sum_{i,j} c_{ij}^2 = 80$
+1	+1	−1	−1	1	
+3	+3	−3	−3	3	

$$R_Q \times F_L = -3(79) + 3(64) + \cdots + 3(280) = -8$$

$$\text{SS}_{R_Q \times F_L} = \frac{(-8)^2}{3(80)} = 0.27$$

For $R_Q \times F_Q$, the cell coefficients are as shown in Table 8.20.

Table 8.20 Quadratic by Quadratic Interaction

	R_Q				
F_Q	+1	−1	−1	+1	
+1	1	−1	−1	+1	
−1	−1	1	1	−1	$\sum_{i,j} c_{ij}^2 = 16$
−1	−1	1	1	−1	
+1	1	−1	−1	1	

$$R_Q \times F_Q = 1(79) - 1(64) + \cdots + 1(280) = -114$$

$$\text{SS}_{R_Q \times F_Q} = \frac{(-114)^2}{3(16)} = 270.75$$

For $R_Q \times F_C$, the cell coefficients are as shown in Table 8.21.

Table 8.21 Quadratic by Cubic Interaction

	R_Q				
F_C	+1	−1	−1	+1	
−1	−1	1	1	−1	
+3	3	−3	−3	3	$\sum_{i,j} c_{ij}^2 = 80$
−3	−3	3	3	−3	
+1	1	−1	−1	1	

$$R_Q \times F_C = -1(79) + 3(64) + \cdots + 1(280) = 594$$

$$\text{SS}_{R_Q \times F_C} = \frac{(594)^2}{3(80)} = 1470.15$$

For $R_C \times F_L$, the cell coefficients are as shown in Table 8.22.

Table 8.22 Cubic by Linear Interaction

F_L	R_C: −1	+3	−3	+1	
−3	3	−9	9	−3	
−1	1	−3	3	−1	$\sum_{i,j} c_{ij}^2 = 400$
+1	−1	3	−3	1	
+3	−3	9	−9	3	

$$R_C \times F_L = 3(79) + 1(64) + \cdots + 3(280) = -1864$$

$$SS_{R_C \times F_L} = \frac{(-1864)^2}{3(400)} = 2895.41$$

For $R_C \times F_Q$, the cell coefficients are as shown in Table 8.23.

Table 8.23 Cubic by Quadratic Interaction

F_Q	R_C: −1	+3	−3	+1	
+1	−1	3	−3	1	
−1	+1	−3	3	−1	$\sum_{i,j} c_{ij}^2 = 80$
−1	1	−3	3	−1	
+1	−1	3	−3	1	

$$R_C \times F_Q = -1(79) + 1(64) + \cdots + 1(280) = 406$$

$$SS_{R_C \times F_Q} = \frac{(406)^2}{3(80)} = 686.82$$

For $R_C \times F_C$, the cell coefficients are as shown in Table 8.24.

Table 8.24 Cubic by Cubic Interaction

F_C	R_C: −1	+3	−3	+1	
−1	1	−3	3	−1	
+3	−3	9	−9	3	$\sum_{i,j} c_{ij}^2 = 400$
−3	3	−9	9	−3	
+1	−1	3	−3	1	

$$R_C \times F_C = 1(79) - 3(64) + \cdots + 1(280) = -38$$

$$SS_{R_C \times F_C} = \frac{(-38)^2}{3(400)} = 1.20$$

If all of these single degree-of-freedom, interaction sums of squares are added, the sum of the overall $R \times F$ interaction sums of squares is

$$\begin{aligned}
SS_{R_L \times F_L} &= 478.80 \\
SS_{R_L \times F_Q} &= 576.60 \\
SS_{R_L \times F_C} &= 94.08 \\
SS_{R_Q \times F_L} &= 0.27 \\
SS_{R_Q \times F_Q} &= 270.75 \\
SS_{R_Q \times F_C} &= 1470.15 \\
SS_{R_C \times F_L} &= 2895.41 \\
SS_{R_C \times F_Q} &= 686.82 \\
SS_{R_C \times F_C} &= 1.20 \\
\hline
SS_{\text{overall } R \times F} &= 6474.08
\end{aligned}$$

Summarizing all of the above results in a complete ANOVA table along with the significance tests, we get Table 8.25.

Table 8.25 Complete ANOVA for Alfalfa Problem

Source	df		SS		MS	F
Radius R_i	3		31,414.42			
R_L		1		27,008.82	27,008.82	81.97***
R_Q		1		4181.33	4181.33	12.69**
R_C		1		224.27	224.27	<1
Force F_j	3		1076.75			
F_L		1		93.75	93.75	<1
F_Q		1		901.33	901.33	2.74
F_C		1		81.67	81.67	<1
$R \times F$ interaction	9		6474.08			
$R_L \times F_L$		1		478.80	478.80	1.45
$R_L \times F_Q$		1		576.60	576.60	1.75
$R_L \times F_C$		1		94.08	94.08	<1
$R_Q \times F_L$		1		0.27	0.27	<1
$R_Q \times F_Q$		1		270.75	270.75	<1
$R_Q \times F_C$		1		1470.15	1470.15	4.46*
$R_C \times F_L$		1		2895.41	2895.41	8.79**
$R_C \times F_Q$		1		686.82	686.82	2.08
$R_C \times F_C$		1		1.20	1.20	<1
Error	32		10,544.00		329.50	
Totals	47		49,509.25			

* One asterisk indicates significance at the 5-percent level; two, the 1-percent level; and three the 0.1-percent level.

The complete analysis of Table 8.25 shows a highly significant linear effect and a significant quadratic effect for the knife-edge radius. No force effects are found to be significant, but there is a highly significant interaction between the cubic component of radius and the linear component of force, and there is also a significant interaction between the quadratic component of radius and the cubic component of force. This $R_C \times F_L$ interaction may mean that linear changes in force produce different cubic trends because of the radius. Also, the $R_Q \times F_C$ may mean that quadratic changes in the radius produce different cubic trends because of the force. Some of these conclusions can be seen in Figure 8.5, a graph of cell totals versus radius for the four force levels.

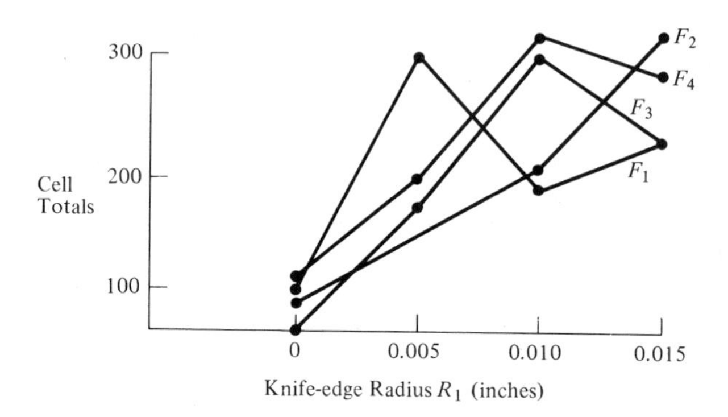

Figure 8.5 Graph I of alfalfa problem.

Figure 8.5 shows the strong linear effect and the overall quadratic effect of radius. The closeness of the lines also shows a lack of force effect. Some sort of interaction may be indicated, as the curves are not parallel. Replotting with cell totals versus force at four radius levels gives Figure 8.6. Figure 8.6

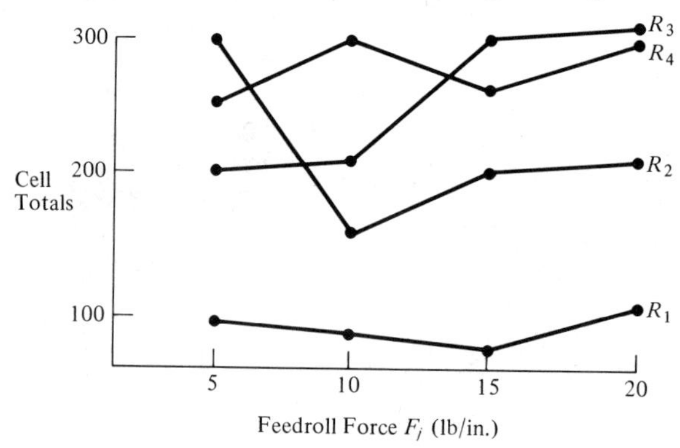

Figure 8.6 Graph II of alfalfa problem.

also shows the strong radius effect, indicated by the gaps between most of the curves. It may also help to show the different curvilinear effects of radius for linear shifts in force.

8.7 SUMMARY

The examples of this chapter may easily be extended to higher order factorials whenever one or more factors are considered at quantitative levels. The use of orthogonal polynomials makes the analysis rather simple, provided the experiment is designed with equispaced quantitative levels. The summary of designs at the end of Chapter 7 has not been changed by this chapter, as these methods can be used on all experiments where quantitative levels are involved.

PROBLEMS

8.1 In Problem 3.1 assume that the five levels of factor A are quantitative and equispaced. Using orthogonal polynomials, determine the sum of squares for linear, quadratic, cubic, and quartic effects of factor A and test for significance.

8.2 Fit the proper polynomial to the data of Problem 8.1 and show how good a predictor it is.

8.3 Determine r^2, η^2, and R^2 for various degree polynomials in Problem 8.2. Discuss.

8.4 In Problem 4.1 assume that the three days are equispaced in time from their date of manufacture and extract the quantitative effects and interactions (or error).

8.5 Data on screen quality for lacquer concentration and standing time effects are as follows:

Standing Time	Lacquer Concentration			
	$\frac{1}{2}$	1	$1\frac{1}{2}$	2
30	16	12	17	13
	14	11	19	11
20	15	14	15	12
	15	17	18	14
10	10	7	10	9
	9	6	14	13

Assuming a completely randomized design, do an ANOVA on these data using the general methods of Chapter 6.

8.6 Since both factors in Problem 8.5 are quantitative and equispaced, set up orthogonal polynomials and pull out all possible effects and their interactions.

8.7 Write prediction equations for predicting screen quality from lacquer concentration and from standing times in Problem 8.5. Explain how these equations are to be used.

8.8 Show graphically the reasonableness of your results in Problem 8.6.

9

3^n-Factorial Experiments

9.1 INTRODUCTION

As 2^n-factorial experiments represent an interesting special case of factorial experimentation, so also do 3^n-factorial experiments. 3^n-factorials consider n factors each at three levels; thus there are 2 df between the levels of each of these factors. If the three levels are quantitative and equispaced, the methods of Chapter 8 may be used to extract linear and quadratic effects and to test these for significance. 3^n factorials also play an important role in more complicated design problems which will be discussed in subsequent chapters. For this chapter, it will be assumed that the design is a completely randomized design and that the levels of the factors considered are fixed levels. Such levels may be either qualitative or quantitative.

9.2 3^2 FACTORIAL

If just two factors are crossed in an experiment and each of the two is set at three levels, there are $3 \times 3 = 9$ treatment combinations. Since each factor is at three levels, the notation of Chapter 7 will no longer suffice. There is now a low, intermediate, and high level for each factor which may be designated as 0, 1, and 2. A model for this arrangement would be

$$Y_{ij} = \mu + A_i + B_j + AB_{ij} + \varepsilon_{ij}$$

where $i = 1, 2, 3$, $j = 1, 2, 3$, and the error term is confounded with the AB interaction unless there are some replications in the nine cells, in which case

$$Y_{ijk} = \mu + A_i + B_j + AB_{ij} + \varepsilon_{k(ij)}$$

and $k = 1, 2, \cdots, r$ for r replications.

To introduce some notation for treatment combinations when three levels are involved, consider the data layout in Figure 9.1. In Figure 9.1, two digits

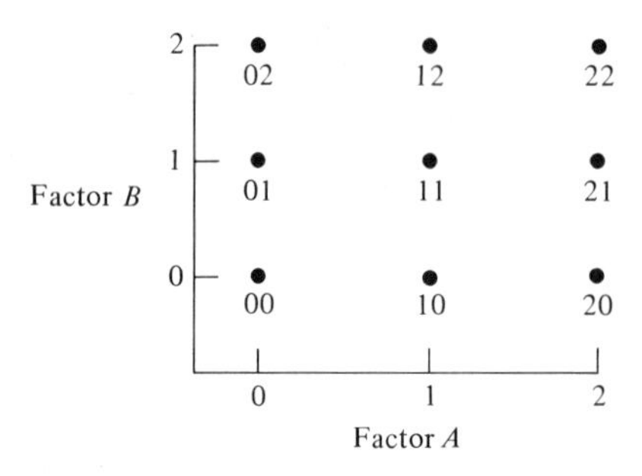

Figure 9.1 3^2 data layout.

are used to describe each of the nine treatment combinations. The first digit indicates the level of Factor A, and the second digit the level of factor B. Thus, 12 means A at its intermediate level and B at its highest level. This notation can easily be extended to more factors and as many levels as are necessary. It could have been used for 2^n factorials as 00, 10, 01, and 11, corresponding respectively to (1), a, b, and ab. The only reason for not using this digital notation on 2^n factorials is that so much of the literature includes this (1), a, b, $\cdots$ notation. By proper choice of coefficients on these treatment combinations, the linear and quadratic effects of both A and B can be determined, as well as their interactions, such as $A_L \times B_L$, $A_L \times B_Q$, $A_Q \times B_L$, and $A_Q \times B_Q$. The methods of analysis will be illustrated on a simple hypothetical example.

Suppose the responses in Table 9.1 were recorded for the treatment combinations indicated in the upper left-hand corner of each cell. Analyzing

Table 9.1 3^2 Factorial with Responses and Totals

	Factor A			
Factor B	0	1	2	$T_{.j}$
0	00 1	10 −2	20 3	2
1	01 0	11 4	21 1	5
2	02 2	12 −1	22 2	3
$T_{i.}$	3	1	6	10

the data of Table 9.1 by the general methods of Chapter 6 gives

$$SS_{\text{total}} = 1^2 + 0^2 + 2^2 + \cdots + 2^2 - \frac{(10)^2}{9} = 28.89$$

$$SS_A = \frac{3^2 + 1^2 + 6^2}{3} - \frac{(10)^2}{9} = 4.22$$

$$SS_B = \frac{2^2 + 5^2 + 3^2}{3} - \frac{(10)^2}{9} = 1.56$$

$$SS_{\text{error}} = 28.89 - 4.22 - 1.56 = 23.11$$

and the ANOVA is given in Table 9.2.

Table 9.2 ANOVA for 3^2 Factorial of Table 9.1

Source	df	SS	MS
A_i	2	4.22	2.11
B_j	2	1.56	0.78
AB_{ij}	4	23.11	5.77
Totals	8	28.89	

A further breakdown of this analysis is now possible by recalling that coefficients of -1, 0, $+1$ applied to the responses at low, intermediate, and high levels of a factor will measure its linear effect, whereas coefficients of $+1$, -2, $+1$ applied to these same responses will measure the quadratic effect of this factor. As in the case of 2^n factorials, products of coefficients will give the proper coefficients for various interactions. This can best be shown by Table 9.3 which indicates the coefficients for each effect to be used with the nine treatment combinations.

Table 9.3 Coefficients for a 3^2 Factorial with Quantitative Levels

	Treatment Combination									
Factor	00	01	02	10	11	12	20	21	22	$\sum c_i^2$
A_L	−1	−1	−1	0	0	0	+1	+1	+1	6
A_Q	+1	+1	+1	−2	−2	−2	+1	+1	+1	18
B_L	−1	0	+1	−1	0	+1	−1	0	+1	6
B_Q	+1	−2	+1	+1	−2	+1	+1	−2	+1	18
A_LB_L	+1	0	−1	0	0	0	−1	0	+1	4
A_LB_Q	−1	+2	−1	0	0	0	+1	−2	+1	12
A_QB_L	−1	0	+1	+2	0	−2	−1	0	+1	12
A_QB_Q	+1	−2	+1	−2	+4	−2	+1	−2	+1	36
Y_{ij}	1	0	2	−2	4	−1	3	1	2	

From Table 9.3, it can be seen that A_L compares all highest levels of $A(+1)$ with all lowest levels of $A(-1)$. A_Q compares the extreme levels with twice the intermediate levels. Both of these effects are taken across *all* levels of B. Now B_L compares the highest versus the lowest level of B at the 0 level of A, then at level 1 of A, then at level 2 of A, reading from left to right across B_L. Similarly, B_Q compares the extreme levels of B with twice the intermediate level at all three levels of A. The coefficients for interaction are found by multiplying corresponding main-effect coefficients. An examination of these coefficients in the light of what interactions there are should make the coefficients seem quite plausible. The sums of squares of the coefficients are given at the right of Table 9.3.

Applying these coefficients to the responses for each treatment combination gives

$$\begin{aligned}
A_L &= -1(1) - 1(0) - 1(2) + 0(-2) + 0(4) \\
&\quad + 0(-1) + 1(3) + 1(1) + 1(2) = 3 \\
A_Q &= +1(1) + 1(0) + 1(2) - 2(-2) - 2(4) \\
&\quad - 2(-1) + 1(3) + 1(1) + 1(2) = 7 \\
B_L &= -1(1) + 0(0) + 1(2) - 1(-2) + 0(4) \\
&\quad + 1(-1) - 1(3) + 0(1) + 1(2) = 1 \\
B_Q &= +1(1) - 2(0) + 1(2) + 1(-2) - 2(4) \\
&\quad + 1(-1) + 1(3) - 2(1) + 1(2) = -5 \\
A_LB_L &= +1(1) + 0(0) - 1(2) + 0(-2) + 0(4) \\
&\quad + 0(-1) - 1(3) + 0(1) + 1(2) = -2 \\
A_LB_Q &= -1(1) + 2(0) - 1(2) + 0(-2) + 0(4) \\
&\quad + 0(-1) + 1(3) - 2(1) + 1(2) = 0 \\
A_QB_L &= -1(1) + 0(0) + 1(2) + 2(-2) + 0(4) \\
&\quad -2(-1) - 1(3) + 0(1) + 1(2) = -2 \\
A_QB_Q &= +1(1) - 2(0) + 1(2) - 2(-2) + 4(4) \\
&\quad - 2(-1) + 1(3) - 2(1) + 1(2) = 28
\end{aligned}$$

The corresponding sums of squares become

$$SS_{A_L} = \frac{(3)^2}{6} = 1.50 \qquad SS_{A_LB_L} = \frac{(-2)^2}{4} = 1.00$$

$$SS_{A_Q} = \frac{(7)^2}{18} = 2.72 \qquad SS_{A_LB_Q} = \frac{0^2}{12} = 0.00$$

$$SS_{B_L} = \frac{(1)^2}{6} = 0.17 \qquad SS_{A_QB_L} = \frac{(-2)^2}{12} = 0.33$$

$$SS_{B_Q} = \frac{(-5)^2}{18} = 1.39 \qquad SS_{A_QB_Q} = \frac{(28)^2}{36} = 21.78$$

Summarizing, we obtain Table 9.4.

Table 9.4 ANOVA Breakdown for 3^2 Factorial

Source	df		SS	
A_i	2		4.22	
A_L		1		1.50
A_Q		1		2.72
B_j	2		1.56	
B_L		1		0.17
B_Q		1		1.39
AB_{ij}	4		23.11	
A_LB_L		1		1.00
A_LB_Q		1		0.00
A_QB_L		1		0.33
A_QB_Q		1		21.78
Totals	8		28.89	

The results of this analysis will not be tested, as there is no separate measure of error, and the interaction effect is obviously large compared to other effects. As the numbers used here are purely hypothetical, the purpose has been only to show how such data can be analyzed and how the notation can be used.

It may be instructive to examine graphically the meaning of this high A_QB_Q interaction since it dwarfs all main effects and other interactions in the example.

If the A_LB_L interaction is graphed using only the extreme levels of each factor, the results are as shown in Figure 9.2. Although the lines do cross, this is a very small interaction. Note also that the average change in response for the lowest level of A to the highest level of A (shown by $\times$) is very small indicating a very small A_L effect as the data show.

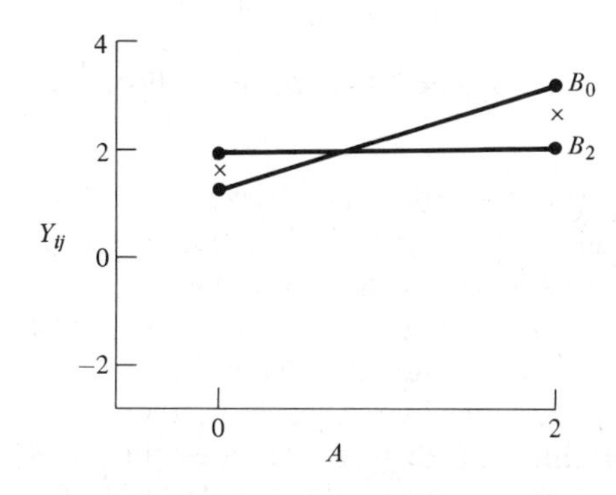

Figure 9.2 A_LB_L interaction.

Plotting $A_L B_Q$ means using all 3 levels of B, giving Figure 9.3, which again shows little interaction.

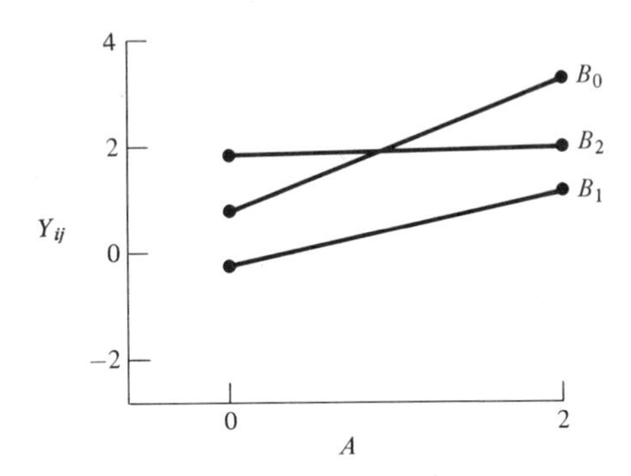

Figure 9.3 $A_L B_Q$ interaction.

For $A_Q B_L$ the results are as shown in Figure 9.4, which once again shows little interaction. This graph does indicate that the quadratic effect of A is more pronounced than its linear effect. This is borne out by the data.

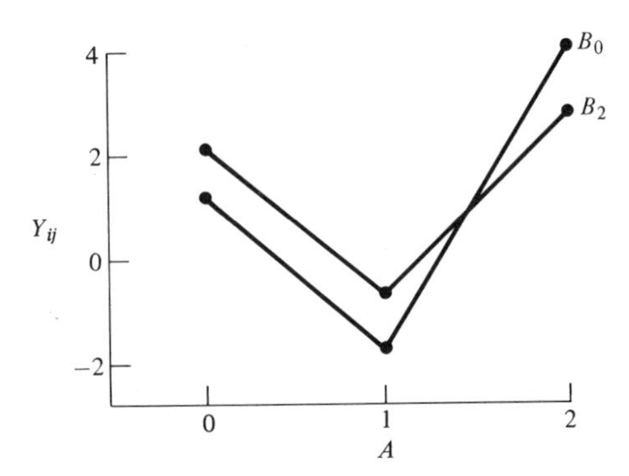

Figure 9.4 $A_Q B_L$ interaction.

Replotting Figure 9.3 with B as absicca gives Figure 9.5 which shows the same lack of serious interaction as in Figure 9.3, but it does show a slight quadratic bend in factor B over the linear effect.

So far no startling results have been seen. Now plot the middle level of B on Figure 9.4 to obtain Figure 9.6. Note the way this "curve" reverses its trend compared to the other two. This shows that not until both factors are considered at all three of their levels does this $A_Q B_Q$ interaction show up. It can also be seen by adding the middle level of A on Figure 9.5 (see Figure 9.7).

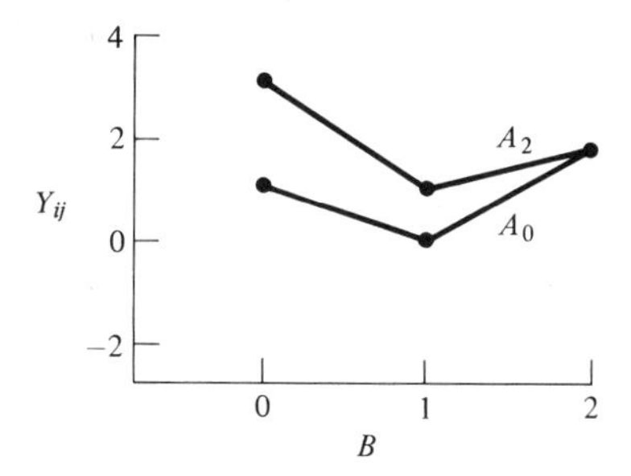

Figure 9.5 $A_L B_Q$ interaction again.

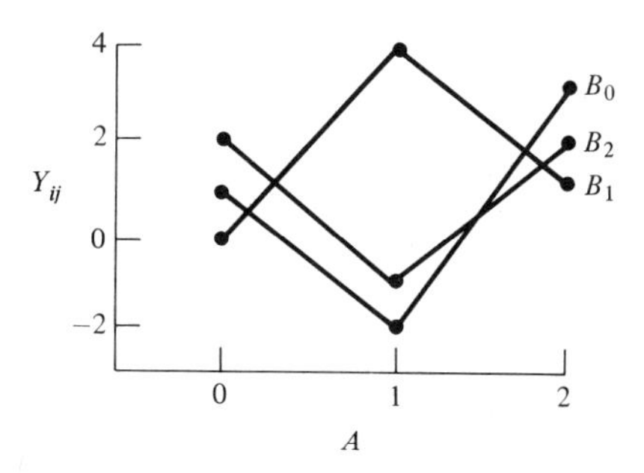

Figure 9.6 $A_Q B_Q$ interaction.

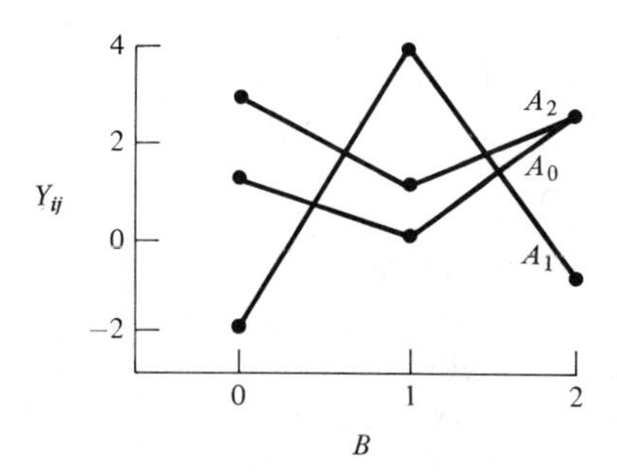

Figure 9.7 $A_Q B_Q$ interaction again.

Before leaving this problem, reconsider the data of Table 9.1. Add the data by diagonals rather than by rows or columns. First consider the diagonals downward from left to right where the main diagonal is $1 + 4 + 2 = 7$, the next one to the right is $-2 + 1 + 2 = 1$, and the last $+2$ is found by repeating the table again to the right of the present one as in Table 9.5.

Table 9.5 Diagonal Computations

	Factor A			Factor A		
Factor B	0	1	2	0	1	2
0	1	−2	3	1	−2	3
1	0	4	1	0	4	1
2	2	−1	2	2	−1	2

Similarly, the next downward diagonal gives $3 + 0 + (-1) = 2$. The sum of squares between these three diagonal terms is then

$$\frac{(7)^2 + (1)^2 + (2)^2}{3} - \frac{(10)^2}{9} = 6.89$$

If the diagonals are now considered downward and to the left, their totals are

$$3 + 4 + 2 = 9$$

$$1 + 1 - 1 = 1$$

$$-2 + 0 + 2 = 0$$

and their sum of squares is

$$\frac{(9)^2 + (1)^2 + (0)^2}{3} - \frac{(10)^2}{9} = 16.22$$

These two somewhat artificial sums of squares of 6.89 and 16.22 are seen to add up to the interaction sum of squares

$$6.89 + 16.22 = 23.11$$

These two components of interaction have no physical significance, but simply illustrate another way to extract two orthogonal components of interaction. Testing each of these separately for significance has no meaning, but this arbitrary breakdown is very useful in more complex designs. Some authors refer to these two components as the I and J components of interaction

$$I(AB) = 6.89 \qquad 2 \text{ df}$$

$$J(AB) = 16.22 \qquad 2 \text{ df}$$

$$\text{total } A \times B = 23.11 \qquad 4 \text{ df}$$

Each such component carries 2 df. These are sometimes referred to as the AB^2 and AB components of $A \times B$ interaction. In this notation, effects can be multiplied together using a modulus of 3, since this is a 3^n factorial. A *modulus* of 3 means that the resultant number is equal to the remainder when the number in the usual base of 10 is divided by 3. Thus 4 = 1 in modulus 3, as 1 is the remainder when 4 is divided by 3. The following associations also hold:

$$\begin{array}{lllll} \text{numbers: } 0 & 3 = 0 & 6 = 0 & 9 = 0 \\ \phantom{\text{numbers: }} 1 & 4 = 1 & 7 = 1 & 10 = 1 \\ \phantom{\text{numbers: }} 2 & 5 = 2 & 8 = 2 & 11 = 2 \\ \phantom{\text{numbers: }} \vdots & & & \end{array}$$

When using the form A^pB^q, it is postulated that the only exponent allowed on the first letter in the expression is a 1. To make it a 1, the expression can be squared and reduced, modulus 3. For example,

$$A^2B = (A^2B)^2 = A^4B^2 = AB^2$$

Hence, AB and AB^2 are the only components of the $A \times B$ interaction with 2 df each. Here the two types of notation are related as follows:

$$I(AB) = AB^2$$

$$J(AB) = AB$$

To summarize this simple experiment, all effects can be expressed with 2 df each, as in Table 9.6.

Table 9.6 3^2 Factorial by 2-df Analysis

Source	df	SS
A_i	2	4.22
B_j	2	1.56
$I(AB) = AB^2$	2	6.89
$J(AB) = AB$	2	16.22
Totals	8	28.89

It will be found very useful to break such an experiment down into 2-df effects when more complex designs are considered. Here, this breakdown is presented merely to show another way to partition the interaction effect.

9.3 3^3 FACTORIAL

If an experimenter has three factors, each at three levels, or a $3 \times 3 \times 3 = 3^3$ factorial, there are several ways to break down the effects of factors A, B, and C and their associated interactions. If the order of experimentation is completely randomized, the model for such an experiment is

$$Y_{ijk} = \mu + A_i + B_j + AB_{ij} + C_k + AC_{ik} + BC_{jk} + ABC_{ijk} + \varepsilon_{ijk}$$

with the last two terms confounded unless there is replication within the cells.

In this model, $i = 1, 2, 3$, $j = 1, 2, 3$, and $k = 1, 2, 3$, making 27 treatment combinations. These 27 treatment combinations may be as shown in Table 9.7.

Table 9.7 3^3 Factorial Treatment Combinations

		Factor A		
Factor B	Factor C	0	1	2
0	0	000	100	200
	1	001	101	201
	2	002	102	202
1	0	010	110	210
	1	011	111	211
	2	012	112	212
2	0	020	120	220
	1	021	121	221
	2	022	122	222

Association of the proper coefficients on these 27 treatment combinations would allow the Table 9.8 breakdown of an ANOVA if all effects were set at quantitative levels.

Table 9.8 3^3 Factorial Analysis for Linear and Quadratic Effects

Source	df		Source	df	
A_i	2		$A_Q C_L$		1
A_L		1	$A_Q C_Q$		1
A_Q		1	BC_{jk}	4	
B_j	2		$B_L C_L$		1
B_L		1	$B_L C_Q$		1
B_Q		1	$B_Q C_L$		1
AB_{ij}	4		$B_Q C_Q$		1
$A_L B_L$		1	ABC_{ijk}	8	
$A_L B_Q$		1	$A_L B_L C_L$		1
$A_Q B_L$		1	$A_L B_L C_Q$		1
$A_Q B_Q$		1	$A_L B_Q C_L$		1
C_k	2		$A_L B_Q C_Q$		1
C_L		1	$A_Q B_L C_L$		1
C_Q		1	$A_Q B_L C_Q$		1
AC_{ik}	4		$A_Q B_Q C_L$		1
$A_L C_L$		1	$A_Q B_Q C_Q$		1
$A_L C_Q$		1			
			Total		26

In an actual problem, these three-way interactions would be hard to explain, and quite often the ABC interaction is left with its 8 df for use as an error term to test the main effects A, B, C and the two-way interactions.

Another possible partitioning of these effects is in terms of 2-df effects using I and J components on AB, AC, and BC interactions. These could be designated as AB, AB^2, AC, AC^2, and BC, BC^2, each with 2 df. However, the three-way interaction with its 8 df may need a further breakdown. Sometimes ABC is broken into four 2-df components called $X(ABC)$, $Y(ABC)$, $Z(ABC)$, and $W(ABC)$, or, using the notation of the last section: AB^2C, ABC^2, ABC, and AB^2C^2. Here again no first letter is squared, and $A^2BC = (A^2BC)^2 = A^4B^2C^2 = AB^2C^2$ modulus 3. Such a partioning would yield Table 9.9.

Table 9.9 3^3 Factorial in 2-df Analyses

Source	df	
A	2	
B	2	
AB	2	4
AB^2	2	
C	2	
AC	2	4
AC^2	2	
BC	2	4
BC^2	2	
ABC	2	8
ABC^2	2	
AB^2C	2	
AB^2C^2	2	
Total	26	

Example 9.1 A problem involving the effect of three factors, each at three levels, was proposed by Professor Burr of Purdue University. Here the measured variable was yield and the factors which might affect this response were days, operators, and concentrations of solvent. Three days, three operators, and three concentrations were chosen. Days and operators were qualitative effects, concentrations were quantitative and set at 0.5, 1.0, 2.0. Although these are not equispaced, the logarithms of these three levels are equispaced, and the logarithms can then be used if a curve fitting is warranted.

For the purposes of this chapter, all levels of all factors will be considered as fixed and the design will be considered as completely randomized. It was decided to take three replications of each of the $3^3 = 27$ treatment combinations. The data, after coding by subtracting 20.0, are presented in Table 9.10.

Table 9.10 Example Data on 3^3 Factorial with Three Replications

	Day D_i								
	5/14			5/15			5/16		
	Operator O_j								
Concentration C_k	A	B	C	A	B	C	A	B	C
0.5	1.0	0.2	0.2	1.0	1.0	1.2	1.7	0.2	0.5
	1.2	0.5	0.0	0.0	0.0	0.0	1.2	0.7	1.0
	1.7	0.7	−0.3	0.5	0.0	0.5	1.2	1.0	1.7
1.0	5.0	3.2	3.5	4.0	3.2	3.7	4.5	3.7	3.7
	4.7	3.7	3.5	3.5	3.0	4.0	5.0	4.0	4.5
	4.2	3.5	3.2	3.5	4.0	4.2	4.7	4.2	3.7
2.0	7.5	6.0	7.2	6.5	5.2	7.0	6.7	7.5	6.2
	6.5	6.2	6.5	6.0	5.7	6.7	7.5	6.0	6.5
	7.7	6.2	6.7	6.2	6.5	6.8	7.0	6.0	7.0

If these data are analyzed on a purely qualitative basis, the methods of Chapter 6, Section 6.4 can be used. The resulting ANOVA is shown in Table 9.11.

Table 9.11 First ANOVA for Example 9.1

Source	df	SS	MS
D_i	2	3.48	1.74**
O_j	2	6.14	3.07**
DO_{ij}	4	4.07	1.02**
C_k	2	468.99	234.49***
DC_{ik}	4	0.59	0.15
OC_{jk}	4	0.89	0.22
DOC_{ijk}	8	1.09	0.14
$\varepsilon_{m(ijk)}$	54	9.98	0.18
Totals	80	495.23	

** Two asterisks indicate significance at the 1-percent level; three, at the 0.1-percent level.

The model for this example is merely $Y_{ijkm} = \mu$ plus the sum of the terms in the Source column in Table 9.11. From this analysis, the concentration effect is tremendous, and the days, operators, and day × operator interaction are all significant at the 1-percent level of significance.

Since concentrations are at quantitative levels, the linear and quadratic effects of concentrations may be computed, as well as the interactions between linear effect of concentration and days, quadratic effect of concentration and days, linear effect of concentration and operators, and quadratic effect of concentration and operators. It is not usually worthwhile to extract three-way interaction in this way. To calculate these quantitative effects, it is usually helpful to construct some two-way tables for the interactions which are being computed. Two of these are shown as Table 9.12(*a*) and Table 9.12(*b*).

Table 9.12 Cell Totals for $D \times C$ and $O \times C$ Interactions

	(*a*) Concentration					(*b*) Concentration			
Day	0.5	1.0	2.0		Operator	0.5	1.0	2.0	
5/14	5.2	34.5	60.5		*A*	9.5	39.1	61.6	
5/15	4.2	33.1	56.6		*B*	4.3	32.5	55.3	
5/16	9.2	38.0	60.4		*C*	4.8	34.0	60.6	
Totals	18.6	105.6	177.5	301.7	Totals	18.6	105.6	177.5	301.7

From Table 9.12(*a*), applying the linear and quadratic coefficients to the concentration totals, we have

Sums of Squares

$$C_L = -1(18.6) + 0(105.6) + 1(177.5) = 158.9 \qquad SS_{C_L} = \frac{(158.9)^2}{27(2)} = 467.58$$

$$C_Q = +1(18.6) - 2(105.6) + 1(177.5) = -15.1 \qquad SS_{C_Q} = \frac{(-15.1)^2}{27(6)} = 1.41$$

$$SS_C = 468.99$$

For the $D \times C$ interactions, consider each level of days separately. At

$$5/14\colon C_L = -1(5.2) + 0(34.5) + 1(60.5) = 55.3$$

$$5/15\colon C_L = -1(4.2) + 0(33.1) + 1(56.6) = 52.4$$

$$5/16\colon C_L = -1(9.2) + 0(38.0) + 1(60.4) = 51.2$$

The $D \times C_L$ $SS_{\text{interaction}}$ is then

$$\frac{(55.3)^2 + (52.4)^2 + (51.2)^2}{9(2)} - \frac{(158.9)^2}{27(2)} = 0.49$$

For quadratic effects, at

$$5/14: C_Q = +1(5.2) - 2(34.5) + 1(60.5) = -3.3$$
$$5/15: C_Q = +1(4.2) - 2(33.1) + 1(56.6) = -5.4$$
$$5/16: C_Q = +1(9.2) - 2(38.0) + 1(60.4) = -6.4$$

The $D \times C_Q$ $SS_{\text{interaction}}$ is then

$$\frac{(-3.3)^2 + (-5.4)^2 + (-6.4)^2}{9(6)} - \frac{(-15.1)^2}{27(6)} = 0.09$$

and

$$SS_{D\times C} = SS_{D\times C_L} + SS_{D\times C_Q} = 0.49 + 0.09 = 0.58$$

If the same procedure is now applied to the data of Table 9.12(*b*), we have

$$SS_{O\times C_L} = \frac{(52.1)^2 + (51.0)^2 + (55.8)^2}{9(2)} - \frac{(158.9)^2}{27(2)} = 0.70$$

$$SS_{O\times C_Q} = \frac{(-7.1)^2 + (-5.4)^2 + (-2.6)^2}{9(6)} - \frac{(-15.1)^2}{27(6)} = 0.19$$

and

$$SS_{O\times C} = SS_{O\times C_L} + SS_{O\times C_Q} = 0.70 + 0.19 = 0.89$$

The resulting ANOVA can now be shown in Table 9.13.

Table 9.13 Second ANOVA for Example 9.1

Source	df	SS	MS
D_i	2	3.48	1.74**
O_j	2	6.14	3.07**
DO_{ij}	4	4.07	1.02**
C_L	1	467.58	467.58***
C_Q	1	1.41	1.41**
$D \times C_L$	2	0.49	0.24
$D \times C_Q$	2	0.09	0.04
$O \times C_L$	2	0.70	0.35
$O \times C_Q$	2	0.19	0.09
DOC_{ijk}	8	1.09	0.14
$\varepsilon_{m(ijk)}$	54	9.98	0.18
Totals	80	495.22	

** Two asterisks indicate significance at the 1-percent level; three, at the 0.1-percent level.

This second analysis shows that the linear effect and the quadratic effect of concentration are extremely significant. Two plots of Figure 9.8 may help in picturing what is really happening in this experiment.

Figure 9.8(*a*) shows the effect of operators, days, and $D \times O$ interaction. Figure 9.8(*b*) indicates that the linear effect of concentration far outweighs the quadratic effect and there is no significant interaction. If a straight line or three straight lines were fit to these data, the logs of the concentrations would be used, as the logs are equispaced.

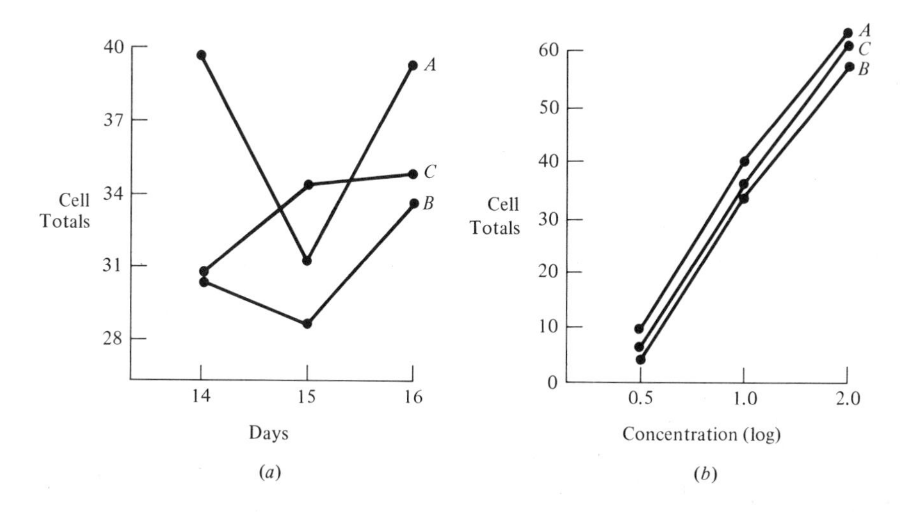

Figure 9.8 Plots of 3^3 example of Table 9.10.

Although this would usually conclude the analysis of this problem, each interaction will be broken down into its diagonal, or *I* and *J*, components in order to illustrate the technique. To compute the two diagonal components of the two-factor interactions, the two parts of Table 9.12 can be used, along with a similar table for the $D \times O$ cells (see Table 9.14).

Table 9.14 Cell Totals for $D \times O$ Interactions

	Operator		
Day	*A*	*B*	*C*
5/14	39.5	30.2	30.5
5/15	31.2	28.6	34.1
5/16	39.5	33.3	34.8

From Table 9.14, the diagonal components of the $D \times O$ interaction are

$$I(D \times O) = \frac{(39.5 + 28.6 + 34.8)^2 + (30.2 + 34.1 + 39.5)^2 + (30.5 + 33.3 + 31.2)^2}{27} - \frac{(301.7)^2}{81} = 1.74$$

Call it DO^2.

$$J(D \times O) = \frac{(30.5 + 28.6 + 39.5)^2 + (96.2)^2 + (106.9)^2}{27} - \frac{(301.7)^2}{81} = 2.33$$

Call it DO. These total $1.74 + 2.33 = 4.07$, the $D \times O$ interaction sum of squares.

Applying the same technique to the two parts of Table 9.12 gives

$$I(DC) = DC^2 = \frac{(98.7)^2 + (102.7)^2 + (100.3)^2}{27} - \frac{(301.7)^2}{81} = 0.30$$

$$J(DC) = DC = \frac{(102.8)^2 + (99.8)^2 + (99.1)^2}{27} - \frac{(301.7)^2}{81} = 0.29$$

$$D \times C = 0.59$$

$$I(OC) = OC^2 = \frac{(102.6)^2 + (99.9)^2 + (99.2)^2}{27} - \frac{(301.7)^2}{81} = 0.24$$

$$J(OC) = OC = \frac{(98.9)^2 + (104.0)^2 + (98.8)^2}{27} - \frac{(301.7)^2}{81} = 0.65$$

$$O \times C = 0.89$$

To break down the 8 df of the $D \times O \times C$ interaction, form an $O \times D$ table showing each of the three levels of concentration C as in Table 9.15.

Table 9.15 Cell Totals for $D \times O$ Interaction at Each Level of Concentration

O_i	D_i at C_1			D_i at C_2			D_i at C_3		
A	3.9	1.5	4.1	13.9	11.0	14.2	21.7	18.7	21.2
B	1.4	1.0	1.9	10.4	10.2	11.9	18.4	17.4	19.5
C	−0.1	1.7	3.2	10.2	11.9	11.9	20.4	20.5	19.7

For each of these concentration levels, find the I and J effect totals, for example, at C_1,

I components are 8.1, 3.3, 7.2

J components are 5.0, 6.1, 7.5

Now form a table with these I and J components at each level of C (see Table 9.16).

Table 9.16 Diagonal Totals for Each Level of Concentration

	$I(DO)$			$J(DO)$		
C_k	i_0	i_1	i_2	j_0	j_1	j_2
0.5	8.1	3.3	7.2	5.0	6.1	7.5
1.0	36.0	33.1	36.5	34.6	33.3	37.7
2.0	58.8	58.6	60.1	59.0	56.8	61.7

Treat each half of Table 9.16 as a simple interaction and compute the I and J components. Thus,

$$DO^2C^2 = I[C \times I(DO)] = \frac{(101.3)^2 + (98.6)^2 + (101.8)^2}{27} - \frac{(301.7)^2}{81} = 0.22$$

$$DO^2C = J[C \times I(DO)] = \frac{(99.1)^2 + (99.4)^2 + (103.2)^2}{27} - \frac{(301.7)^2}{81} = 0.39$$

$$DOC^2 = I[C \times J(DO)] = \frac{(100.0)^2 + (102.8)^2 + (98.9)^2}{27} - \frac{(301.7)^2}{81} = 0.30$$

$$DOC = J[C \times J(DO)] = \frac{(99.8)^2 + (102.4)^2 + (99.5)^2}{27} - \frac{(301.7)^2}{81} = 0.19$$

$$\text{total } D \times O \times C = 1.10$$

compared to 1.09 in Table 9.13. This last breakdown into four parts could also have been accomplished by considering the $C \times O$ interaction at three levels of D_i, or the $C \times D$ interaction at three levels of O_j.

The resulting analysis is summarized in Table 9.17. This analysis is in substantial agreement with Tables 9.11 and 9.13. No new tests would be performed on the data in Table 9.17, as they represent only an arbitrary breakdown of the interactions into 2-df components. The purpose of such a breakdown will be discussed in subsequent chapters. For testing hypotheses on interaction, these components are added together again.

Table 9.17 Third ANOVA for 3^3 Experiment

Source	df	SS
D_i	2	3.48
O_j	2	6.14
DO	2	2.33
DO^2	2	1.74
C_k	2	468.99
DC	2	0.29
DC^2	2	0.30
OC	2	0.65
OC^2	2	0.24
DOC	2	0.19
DOC^2	2	0.30
DO^2C	2	0.39
DO^2C^2	2	0.22
$\varepsilon_{m(ijk)}$	54	9.98
Totals	80	495.24

9.4 SUMMARY

The summary at the end of Chapter 7 may now be extended for Part II.

Experiment	*Design*	*Analysis*
II. Two or more factors *A*. Factorial (crossed)		
	1. Completely randomized $Y_{ijk} = \mu + A_i + B_j + AB_{ij} + \varepsilon_{k(ij)} \cdots$ for more factors	1.
	a. General case	*a*. ANOVA with interactions
	b. 2^n case	*b*. Yates method or general ANOVA; use (1), *a*, *b*, *ab*, $\cdots$
	c. 3^n case	*c*. General ANOVA; use 00, 10, 20, 01, 11, $\cdots$ and $A \times B = AB + AB^2, \cdots$ for interaction

PROBLEMS

9.1 Pull-off force in pounds on glued boxes at three temperatures and three humidities with two observations per treatment combination in a completely randomized experiment gives

	Temperature A		
Humidity B	Cold	Ambient	Hot
50%	0.8	1.5	2.5
	2.8	3.2	4.2
70%	1.0	1.6	1.8
	1.6	1.8	1.0
90%	2.0	1.5	2.5
	2.2	0.8	4.0

Do a complete analysis of this problem by the general methods of Chapter 6.

9.2 Assuming the temperatures in Problem 9.1 are equispaced, extract linear and quadratic effects of both temperature and humidity as well as all components of interaction.

9.3 From Problem 9.1, extract the AB and AB^2 components of interaction.

9.4 Develop a "Yates method" for this 3^2 experiment and check the results with those above.

9.5 A behavior variable on concrete pavements was measured for three surface thicknesses: 3 in., 4 in., and 5 in.; three base thicknesses: 0 in., 3 in., and 6 in., and three subbase thicknesses: 4 in., 8 in., and 12 in. Two observations were made under each of the 27 pavement conditions and complete randomization performed. The results were

	Surface Thickness (inches)								
	3			4			5		
Subbase Thickness (inches)	Base Thickness (inches)			Base Thickness (inches)			Base Thickness (inches)		
	0	3	6	0	3	6	0	3	6
4	2.8	4.3	5.7	4.1	5.4	6.7	6.0	6.3	7.1
	2.6	4.5	5.3	4.4	5.5	6.9	6.2	6.5	6.9
8	4.1	5.7	6.9	5.3	6.5	7.7	6.1	7.2	8.1
	4.4	5.8	7.1	5.1	6.7	7.4	5.8	7.1	8.4
12	5.5	7.0	8.1	6.5	7.7	8.8	7.0	8.0	9.1
	5.3	6.8	8.3	6.7	7.5	9.1	7.2	8.3	9.0

Do a complete ANOVA of this experiment by the methods of Chapter 6.

9.6 Since all three factors are quantitative and equispaced, determine linear and quadratic effects for each factor and all interaction breakdowns. Test for significance.

9.7 Break down the interactions of Problem 9.5 into 2-df components such as AB, AB^2, ABC, AB^2C, and so on.

9.8 Use a Yates method to solve Problem 9.5 and check the results.

9.9 Plot any significant results of Problem 9.5.

9.10 From Example 9.1, fit the proper degree regression equation for the quantitative variable and express it in terms of the original variable.

9.11 The following data are on the wet-film thickness (in mils) of lacquer. The factors studied were: type of resin (two types), gate-blade setting in mils (three settings), and weight fraction of nonvolatile material in the lacquer (three fractions).

	Resin Type					
	1			2		
Gate	Weight Fraction			Weight Fraction		
Setting	0.20	0.25	0.30	0.20	0.25	0.30
2	1.6	1.5	1.5	1.5	1.4	1.6
	1.5	1.3	1.3	1.4	1.3	1.4
4	2.7	2.5	2.4	2.4	2.6	2.2
	2.7	2.5	2.3	2.3	2.4	2.1
6	4.0	3.6	3.5	4.0	3.7	3.4
	3.9	3.8	3.4	4.0	3.6	3.3

Do an ANOVA of the above data and pull out any quantitative effects in terms of their components.

9.12 Using $\alpha = 0.01$, write prediction equations where possible and check them (Problem 9.11).

9.13 If the gate setting is 5 mils and the weight fraction 0.20, what is the best prediction of wet-film thickness? If interaction is ignored, what is the prediction? How much difference does it make? (Problem 9.12).

9.14 Plot results indicated in Problem 9.12.

10
Fixed, Random, and Mixed Models

10.1 INTRODUCTION

In Chapter 1 it was pointed out that, in the planning stages of an experiment, the experimenter must decide whether the levels of factors considered are to be set at fixed values or are to be chosen at random from many possible levels. In the intervening chapters it has always been assumed that the factor levels were fixed. In practice it may be desirable to choose the levels of some factors at random, depending on the objectives of the experiment. Are the results to be judged for these levels alone or are they to be extended to more levels of which those in the experiment are but a random sample? In the case of some factors such as temperature, time, or pressure, it is usually desirable to pick fixed levels, often near the extremes and at some intermediate points, because a random choice might not cover the range in which the experimenter is interested. In such cases of fixed quantitative levels, we often feel safe in interpolating between the fixed levels chosen. Other factors such as operators, days, or batches may often be only a small sample of all possible operators, days, or batches. In such cases, the particular operator, day, or batch may not be very important but only whether or not operators, days, or batches in general increase the variability of the experiment.

It is not reasonable to decide after the data have been collected whether the levels are to be considered fixed or random. This decision must be made prior to the running of the experiment, and if random levels are to be used, they must be chosen from all possible levels by a random process. In the case of random levels, it will be assumed that the levels are chosen from an infinite population of possible levels. Bennett and Franklin [3] discuss a case in which the levels chosen are from a finite set of possible levels.

When all levels are fixed, the mathematical model of the experiment is called a *fixed model*. When all levels are chosen at random, the model is called a *random model*. When several factors are involved, some at fixed levels and others at random levels, the model is called a *mixed model*.

10.2 SINGLE-FACTOR MODELS

In the case of a single-factor experiment, the factor may be referred to as a *treatment effect*, as in Chapter 3; and if the design is completely randomized, the model is

$$Y_{ij} = \mu + \tau_j + \varepsilon_{ij} \qquad (10.1)$$

Whether the treatment levels are fixed or random, it is assumed in this model that μ is a fixed constant and the errors are normally and independently distributed with a zero mean and the same variance, that is, ε_{ij} are NID $(0, \sigma_\varepsilon^2)$. The decision as to whether the levels of the treatments are fixed or random will affect the assumptions about the treatment term τ_j. The different assumptions and other differences will be compared in parallel columns.

Fixed Model

1. Assumptions: τ_j's are fixed constants.

$$\sum_{j=1}^{k} \tau_j = \sum_{j=1}^{k} (\mu_{.j} - \mu) = 0$$

(These add to zero as they are the only treatment means being considered.)

2. Analysis: Procedures as given in Chapter 3 for computing SS.
3. EMS:

Source	df	EMS
τ_j	$k - 1$	$\sigma_\varepsilon^2 + n\phi_\tau$
ε_{ij}	$k(n - 1)$	σ_ε^2

4. Hypothesis tested:

$$H_0: \tau_j = 0 \text{ (for all } j)$$

Random Model

1. Assumptions: τ_j's are random variables and are

$$\text{NID } (0, \sigma_\tau^2)$$

(Here σ_τ^2 represents the variance among the τ_j's or among the true treatment means $\mu_{.j}$. The τ_j average to zero when averaged over all possible levels, but for the k levels of the experiment they usually will not average 0.)

2. Analysis: Same as for fixed model.
3. EMS:

Source	df	EMS
τ_j	$k - 1$	$\sigma_\varepsilon^2 + n\sigma_\tau^2$
ε_{ij}	$k(n - 1)$	σ_ε^2

4. Hypothesis tested:

$$H_0: \sigma_\tau^2 = 0$$

The expected mean square (EMS) column turns out to be extremely important in more complex experiments as an aid in deciding how to set up an F test for significance. The EMS for any term in the model is the long-range average of the calculated mean square when the Y_{ij} from the model is substituted in algebraic form into the mean square computation. The derivation of these EMS values is often complicated, but those for the single-factor model were derived in Chapter 3 and those for two-factor models will be derived in a later section of this chapter.

For the fixed model, if the hypothesis is true that $\tau_j = 0$ for all j, that is, all the k fixed treatment means are equal, then $\sum_j \tau_j^2 = 0$ and the EMS for τ_j and ε_{ij} are both σ_ε^2. Hence the observed mean squares for treatments and error mean square are both estimates of the error variance, and they can be compared by means of an F test. If this F test shows a significantly high value, it must mean that $n \sum_j \tau_j^2/(k - 1) = \phi_\tau$ is not zero and the hypothesis is to be rejected.

For the random model, if the hypothesis is true that $\sigma_\tau^2 = 0$, that is, the variance among all treatment means is zero, then again each mean square is an estimate of the error variance. Again an F test between the two mean squares is appropriate.

From the two tables in step 3 above, it is seen that for a single-factor experiment there is no difference in the test to be made after the analysis, and the only difference is in the generality of the conclusions. If H_0 is rejected, there is probably a difference between the k fixed treatment means for the fixed model; for the random model there is a difference between all treatments of which the k examined are but a random sample.

10.3 TWO-FACTOR MODELS

For two factors A and B, the model in the general case is

$$Y_{ijk} = \mu + A_i + B_j + AB_{ij} + \varepsilon_{k(ij)}$$

with

$$i = 1, 2, \cdots, a \qquad j = 1, 2, \cdots, b \qquad k = 1, 2, \cdots, n$$

provided the design is completely randomized. In this model, it is again assumed that μ is a fixed constant and $\varepsilon_{k(ij)}$'s are NID $(0, \sigma_\varepsilon^2)$. If both A and B are at fixed levels, the model is a fixed model. If both are at random levels, the model is a random model, and if one is at fixed levels and the other at random levels, the model is a mixed model. Comparing each of these models gives

Fixed	*Random*	*Mixed*
1. Assumptions: A_i's are fixed constants and $\sum_{i=1}^{a} A_i = 0$	1. Assumptions: A_i's are NID $(0, \sigma_A^2)$	1. Assumptions: A_i's are fixed and $\sum_{i}^{a} A_i = 0$

Fixed	Random	Mixed
B_j's are fixed constants and $\sum_{j=1}^{b} B_j = 0$	B_j's are NID $(0, \sigma_B^2)$	B_j's are NID $(0, \sigma_B^2)$
AB_{ij}'s are fixed constants and $\sum_{i}^{a} \sum_{j}^{b} AB_{ij} = 0$	AB_{ij}'s are NID $(0, \sigma_{AB}^2)$	AB_{ij}'s are NID $(0, \sigma_{AB}^2)$ but $\sum_{i}^{a} AB_{ij} = 0$, $\sum_{j}^{b} AB_{ij} \neq 0$ (for A fixed, B random)
2. Analysis: Procedures of Chapter 6 for sums of squares	2. Analysis: Same	2. Analysis: Same
3. EMS:	3. EMS:	3. EMS:

Source	df	EMS (Fixed)	EMS (Random)	EMS (Mixed)
A_i	$a - 1$	$\sigma_\varepsilon^2 + nb\phi_A$	$\sigma_\varepsilon^2 + n\sigma_{AB}^2 + nb\sigma_A^2$	$\sigma_\varepsilon^2 + n\sigma_{AB}^2 + nb\phi_A$
B_j	$b - 1$	$\sigma_\varepsilon^2 + na\phi_B$	$\sigma_\varepsilon^2 + n\sigma_{AB}^2 + na\sigma_B^2$	$\sigma_\varepsilon^2 + na\sigma_B^2$
AB_{ij}	$(a - 1)(b - 1)$	$\sigma_\varepsilon^2 + n\phi_{AB}$	$\sigma_\varepsilon^2 + n\sigma_{AB}^2$	$\sigma_\varepsilon^2 + n\sigma_{AB}^2$
$\varepsilon_{k(ij)}$	$ab(n - 1)$	σ_ε^2	σ_ε^2	σ_ε^2

Fixed	Random	Mixed
4. Hypotheses tested:	4. Hypotheses tested:	4. Hypotheses tested:
H_1: $A_i = 0$ for all i	H_1: $\sigma_A^2 = 0$	H_1: $A_i = 0$ for all i
H_2: $B_j = 0$ for all j	H_2: $\sigma_B^2 = 0$	H_2: $\sigma_B^2 = 0$
H_3: $AB_{ij} = 0$ for all i and j	H_3: $\sigma_{AB}^2 = 0$	H_3: $\sigma_{AB}^2 = 0$

In the assumptions for the mixed model, the fact that summing the interaction term over the fixed factor ($\sum_i$) is zero but summing it over the random factor ($\sum_j$) is not zero affects the expected mean squares, as seen in item 3 above.

For the fixed model, the mean squares for A, B, and AB are each compared to the error mean square to test the respective hypotheses, as should be clear from an examination of the EMS column when the hypotheses are true.

For the random model, the third hypothesis of no interaction is tested by comparing the mean square for interaction to the mean square for error, but the first and second hypotheses are each tested by comparing the mean square for the main effect (A_i or B_j) with the mean square for the interaction as seen by their expected mean square values. For a mixed model, the interaction hypothesis is tested by comparing the interaction mean square with the error mean square. The random effect B_j is also tested by comparing its mean square with the error mean square. The fixed effect (A_i), however, is tested by comparing its mean square with the interaction mean square.

From these observations on a two-factor experiment, the importance of the EMS column is evident, as this column can be used to see how the tests of hypotheses should be run. It is also important to note that these EMS expressions can be determined prior to the running of the experiment. This will indicate whether or not a good test of a hypothesis exists. In some cases, the proper test indicated by the EMS column will have insufficient degrees of freedom to be sufficiently sensitive; in which case, the investigator might wish to change the experiment. This would involve such changes as a choice of more levels of some factors, or changing from random to fixed levels of some factors.

10.4 EMS RULES

The two examples above have shown the importance of the EMS column in determining what tests of significance are to be run after the analysis is completed. Because of the importance of this EMS column in these and more complex models, it is often useful to have some simple method of determining these values from the model for the given experiment. A set of rules can be stated which will determine the EMS column very rapidly, without recourse to their derivation. The rules will be illustrated on the two-factor mixed model of Section 10.3. To determine the EMS column for any model:

1. Write the variable terms in the model as row headings in a two-way table.

A_i
B_j
AB_{ij}
$\varepsilon_{k(ij)}$

2. Write the subscripts in the model as column headings; over each subscript write F if the factor levels are fixed, R if random. Also write the number of observations each subscript is to cover.

	a	b	n
	F	R	R
	i	j	k
A_i			
B_j			
AB_{ij}			
$\varepsilon_{k(ij)}$			

3. For each row (each term in the model) copy the number of observations under each subscript, providing the subscript does not appear in the row heading.

	a	b	n
	F	R	R
	i	j	k
A_i		b	n
B_j	a		n
AB_{ij}			n
$\varepsilon_{k(ij)}$			

4. For any bracketed subscripts in the model, place a 1 under those subscripts which are inside the brackets.

	a	b	n
	F	R	R
	i	j	k
A_i		b	n
B_j	a		n
AB_{ij}			n
$\varepsilon_{k(ij)}$	1	1	

5. Fill the remaining cells with a 0 or a 1, depending upon whether the subscript represents a fixed F or a random R factor.

	a	b	n
	F	R	R
	i	j	k
A_i	0	b	n
B_j	a	1	n
AB_{ij}	0	1	n
$\varepsilon_{k(ij)}$	1	1	1

6. To find the expected mean square for any term in the model:
   a. Cover the entries in the column (or columns) which contain non-bracketed subscript letters in this term in the model (for example, for A_i, cover column i; for $\varepsilon_{k(ij)}$, cover column k).
   b. Multiply the remaining numbers in each row. Each of these products is the coefficient for its corresponding term in the model, provided the subscript on the term is also a subscript on the term whose expected mean square is being determined. The sum of these coefficients multiplied by the variance of their corresponding terms (ϕ_τ or σ_τ^2) is the EMS of the term being considered (for example, for A_i, cover column i). The products of the remaining coefficients are bn, n, n, and 1, but the first n is not used, as there is no i in its term (B_j). The resulting EMS is then $bn\phi_A + n\sigma_{AB}^2 + 1 \cdot \sigma_\varepsilon^2$). For all terms, these rules give

	a F i	b R j	n R k	EMS
A_i	0	b	n	$\sigma_\varepsilon^2 + n\sigma_{AB}^2 + nb\phi_A$
B_j	a	1	n	$\sigma_\varepsilon^2 + na\sigma_B^2$
AB_{ij}	0	1	n	$\sigma_\varepsilon^2 + n\sigma_{AB}^2$
$\varepsilon_{k(ij)}$	1	1	1	σ_ε^2

These results are seen to be in agreement with the EMS values for the mixed model in Section 10.3. Here ϕ_A is, of course, a fixed type of variance

$$\phi_A = \frac{\sum_i A_i^2}{a - 1}$$

Although the rules seem rather involved, they become very easy to use with a bit of practice. A more complex example will illustrate the point.

Example 10.1 An industrial engineering student wished to determine the effect of five different clearances on the time required to position and assemble mating parts. As all such experiments involve operators, it was natural to consider a random sample of operators to perform the experiment. He also decided the part should be assembled directly in front of the operator and at arm's length from the operator. He also tried four different angles, from 0° directly in front of the operator through 30°, 60°, and 90° from this position. Thus, four factors were involved, any one of which might affect the time required to position and assemble the part. The experimenter decided to replicate each setup six times and to completely randomize the order of experimentation. Here operators O_i were at random levels (six being chosen), angles A_j at four fixed levels (0°, 30°, 60°, 90°), clearances C_k at five fixed levels and locations L_m fixed either in front of or at arm's length from the operator.

This is a 6 × 4 × 5 × 2 factorial experiment with six replications, run in a completely randomized design. The expected mean square values can be determined from the rules given in Section 10.4 as shown in Table 10.1.

Table 10.1 EMS for Clearance Problem

Source	6 R i	4 F j	5 F k	2 F m	6 R q	EMS
O_i	1	4	5	2	6	$\sigma_\varepsilon^2 + 240\sigma_O^2$
A_j	6	0	5	2	6	$\sigma_\varepsilon^2 + 60\sigma_{OA}^2 + 360\phi_A$
OA_{ij}	1	0	5	2	6	$\sigma_\varepsilon^2 + 60\sigma_{OA}^2$
C_k	6	4	0	2	6	$\sigma_\varepsilon^2 + 48\sigma_{OC}^2 + 288\phi_C$
OC_{ik}	1	4	0	2	6	$\sigma_\varepsilon^2 + 48\sigma_{OC}^2$
AC_{jk}	6	0	0	2	6	$\sigma_\varepsilon^2 + 12\sigma_{OAC}^2 + 72\phi_{AC}$
OAC_{ijk}	1	0	0	2	6	$\sigma_\varepsilon^2 + 12\sigma_{OAC}^2$
L_m	6	4	5	0	6	$\sigma_\varepsilon^2 + 120\sigma_{OL}^2 + 720\phi_L$
OL_{im}	1	4	5	0	6	$\sigma_\varepsilon^2 + 120\sigma_{OL}^2$
AL_{jm}	6	0	5	0	6	$\sigma_\varepsilon^2 + 30\sigma_{OAL}^2 + 180\phi_{AL}$
OAL_{ijm}	1	0	5	0	6	$\sigma_\varepsilon^2 + 30\sigma_{OAL}^2$
CL_{km}	6	4	0	0	6	$\sigma_\varepsilon^2 + 24\sigma_{OCL}^2 + 144\phi_{CL}$
OCL_{ikm}	1	4	0	0	6	$\sigma_\varepsilon^2 + 24\sigma_{OCL}^2$
ACL_{jkm}	6	0	0	0	6	$\sigma_\varepsilon^2 + 6\sigma_{OACL}^2 + 36\phi_{ACL}$
$OACL_{ijkm}$	1	0	0	0	6	$\sigma_\varepsilon^2 + 6\sigma_{OACL}^2$
$\varepsilon_{q(ijkm)}$	1	1	1	1	1	σ_ε^2

From this table it is easily seen that all interactions involving operators and the operator main effect are tested against the error mean square at the bottom of the table. All interactions and main effects involving fixed factors are tested by the mean square just below them in the table.

The rules given in this section are general enough to be applied to the most complex designs, as will be seen in later chapters.

10.5 EMS DERIVATIONS

Single-Factor Experiment

The EMS expressions for a single-factor experiment were derived in Chapter 3, Section 3.5.

Two-Factor Experiment

Using the definitions of expected values given in Chapter 2 and the procedures of Section 3.5, the EMS expressions may be derived for the two-factor experiment.

For a two-factor experiment, the model is

$$Y_{ijk} = \mu + A_i + B_j + AB_{ij} + \varepsilon_{k(ij)} \tag{10.2}$$

with

$$i = 1, 2, \cdots, a \qquad j = 1, 2, \cdots, b \qquad k = 1, 2, \cdots, n$$

The sum of squares for factor A is

$$\text{SS}_A = \sum_{i=1}^{a} nb(\overline{Y}_{i..} - \overline{Y}_{...})^2$$

From the model in Equation (10.2)

$$\overline{Y}_{i..} = \sum_{j}^{b}\sum_{k}^{n} Y_{ijk}/bn = \sum_{j}^{b}\sum_{k}^{n} (\mu + A_i + B_j + AB_{ij} + \varepsilon_{k(ij)})/bn$$

$$\overline{Y}_{i..} = \mu + A_i + \sum_{j}^{b} B_j/b + \sum_{j}^{b} AB_{ij}/b + \sum_{j}^{b}\sum_{k}^{n} \varepsilon_{k(ij)})/bn$$

$$\overline{Y}_{...} = \sum_{i}^{a}\sum_{j}^{b}\sum_{k}^{n} Y_{ijk}/abn$$

$$= \sum_{i}^{a}\sum_{j}^{b}\sum_{k}^{n} (\mu + A_i + B_j + AB_{ij} + \varepsilon_{k(ij)})/abn$$

$$\overline{Y}_{...} = \mu + \sum_{i}^{a} A_i/a + \sum_{j}^{b} B_j/b + \sum_{i}^{a}\sum_{j}^{b} AB_{ij}/ab$$

$$+ \sum_{i}^{a}\sum_{j}^{b}\sum_{k}^{n} \varepsilon_{k(ij)}/nab$$

Subtracting gives

$$\overline{Y}_{i..} - \overline{Y}_{...} = \left(A_i - \sum_{i}^{a} A_i/a\right) + \left(\sum_{j}^{b} AB_{ij}/b - \sum_{i}^{a}\sum_{j}^{b} AB_i/ab\right)$$

$$+ \left(\sum_{j}^{b}\sum_{k}^{n} \varepsilon_{k(ij)}/bn - \sum_{i}^{a}\sum_{j}^{b}\sum_{k}^{n} \varepsilon_{k(ij)}/abn\right)$$

Note that the B effect cancels out of the A sum of squares as it should, since A and B are orthogonal effects in a factorial experiment. Squaring and adding gives

$$\text{SS}_A = nb \sum_{i=1}^{a} \left(A_i - \sum_{i}^{a} A_i/a\right)^2$$

$$+ nb \sum_{i=1}^{a} \left(\sum_{j}^{b} AB_{ij}/b - \sum_{i}^{a}\sum_{j}^{b} AB_{ij}/ab\right)^2$$

$$+ nb \sum_{i=1}^{a} \left(\sum_{j}^{b}\sum_{k}^{n} \varepsilon_{k(ij)}/bn - \sum_{i}^{a}\sum_{j}^{b}\sum_{k}^{n} \varepsilon_{k(ij)}/abn\right)^2$$

+ (cross product terms)

Taking the expected value for a fixed model where

$$\sum_i A_i = 0 \qquad \sum_{i \text{ or } j} AB_{ij} = 0$$

the result is

$$E(\text{SS}_A) = nb \sum_i^a A_i^2 + 0 + \frac{nb}{n^2b^2}(abn - bn)\sigma_\varepsilon^2$$

$$E(\text{MS}_A) = E[\text{SS}_A/(a-1)] = nb \sum_{i=1}^a A_i^2/(a-1) + \sigma_\varepsilon^2 = nb\phi_A + \sigma_\varepsilon^2$$

which agrees with Section 10.3 for the fixed model.

If now the levels of A and B are random,

$$\sum_i A_i \neq 0 \qquad \sum_{i \text{ or } j} AB_{ij} \neq 0$$

$$E(\text{SS}_A) = nb(a-1)\sigma_A^2 + \frac{nb}{b^2}(ab - b)\sigma_{AB}^2 + \frac{nb}{n^2b^2}(abn - bn)\sigma_\varepsilon^2$$

$$E(\text{MS}_A) = nb\sigma_A^2 + n\sigma_{AB}^2 + \sigma_\varepsilon^2$$

as stated in Section 10.3 for a random model.

If the model is mixed with A fixed and B random,

$$\sum_i^a A_i = 0 \qquad \sum_i^a AB_{ij} = 0$$

but

$$\sum_j^b AB_{ij} \neq 0$$

then

$$E(\text{SS}_A) = nb \sum_i^a A_i^2 + \frac{nb}{b^2}(ab - b)\sigma_{AB}^2 + \frac{nb}{n^2b^2}(nab - nb)\sigma_\varepsilon^2$$

$$E(\text{MS}_A) = nb \sum_i^a A_i^2/(a-1) + n\sigma_{AB}^2 + \sigma_\varepsilon^2 = nb\phi_A + n\sigma_{AB}^2 + \sigma_\varepsilon^2$$

which agrees with the value stated in Section 10.3 for a mixed model.

Using these expected value methods, one can derive all EMS values given in Section 10.3. It might be noted that if A were random in the mixed model

$$\sum_j^a AB_{ij} = 0$$

then the interaction term would not appear in the factor A sum of squares. This is true of B in the mixed model of Section 10.3.

These few derivations should be sufficient to show the general method of derivation and to demonstrate the advantages of the simple rules in Section 10.4 in determining these EMS values.

10.6 THE PSEUDO F TEST

Occasionally the EMS column for a given experiment indicates that there is no exact F test for one or more factors in the design model. Consider the following example.

Example 10.2 Two days in a given month were randomly selected in which to run an experiment. Three operators were also selected at random from a large pool of available operators. The experiment consisted of measuring the dry-film thickness of varnish in mils for three different gate settings: 2, 4, and 6 mils. Two determinations were made by each operator each day and at each of the three gate settings. Results are shown in Table 10.2.

Table 10.2 Dry-Film Thickness Experiment

	Day					
	1			2		
Gate	Operator			Operator		
Setting	*A*	*B*	*C*	*A*	*B*	*C*
2	0.38	0.39	0.45	0.40	0.39	0.41
	0.40	0.41	0.40	0.40	0.43	0.40
4	0.63	0.72	0.78	0.68	0.77	0.85
	0.59	0.70	0.79	0.66	0.76	0.84
6	0.76	0.95	1.03	0.86	0.86	1.01
	0.78	0.96	1.06	0.82	0.85	0.98

Assuming that days and operators are random effects, gate settings are fixed, and the design is completely randomized, the analysis yields Table 10.3.

Table 10.3 Analysis of Dry-Film Thickness Experiment

Source	df	SS	MS	EMS
Days D	1	0.0010	0.0010	$\sigma_\varepsilon^2 + 6\sigma_{DO}^2 + 18\sigma_D^2$
Operators O	2	0.1121	0.0560	$\sigma_\varepsilon^2 + 6\sigma_{DO}^2 + 12\sigma_O^2$
$D \times O$ interaction	2	0.0060	0.0030	$\sigma_\varepsilon^2 + 6\sigma_{DO}^2$
Gate setting G	2	1.5732	0.7866	$\sigma_\varepsilon^2 + 2\sigma_{DOG}^2 + 4\sigma_{OG}^2 + 6\sigma_{DG}^2 + 12\phi_G$
$D \times G$ interaction	2	0.0113	0.0056	$\sigma_\varepsilon^2 + 2\sigma_{DOG}^2 + 6\sigma_{DG}^2$
$O \times G$ interaction	4	0.0428	0.0107	$\sigma_\varepsilon^2 + 2\sigma_{DOG}^2 + 4\sigma_{OG}^2$
$D \times O \times G$ interaction	4	0.0099	0.0025**	$\sigma_\varepsilon^2 + 2\sigma_{DOG}^2$
Error	18	0.0059	0.0003	σ_ε^2
Total	35	1.7622		

** Two asterisks indicate significance at the 1-percent level.

All F tests are clear from the EMS column except for the test on gate setting which is probably the most important factor in the experiment. Only the three-way interaction shows significance. It is obvious from the results that gate setting is the most important factor, but how can it be tested? If the $D \times G$ interaction is assumed to be zero, then the gate setting can be tested against the $O \times G$ interaction term. On the other hand, if the $O \times G$ interaction is assumed to be zero, gate setting can be tested against the $D \times G$ interaction term. Although neither of these interactions is significant at the 5-percent level, both are numerically larger than the $D \times O \times G$ interaction against which they are tested. In this case any test on G is contingent upon these tests on interaction. One method for testing hypotheses in such situations was developed by Satterthwaite and is given in Bennett and Franklin [3, pp. 367–368].

The scheme consists of constructing a mean square as a linear combination of the mean squares in the experiment, where the EMS for this mean square includes the same terms as in the EMS of the term being tested, except for the variance of that term. For example, to test the gate-setting effect G in Table 10.3, a mean square is to be constructed whose expected value is

$$\sigma_\varepsilon^2 + 2\sigma_{DOG}^2 + 4\sigma_{OG}^2 + 6\sigma_{DG}^2$$

This can be found by the linear combination

$$\text{MS} = \text{MS}_{DG} + \text{MS}_{OG} - \text{MS}_{DOG}$$

as its expected value is

$$\begin{aligned} E(\text{MS}) &= \sigma_\varepsilon^2 + 2\sigma_{DOG}^2 + 4\sigma_{OG}^2 + \sigma_\varepsilon^2 + 2\sigma_{DOG}^2 + 6\sigma_{DG}^2 - \sigma_\varepsilon^2 - 2\sigma_{DOG}^2 \\ &= \sigma_\varepsilon^2 + 2\sigma_{DOG}^2 + 4\sigma_{OG}^2 + 6\sigma_{DG}^2 \end{aligned}$$

An F test can now be constructed using the mean square for gate setting as the numerator and this mean square as the denominator. Such a test is called a pseudo-F or F' test. The real problem here is to determine the degrees of freedom for the denominator mean square. According to Bennett and Franklin, if

$$\text{MS} = a_1(\text{MS})_1 + a_2(\text{MS})_2 + \cdots$$

and $(\text{MS})_1$ is based on v_1 df, $(\text{MS})_2$ is based on v_2 df, and so on, then the degrees of freedom for MS are

$$v = \frac{(\text{MS})^2}{a_1^2[(\text{MS})_1^2/v_1] + a_2^2[(\text{MS})_2^2/v_2] + \cdots}$$

In the case of testing for the gate-setting effect above, $a_1 = 1$, $a_2 = 1$, and $a_3 = -1$ and the degrees of freedom are $v_1 = 4$, $v_2 = 2$, and $v_3 = 4$. Here

MS = 0.0107 + 0.0056 − 0.0025 = 0.0138 and its df is

$$v = \frac{(0.0138)^2}{(1)^2[(0.0107)^2/4] + (1)^2[(0.0056)^2/2] + (-1)^2[(0.0025)^2/4]}$$

$$= \frac{1.9044 \times 10^{-4}}{0.4586 \times 10^{-4}} = 4.2$$

Hence the F' test is

$$F' = \frac{\text{MS}_G}{\text{MS}} = \frac{0.7866}{0.0138} = 57.0$$

with 2 and 4.2 df, which is significant at the 1-percent level of significance based on F with 2 and 4 or 2 and 5 df.

10.7 REMARKS

The examples in this chapter should be sufficient to show the importance of the EMS column in deciding just what mean squares should be compared in an F test of a given hypothesis. This EMS column is also useful (usually in random models) to solve for components of variance as illustrated in Chapter 3, Section 3.5.

One special case is of interest. In a two-factor factorial when there is but one observation per cell ($k = 1$), the EMS columns of Section 10.3 reduce to those in Table 10.4.

Table 10.4 EMS for One Observation per Cell

Source	EMS (Fixed)	EMS (Random)	EMS (Mixed)
A_i	$\sigma_\varepsilon^2 + b\phi_A$	$\sigma_\varepsilon^2 + \sigma_{AB}^2 + b\sigma_A^2$	$\sigma_\varepsilon^2 + \sigma_{AB}^2 + b\phi_A$
B_j	$\sigma_\varepsilon^2 + a\phi_B$	$\sigma_\varepsilon^2 + \sigma_{AB}^2 + a\sigma_B^2$	$\sigma_\varepsilon^2 + a\sigma_B^2$
AB_{ij} or ε_{ij}	$\sigma_\varepsilon^2 + \phi_{AB}$	$\sigma_\varepsilon^2 + \sigma_{AB}^2$	$\sigma_\varepsilon^2 + \sigma_{AB}^2$

A glance at these EMS values will show that there is no test for the main effects A and B in a fixed model, as interaction and error are hopelessly confounded. The only test possible is to assume that there is no interaction; then $\phi_{AB} = 0$, and the main effects are tested against the error. If a no-interaction assumption is not reasonable from information outside the experiment, the investigator should not run one observation per cell but should replicate the data in a fixed model.

For a random model, both main effects can be tested whether interaction is present or not. For a mixed model, there is a test for the fixed effect A but no test for the random effect B. This may not be a serious drawback, since the fixed effect is often the most important; the B effect is included chiefly for reduction of the error term. Such a situation is seen in a randomized block design where treatments are fixed, but blocks may be chosen at random.

In the discussion of the single-factor experiment, it was assumed that there were equal sample sizes n for each treatment. If this is not the case, it can be shown that the expected treatment mean square is

$$\sigma_\varepsilon^2 + n_0\sigma_\tau^2 \quad \text{or} \quad \sigma_\varepsilon^2 + n_0\phi_\tau$$

and

$$n_0 = \frac{N^2 - \sum_{j=1}^k n_j^2}{(k-1)N}$$

where

$$N = \sum_{j=1}^{k} n_j$$

The test for treatment effect is to compare the treatment mean square to the error mean square; the use of n_0 is primarily for computing components of variance.

PROBLEMS

10.1 An experiment is run on the effects of three randomly selected operators and five fixed aluminizers on the aluminum thickness of a TV tube. Two readings are made for each operator–aluminizer combination. The following ANOVA table is compiled:

Source	df	SS	MS
Operators	2	107,540	53,770
Aluminizers	4	139,805	34,951
$O \times A$ interaction	8	84,785	10,598
Error	15	230,900	15,393
Totals	29	563,030	

Assuming complete randomization, determine the EMS column for this problem and make the indicated significance tests.

10.2 Consider a three-factor experiment where factor A is at a levels, factor B at b levels, and factor C at c levels. The experiment is to be run in a completely randomized manner with n observations for each treatment combination. If factor A is run at a random levels and both B and C at fixed levels, determine the EMS column and indicate what tests would be made after the analysis.

10.3 Repeat Problem 10.2 with A and B at random levels, but C at fixed levels.

10.4 Repeat Problem 10.2 with all three factors at random levels.

10.5 Consider the completely randomized design of a four-factor experiment similar to Example 10.1. If factors A and B are at fixed levels and C and D are at random levels, set up the EMS column and indicate the tests to be made.

10.6 If three factors are at random levels and one is at fixed levels in Problem 10.5, work out the EMS column and the tests to be run.

10.7 Determine the EMS column for Problem 3.6 and solve for components of variance.

10.8 Derive the expression for n_0 in the EMS column of a single-factor completely randomized experiment where the n's are unequal.

10.9 Verify the numerical results of Table 10.3 using the methods of Chapter 6.

10.10 Verify the EMS column of Table 10.3 by the method of Section 10.4.

10.11 Set up F' tests for Problem 10.4 and explain how the df would be determined.

10.12 Set up F' tests for Problem 10.5 and explain how the df would be determined.

11
Nested and Nested-Factorial Experiments

11.1 INTRODUCTION

In a recent in-plant training course, the members of the class were assigned a final problem. Each class member was to go into the plant and set up an experiment using the techniques that had been discussed in class. One engineer wanted to study the strain readings of glass cathode supports from five different machines. Each machine had four "heads" on which the glass was formed, and he decided to take four samples from each head. He treated this experiment as a 5 × 4 factorial with four replications per cell. Complete randomization of the testing for strain readings presented no problem. His model was

$$Y_{ijk} = \mu + M_i + H_j + MH_{ij} + \varepsilon_{k(ij)}$$

with

$$i = 1, 2, \cdots, 5 \qquad j = 1, \cdots, 4 \qquad k = 1, \cdots, 4$$

His data and analysis appear in Table 11.1. In this model, he assumed that both machines and heads were fixed, and he used the 10-percent significance level. The results indicated no significant machine or head effect on strain readings, but there was a significant interaction at the 10-percent level of significance.

The question was raised whether the four heads were actually removed from machine A and mounted on machine B, then on C, and so on. Of course, the answer was "no" as each machine had its own four heads. So machines and heads did not form a factorial experiment, as the heads on each machine were unique for that particular machine. In such a case, the experiment is called a *nested experiment*: levels of one factor are nested within, or are subsamples of, levels of another factor. Such experiments are also sometimes called *hierarchical* experiments.

Table 11.1 Data and ANOVA for Strain-Reading Problem

	Machine				
Head	*A*	*B*	*C*	*D*	*E*
1	6	10	0	11	1
	2	9	0	0	4
	0	7	5	6	7
	8	12	5	4	9
2	13	2	10	5	6
	3	1	11	10	7
	9	1	6	8	0
	8	10	7	3	3
3	1	4	8	1	3
	10	1	5	8	0
	0	7	0	9	2
	6	9	7	4	2
4	7	0	7	0	3
	4	3	2	8	7
	7	4	5	6	4
	9	1	4	5	0

Source	df	SS	MS	EMS	F	$F_{0.90}$
M_i	4	45.08	11.27	$\sigma_\varepsilon^2 + 16\phi_M$	1.05	2.04
H_j	3	46.25	15.48	$\sigma_\varepsilon^2 + 20\phi_H$	1.45	2.18
MH_{ij}	12	236.42	19.70	$\sigma_\varepsilon^2 + 4\phi_{MH}$	1.84	1.66
$\varepsilon_{k(ij)}$	60	642.00	10.70			
Totals	79	969.95				

11.2 NESTED EXPERIMENTS

The above example can now be reanalyzed by treating it as a nested experiment, since heads are nested within machines. Such a factor may be represented in the model as $H_{j(i)}$, where j covers all levels 1, 2, $\cdots$ within the ith level of M_i. The number of levels of the nested factor need not be the same for all levels of the other factor. They are all equal in this problem, that is, $j = 1, 2, 3, 4$ for all i. The errors, in turn, are nested within the levels of i and j; $\varepsilon_{k(ij)}$ and $k = 1, 2, 3, 4$ for all i and j.

In order to emphasize the fact that the heads on each machine are different heads, the data layout in Table 11.2 shows heads 1, 2, 3, and 4 on machine A; heads 5, 6, 7, and 8 on machine B; and so on.

Table 11.2 Data for Strain-Reading Problem in a Nested Arrangement

Machine	*A*				*B*				*C*				*D*				*E*			
Head	1	2	3	4	5	6	7	8	9	10	11	12	13	14	15	16	17	18	19	20
	6	13	1	7	10	2	4	0	0	10	8	7	11	5	1	0	1	6	3	3
	2	3	10	4	9	1	1	3	0	11	5	2	0	10	8	8	4	7	0	7
	0	9	0	7	7	1	7	4	5	6	0	5	6	8	9	6	7	0	2	4
	8	8	6	9	12	10	9	1	5	7	7	4	4	3	4	5	9	3	2	0
Head totals	16	33	17	27	38	14	21	8	10	34	20	18	21	26	22	19	21	16	7	14
Machine totals	93				81				82				88				58			

As the heads which are mounted on the machine can be chosen from many possible heads, we might consider the four heads as a random sample of heads that might be used on a given machine. If such heads are selected at random for the machines, the model would be

$$Y_{ijk} = \mu + M_i + H_{j(i)} + \varepsilon_{k(ij)}$$

with

$$i = 1, \cdots, 5 \qquad j = 1, \cdots, 4 \qquad k = 1, \cdots, 4$$

This nested model has no interaction present, as the heads are not crossed with the five machines. If heads are considered random and machines fixed, the proper EMS values can be determined by the rules given in Chapter 10, as shown in Table 11.3.

Table 11.3 EMS for Nested Experiment

	5	4	4	
	F	*R*	*R*	
Source	*i*	*j*	*k*	EMS
M_i	0	4	4	$\sigma_\varepsilon^2 + 4\sigma_H^2 + 16\phi_M$
$H_{j(i)}$	1	1	4	$\sigma_\varepsilon^2 + 4\sigma_H^2$
$\varepsilon_{k(ij)}$	1	1	1	σ_ε^2

This breakdown shows that the head effect is to be tested against the error, and the machine effect is to be tested against the heads-within-machines effect.

To analyze the data for a nested design, first determine the total sum of squares

$$SS_{\text{total}} = 6^2 + 2^2 + \cdots + 4^2 + 0^2 - \frac{(402)^2}{80} = 969.95$$

and for machines

$$SS_M = \frac{(93)^2 + (81)^2 + (82)^2 + (88)^2 + (58)^2}{16} - \frac{(402)^2}{80} = 45.08$$

To determine the sum of squares between heads within machines, consider each machine separately:

machine A:

$$SS_H = \frac{(16)^2 + (33)^2 + (17)^2 + (27)^2}{4} - \frac{(93)^2}{16} = 50.19$$

machine B:

$$SS_H = \frac{(38)^2 + (14)^2 + (21)^2 + (8)^2}{4} - \frac{(81)^2}{16} = 126.18$$

machine C:

$$SS_H = \frac{(10)^2 + (34)^2 + (20)^2 + (18)^2}{4} - \frac{(82)^2}{16} = 74.75$$

machine D:

$$SS_H = \frac{(21)^2 + (26)^2 + (22)^2 + (19)^2}{4} - \frac{(88)^2}{16} = 6.50$$

machine E:

$$SS_H = \frac{(21)^2 + (16)^2 + (7)^2 + (14)^2}{4} - \frac{(58)^2}{16} = 25.25$$

$$\text{total } SS_H = 282.87$$

The error sum of squares by subtraction is then

$$969.95 - 45.08 - 282.87 = 642.00$$

The degrees of freedom between heads within machine A are $4 - 1 = 3$; for all five machines, the degrees of freedom will be $5 \times 3 = 15$. The analysis follows in Table 11.4.

From this analysis, machines appear to have no significant effect on strain readings, but there is a slightly significant (10-percent level) effect of heads within machines on the strain readings. Note that what the experimenter took as head effect (3 df) and interaction effect (12 df) is really heads-within-machines effect (15 df). These results might suggest a more careful

Table 11.4 ANOVA for Nested Strain-Reading Problem

Source	df	SS	MS	EMS	F	$F_{0.90}$
M_i	4	45.08	11.27	$\sigma_\varepsilon^2 + 4\sigma_H^2 + 16\phi_M$	<1	2.36
$H_{j(i)}$	15	282.87	18.85	$\sigma_\varepsilon^2 + 4\sigma_H^2$	1.76	1.60
$\varepsilon_{k(ij)}$	60	642.00	10.70	σ_ε^2		
Totals	79	969.95				

adjustment between heads on the same machine. In a nested model it is also seen that the 5 SS between heads within each machine are "pooled," or added, to give 282.87. This assumes that these sums of squares (which are proportional to variances) within each machine are of about the same magnitude. This might be questioned, as these sums of squares are 50.19, 126.18, 74.75, 6.50, and 25.25. If these five are really different, it appears as if the greatest variability is in machine B. This should indicate the need for work on each machine with an aim toward more homogeneous strain readings between heads on each machine. This example shows the importance of recognizing the difference between a nested experiment and a factorial experiment.

11.3 ANOVA RATIONALE

To see that the sums of squares computed in Section 11.2 were correct, consider the nested model

$$Y_{ijk} = \mu + A_i + B_{j(i)} + \varepsilon_{k(ij)}$$

or

$$Y_{ijk} \equiv \mu + (\mu_{i..} - \mu) + (\mu_{ij.} - \mu_{i..}) + (Y_{ijk} - \mu_{ij.})$$

which is an identity.

Using the best estimates of these population means from the sample data, the sample model is

$$Y_{ijk} \equiv \bar{Y}_{...} + (\bar{Y}_{i..} - \bar{Y}_{...}) + (\bar{Y}_{ij.} - \bar{Y}_{i..}) + (Y_{ijk} - \bar{Y}_{ij.})$$

with

$$i = 1, 2, \cdots, a \qquad j = 1, 2, \cdots, b \qquad k = 1, 2, \cdots, n$$

Transposing $\bar{Y}_{...}$ to the left of this expression, squaring both sides, and adding over i, j, and k gives

$$\sum_i^a \sum_j^b \sum_k^n (Y_{ijk} - \bar{Y}_{...})^2 = \sum_{i=1}^a nb(\bar{Y}_{i..} - \bar{Y}_{...})^2 + \sum_i^a \sum_j^b n(\bar{Y}_{ij.} - \bar{Y}_{i..})^2$$
$$+ \sum_i^a \sum_j^b \sum_k^n (Y_{ijk} - \bar{Y}_{ij.})^2$$

as the sums of cross products equal zero. This expresses the idea that the total sum of squares is equal to the sum of squares between levels of A, plus the sum of squares between levels of B within each level of A, plus the sum of the squares of the errors. The degrees of freedom are

$$(abn - 1) \equiv (a - 1) + a(b - 1) + ab(n - 1)$$

Dividing each independent sum of squares by its corresponding degrees of freedom gives estimates of population variance as usual. For computing purposes, the sum of squares as given above should be expanded in terms of totals, with the general results shown in Table 11.5.

Table 11.5 General ANOVA for a Nested Experiment

Source	df	SS	MS
A_i	$a - 1$	$\sum_i^a \frac{T_{i..}^2}{nb} - \frac{T_{...}^2}{nab}$	$\frac{SS_A}{a - 1}$
$B_{j(i)}$	$a(b - 1)$	$\sum_i^a \sum_j^b \frac{T_{ij.}^2}{n} - \sum_i^a \frac{T_{i..}^2}{nb}$	$\frac{SS_B}{a(b - 1)}$
$\varepsilon_{k(ij)}$	$ab(n - 1)$	$\sum_i^a \sum_j^b \sum_k^n Y_{ijk}^2 - \sum_i^a \sum_j^b \frac{T_{ij.}^2}{n}$	$\frac{SS_\varepsilon}{ab(n - 1)}$
Totals	$abn - 1$	$\sum_i^a \sum_j^b \sum_k^n Y_{ijk}^2 - \frac{T_{...}^2}{nab}$	

This is essentially the form followed in the problem of the last section. Note that the

$$SS_{B_{j(i)}} = \sum_i^a \sum_j^b \frac{T_{ij.}^2}{n} - \sum_i^a \frac{T_{i..}^2}{nb}$$

$$= \sum_i^a \left(\sum_j^b \frac{T_{ij.}^2}{n} - \frac{T_{i..}^2}{nb} \right)$$

which shows the way in which the sum of squares was calculated in the last section: by getting the sum of squares between levels of B for each level of A and then pooling over all levels of A.

11.4 NESTED-FACTORIAL EXPERIMENTS

In many experiments where several factors are involved, some may be factorial or crossed with others; some may be nested within levels of the others. When both factorial and nested factors appear in the same experiment, it is known as a *nested-factorial experiment*. The analysis of such an experiment is simply an extension of the methods of Chapter 6 and this chapter. Care must be exercised, however, in computing some of the interactions. Levels of both factorial and nested factors may be either fixed or random. The methods of Chapter 10 can be used to determine the EMS values and the proper tests to be run.

The nested-factorial experiment is best explained by an example. An investigator wished to improve the number of rounds per minute that could be fired from a Navy gun. He devised a new loading method which he hoped would increase the number of rounds per minute when compared to the existing method of loading. To test this hypothesis, he needed teams of men to operate the equipment. As the general physique of a man might affect the speed with which he could handle the loading of the gun, he chose teams of men in three general groupings—slight, average, and heavy or rugged men. The classification of such men was on the basis of an Armed Services classification table. He chose three teams at random to represent each of the three physique groupings. Each team was presented with the two methods of gun loading in a random order and each team used each method twice. The model for this experiment was

$$Y_{ijkm} = \mu + M_i + G_j + MG_{ij} + T_{k(j)} + MT_{ik(j)} + \varepsilon_{m(ijk)}$$

where

$$M_i = \text{methods}, i = 1, 2$$
$$G_j = \text{groups}, j = 1, 2, 3$$
$$T_{k(j)} = \text{teams within groups}, k = 1, 2, 3 \text{ for all } j$$
$$\varepsilon_{m(ijk)} = \text{random error}, m = 1, 2 \text{ for all } i, j, k$$

The EMS values are shown in Table 11.6 which indicates the proper F tests to run.

Table 11.6 EMS for Gun-Loading Problem

	2	3	3	2	
	F	F	R	R	
Source	i	j	k	m	EMS
M_i	0	3	3	2	$\sigma_\varepsilon^2 + 2\sigma_{MT}^2 + 18\phi_M$
G_j	2	0	3	2	$\sigma_\varepsilon^2 + 4\sigma_T^2 + 12\phi_G$
MG_{ij}	0	0	3	2	$\sigma_\varepsilon^2 + 2\sigma_{MT}^2 + 6\phi_{MG}$
$T_{k(j)}$	2	1	1	2	$\sigma_\varepsilon^2 + 4\sigma_T^2$
$MT_{ik(j)}$	0	1	1	2	$\sigma_\varepsilon^2 + 2\sigma_{MT}^2$
$\varepsilon_{m(ijk)}$	1	1	1	1	σ_ε^2

The data and analysis of this experiment appear in Table 11.7.

Table 11.7 Data and ANOVA for Gun-Loading Problem

Group	I			II			III		
Team	1	2	3	4	5	6	7	8	9
Method I	20.2	26.2	23.8	22.0	22.6	22.9	23.1	22.9	21.8
	24.1	26.9	24.9	23.5	24.6	25.0	22.9	23.7	23.5
Method II	14.2	18.0	12.5	14.1	14.0	13.7	14.1	12.2	12.7
	16.2	19.1	15.4	16.1	18.1	16.0	16.1	13.8	15.1

Source	df	SS	MS	EMS
M_i	1	651.95	651.95	$\sigma_\varepsilon^2 + 2\sigma_{MT}^2 + 18\phi_M$
G_j	2	16.05	8.02	$\sigma_\varepsilon^2 + 4\sigma_T^2 + 12\phi_G$
MG_{ij}	2	1.19	0.60	$\sigma_\varepsilon^2 + 2\sigma_{MT}^2 + 6\phi_{MG}$
$T_{k(j)}$	6	39.23	6.54	$\sigma_\varepsilon^2 + 4\sigma_T^2$
$MT_{ik(j)}$	6	10.75	1.79	$\sigma_\varepsilon^2 + 2\sigma_{MT}^2$
$\varepsilon_{m(ijk)}$	18	41.59	2.31	σ_ε^2
Totals	35	760.76		

It would be well for the reader to verify that the sums of squares of this table are correct. The only term that is somewhat different in this model than those previously handled is $MT_{ik(j)}$, that is, the interaction between methods and teams within groups. The safest way to compute this term is to compute the $M \times T$ interaction sums of squares within each of the three groups separately and then pool these sums of squares. (See Tables 11.8 through 11.10.)

Table 11.8 Data on Gun-Loading Problem for Group I

Team	1		2		3		Method Totals
Method I	20.2		26.2		23.8		146.1
	24.1		26.9		24.9		
		44.3		53.1		48.7	
Method II	14.2		18.0		12.5		95.4
	16.2		19.1		15.4		
		30.4		37.1		27.9	
Team totals		74.7		90.2		76.6	241.5

To compute the $M \times T$ interaction for group I, we have

$$SS_{\text{cell}} = \frac{(44.3)^2 + (53.1)^2 + (48.7)^2 + (30.4)^2 + (37.1)^2 + (27.9)^2}{2}$$

$$- \frac{(241.5)^2}{12} = 5116.39 - 4860.21 = 256.18$$

$$SS_{\text{method}} = \frac{(146.1)^2 + (95.4)^2}{6} - 4860.21 = 214.19$$

$$SS_{\text{team}} = \frac{(74.7)^2 + (90.2)^2 + (76.6)^2}{4} - 4860.21 = 35.71$$

$$SS_{M \times T \text{ interaction}} = 256.18 - 214.19 - 35.71 = 6.28$$

Table 11.9 Data on Gun-Loading Problem for Group II

Team	4		5		6		Method Totals
Method I	22.0		22.6		22.9		
	23.5		24.6		25.0		
		45.5		47.2		47.9	140.6
Method II	14.1		14.0		13.7		
	16.1		18.1		16.0		
		30.2		32.1		29.7	92.0
Team totals		75.7		79.3		77.6	232.6

$$SS_{\text{cell}} = \frac{(45.5)^2 + (47.2)^2 + (47.9)^2 + (30.2)^2 + (32.1)^2 + (29.7)^2}{2}$$

$$- \frac{(232.6)^2}{12} = 199.96$$

$$SS_{\text{method}} = \frac{(140.6)^2 + (92.0)^2}{6} - 4508.56 = 196.83$$

$$SS_{\text{team}} = \frac{(75.7)^2 + (79.3)^2 + (77.6)^2}{4} - 4508.56 = 1.62$$

$$SS_{M \times T \text{ interaction}} = 199.96 - 196.83 - 1.62 = 1.51$$

Table 11.10 Data on Gun-Loading Problem for Group III

Team	7		8		9		Method Totals
Method I	23.1		22.9		21.8		
	22.9		23.7		23.5		
		46.0		46.6		45.3	137.9
Method II	14.1		12.2		12.7		
	16.1		13.8		15.1		
		30.2		26.0		27.8	84.0
Team totals		76.2		72.6		73.1	221.9

$$SS_{\text{cell}} = \frac{(46.0)^2 + (46.6)^2 + (45.3)^2 + (30.2)^2 + (26.0)^2 + (27.8)^2}{2} - \frac{(221.9)^2}{12} = 246.96$$

$$SS_{\text{method}} = \frac{(137.9)^2 + (84.0)^2}{6} - 4103.30 = 242.10$$

$$SS_{\text{team}} = \frac{(76.2)^2 + (72.6)^2 + (73.1)^2}{4} - 4103.30 = 1.90$$

$$SS_{M \times T\ \text{interaction}} = 246.96 - 242.10 - 1.90 = 2.96$$

Pooling for all three groups gives

$$SS_{M \times T} = 6.28 + 1.51 + 2.96 = 10.75$$

which is recorded in Table 11.7.

The results of this experiment show a very significant method effect (the new method averaged 23.58 rounds per minute, and the old method averaged only 15.08 rounds per minute). The results also show a significant difference between teams within groups at the 5-percent significance level. This points up individual differences in the men. No other effects or interactions were significant.

11.5 SUMMARY

The summary at the end of Chapter 7 may now be extended for Part II.

Experiment	*Design*	*Analysis*
II. Two or more factors		
A. Factorial (crossed)		
	1. Completely randomized $Y_{ijk} = \mu + A_i + B_j + AB_{ij} + \varepsilon_{k(ij)}, \cdots$ for more factors	1.
	a. General case	*a.* ANOVA with interactions
	b. 2^n case	*b.* Yates method or general ANOVA; use (1), *a*, *b*, *ab*, $\cdots$
	c. 3^n case	*c.* General ANOVA; use 00, 10, 20, 01, 11, $\cdots$, and $A \times B = AB + AB^2, \cdots$ for interaction
B. Nested (hierarchical)		
	1. Completely randomized $Y_{ijk} = \mu + A_i + B_{j(i)} + \varepsilon_{k(ij)}$	1. Nested ANOVA
C. Nested factorial		
	1. Completely randomized $Y_{ijkm} = \mu + A_i + B_{j(i)} + C_k + AC_{ik} + BC_{jk(i)} + \varepsilon_{m(ijk)}$	1. Nested factorial ANOVA

PROBLEMS

11.1 Porosity readings on condenser paper were recorded for paper from four rolls taken at random from each of three lots. The results were as follows. Analyze these data, assuming lots are fixed and rolls random.

Lot	I				II				III			
Roll	1	2	3	4	5	6	7	8	9	10	11	12
	1.5	1.5	2.7	3.0	1.9	2.3	1.8	1.9	2.5	3.2	1.4	7.8
	1.7	1.6	1.9	2.4	1.5	2.4	2.9	3.5	2.9	5.5	1.5	5.2
	1.6	1.7	2.0	2.6	2.1	2.4	4.7	2.8	3.3	7.1	3.4	5.0

11.2 In Problem 11.1, how would the results change (if they do) if the lots were chosen at random?

11.3 Set up the EMS column and indicate the proper tests to make if A is a fixed factor at five levels, B is nested within A at four random levels for each level of A, C is nested within B at three random levels and two observations are made in each "cell."

11.4 Repeat Problem 11.3 for A and B crossed or factorial and C nested within the A, B cells.

11.5 Two types of machines are used to wind coils. One type is hand operated; two machines of this type are available. The other type is power operated; two machines of this type are available. Three coils are wound on each machine from two different wiring stocks. Each coil is then measured for the outside diameter of the wire at a middle position on the coil. The results were as follows. (Units are 10^{-5} inch.)

Machine Type	Hand		Power	
Machine Number	2	3	5	8
Stock 1	3279	3527	1904	2464
	3262	3136	2166	2595
	3246	3253	2058	2303
Stock 2	3294	3440	2188	2429
	2974	3356	2105	2410
	3157	3240	2379	2685

Set up the model for this problem and determine what tests can be run.

11.6 Do a complete analysis of Problem 11.5.

11.7 If the outside diameter readings in Problem 11.5 were taken at three fixed positions on the coil (tip, middle, and end), what model would now be appropriate and what tests could be run?

11.8 Some research on the abrasive resistance of filled epoxy plastics was carried out by molding blocks of plastic using five fillers: iron oxide, iron filings, copper, alumina, and graphite. Two concentration ratios were used for the ratio of filler to epoxy resin: $\frac{1}{2}:1$ and $1:1$. Three sample blocks were made up of each of the ten combinations above. These blocks were then subjected to a reciprocating motion where gritcloth was used as the abrasive material in all tests. After 10,000 cycles, each sample block was measured in three places at each of three fixed positions on the plates: I, II, and III. These blocks had been measured before cycling so that the difference in thickness was used as the measured variable. Measurements were made to the nearest 0.0001 inch. Assuming complete randomization of the order of testing of the 30 blocks, set up a mathematical model for this situation and set up the ANOVA table with an EMS column. We are interested in the effect of concentration ratio, fillers, and positions on the block. The three observations at each position will be considered as error, and there may well be differences between the average of the three sample blocks within each treatment combination.

11.9 Data for Problem 11.8 were found to be as shown on the following two tables on pages 202 and 203. Complete an ANOVA for these data and state your conclusions.

11.10 From the results of Problem 11.9, what filler–concentration combination would you recommend if you were interested in a minimum amount of wear? The 1 : 1 ratio is cheaper than $\frac{1}{2}:1$.

11.11 In a study made of the characteristics associated with guidance competence versus counseling competence, 144 students were divided into nine groups of 16 each. These nine groups represented all combinations of three levels of guidance ranking (high, medium, low) and three levels of counseling ranking (high, medium, low). All subjects were then given nine subtests. Assuming the rankings as two fixed factors, the subtests as fixed, and the subjects within the nine groups as random, set up a data layout and mathematical model for this experiment.

11.12 Determine the ANOVA layout and the EMS column for Problem 11.11 and indicate the proper tests which can be made.

11.13 Three days of sampling where each sample was subjected to two types of size graders gave the following results, coded by subtracting 4 percent moisture and multiplying by 10.

Day	1		2		3	
Grader	*A*	*B*	*A*	*B*	*A*	*B*
Sample 1	4	11	5	11	0	6
2	6	7	17	13	−1	−2
3	6	10	8	15	2	5
4	13	11	3	14	8	2
5	7	10	14	20	8	6
6	7	11	11	19	4	10
7	14	16	6	11	5	18
8	12	10	11	17	10	13
9	9	12	16	4	16	17
10	6	9	−1	9	8	15
11	8	13	3	14	7	11

Asusming graders fixed, days random, and samples within days random, set up a mathematical model for this experiment and determine the EMS column.

11.14 Complete the ANOVA for Problem 11.13 and comment on the results.

11.15 In a filling process, two random runs are made. For each run six hoppers are used, but they may not be the same hoppers on the second run. Assume that hoppers are a random sample of possible hoppers. Data are taken from the bottom, middle, and top of each hopper in order to check on a possible position effect on filling. Two observations are made within positions and runs on all hoppers. Set up a mathematical model for this problem and the associated EMS column.

11.16 The data below are from the experiment described in Problem 11.15.

	Run 1						Run 2					
Head	Bottom		Middle		Top		Bottom		Middle		Top	
1	19	0	31	25	25	13	6	6	0	6	0	0
2	13	0	19	38	31	25	0	13	0	31	6	0
3	13	0	31	19	13	0	28	6	25	19	13	28
4	13	0	19	31	25	13	28	0	6	0	28	25
5	19	19	19	38	19	25	16	13	13	19	0	0
6	6	0	0	0	0	0	13	6	6	19	0	19

Do a complete analysis of these data.

	Alumina						Graphite					
Concentration	½ : 1			1 : 1			½ : 1			1 : 1		
Position	I	II	III	I	II	III	I	II	III	I	II	III
Sample 1	7.0	5.4	6.4	7.1	5.0	6.4	8.8	5.5	9.0	18.0	14.4	18.5
	7.3	5.1	5.7	6.9	5.7	6.6	8.3	5.9	11.0	17.6	12.4	19.2
	6.6	5.0	7.7	6.5	4.0	6.2	6.9	4.7	11.3	18.4	12.6	19.4
	20.9	15.5	19.8	20.5	14.7	19.2	24.0	16.1	31.3	54.0	39.4	57.1
Sample 2	6.8	5.1	4.6	6.8	4.0	7.5	7.7	6.8	10.1	15.6	11.6	17.6
	6.9	5.5	5.8	7.2	5.2	7.8	6.5	7.0	11.2	14.6	12.8	18.7
	6.2	4.8	5.8	4.8	4.4	6.0	5.9	7.2	10.8	15.1	13.4	19.3
	19.9	15.4	16.2	18.8	13.6	21.3	20.1	21.0	32.1	45.3	37.8	55.6
Sample 3	7.3	6.3	5.1	6.2	4.4	5.7	7.2	7.0	7.7	13.3	11.7	15.5
	7.6	5.3	6.5	6.6	4.8	6.6	6.8	6.9	9.3	13.8	13.4	18.0
	5.6	4.8	7.1	4.4	3.8	5.6	6.2	5.2	9.8	13.6	12.1	19.6
	20.5	16.4	18.7	17.2	13.0	17.9	20.2	19.1	26.8	40.7	37.2	53.1
Totals		163.3			156.2			210.7			420.2	

	Iron Filings						Iron Oxide						Copper					
Concentration	$\frac{1}{2}:1$			1 : 1			$\frac{1}{2}:1$			1 : 1			$\frac{1}{2}:1$			1 : 1		
Position	I	II	III	I	II	III	I	II	III	I	II	III	I	II	III	I	II	III
Sample 1	2.1	1.1	1.3	3.6	1.6	2.3	3.0	1.1	3.2	1.8	1.3	2.4	2.7	1.4	2.8	2.8	1.5	2.4
	2.1	1.1	1.7	3.8	1.0	2.5	3.8	1.8	4.1	2.1	1.0	1.9	2.9	2.2	3.8	2.6	1.1	2.1
	1.0	0.9	1.7	2.9	1.6	2.4	3.0	1.1	3.3	1.9	1.6	1.8	3.0	1.8	3.4	1.9	1.4	2.0
	5.2	3.1	4.7	10.3	4.2	7.2	9.8	4.0	10.6	5.8	3.9	6.1	8.6	5.4	10.0	7.3	4.0	6.5
Sample 2	1.7	1.1	1.7	2.1	1.1	1.7	3.1	1.6	2.2	2.1	1.0	1.5	2.5	1.8	2.4	2.1	1.0	1.5
	1.8	1.0	2.0	2.6	1.1	1.0	3.0	1.7	3.0	2.0	0.8	2.0	3.0	2.6	3.3	2.5	1.0	1.6
	1.3	0.8	2.0	1.6	0.9	1.4	2.7	1.2	2.7	2.3	1.0	2.0	2.0	2.1	2.4	1.5	1.2	1.7
	4.8	2.9	5.7	6.3	3.1	4.1	8.8	4.5	7.9	6.4	2.8	5.5	7.5	6.5	8.1	6.1	3.2	4.8
Sample 3	3.2	0.8	1.4	2.3	1.1	1.8	2.1	1.5	2.0	1.6	1.2	1.6	3.3	2.1	3.4	2.6	1.3	2.6
	2.8	0.8	2.0	2.6	1.2	2.0	2.9	1.9	2.5	1.6	1.1	1.4	3.2	2.0	4.1	2.9	1.5	3.3
	2.0	0.5	1.9	2.1	0.7	2.4	1.6	2.5	2.4	1.9	1.2	2.1	3.2	2.0	3.8	2.8	1.2	2.7
	8.0	2.1	5.3	7.0	3.0	6.2	6.6	5.9	6.9	5.1	3.5	5.1	9.7	6.1	11.3	8.3	4.0	8.6
Totals	41.8			51.4			65.0			44.2			73.2			52.8		

12

Experiments of Two or More Factors—Restrictions on Randomization

12.1 INTRODUCTION

In the discussion of factorial and nested experiments, it was assumed that the whole experiment was performed in a completely randomized manner. In practice, however, it may not be feasible to run the whole experiment several times in one day, or by one experimenter, and so forth. It then becomes necessary to restrict this complete randomization and block the experiment in the same manner that a single-factor experiment was blocked in Chapter 4. Instead of running several replications of the experiment all at one time, it may be possible to run one complete replication on one day, a second complete replication on another day, a third replication on a third day, and so on. In this case, each replication is a block, and the design is a randomized block design with a complete factorial or nested experiment randomized within each block.

Occasionally a second restriction on randomization is necessary, in which case a Latin square design might be used, with the treatments in the square representing a complete factorial or nested experiment.

12.2 FACTORIAL EXPERIMENT IN A RANDOMIZED BLOCK DESIGN

Just as the factorial arrangements were extracted from the treatment effect in Equations (6.1) and (6.2), so can these arrangements be extracted from the treatment effect in the randomized complete block design of Chapter 4 (Section 4.2). For the completely randomized design, you will recall that

$$Y_{ij} = \mu + \tau_j + \varepsilon_{ij}$$

can be subdivided into

$$Y_{ijk} = \mu + A_i + B_j + AB_{ij} + \varepsilon_{k(ij)}$$

where $A_i + B_j + AB_{ij} = \tau_j$ of the first model.

When a complete experiment is replicated several times and R_k represents the blocks or replications, the randomized block model would be

$$Y_{km} = \mu + R_k + \tau_m + \varepsilon_{km}$$

where R_k represents the blocks or replications and τ_m represents the treatments. If the treatments are formed from a two-factor factorial experiment, then

$$\tau_m = A_i + B_j + AB_{ij}$$

and the complete model is

$$Y_{ijk} = \mu + R_k + A_i + B_j + AB_{ij} + \varepsilon_{ijk} \qquad (12.1)$$

Here it is assumed that there is no interaction between replications (blocks) and treatments, which is the usual assumption underlying a randomized block design. Any such interaction is confounded in the error term ε_{ijk}.

To get a more concrete picture of this type of design, consider three levels of factor A, two levels of B, and four replications. The treatment combinations may be written as A_1B_1, A_1B_2, A_2B_1, A_2B_2, A_3B_1, and A_3B_2. Such a design assumes that all six of these treatment combinations can be run on a given day, if days are the replications or blocks. A layout in which these six treatment combinations are randomized within each replication might be

Replication I: A_1B_2 A_3B_1 A_3B_2 A_2B_1 A_1B_1 A_2B_2

Replication II: A_2B_2 A_1B_1 A_3B_2 A_2B_1 A_1B_2 A_3B_1

Replication III: A_2B_1 A_3B_2 A_1B_2 A_3B_1 A_2B_2 A_1B_1

Replication IV: A_1B_1 A_3B_1 A_3B_2 A_2B_1 A_1B_2 A_2B_2

An analysis breakdown would be that shown in Table 12.1.

Table 12.1 Two-Factor Experiment in a Randomized Block Design

Source	df	
Replications R_k	3	
Treatments τ_m	5	
A_i		2
B_j		1
AB_{ij}		2
Error ε_{ijk}	15	
Total	23	

To see how the interaction of replications and treatments is used as error, consider the expanded model and the corresponding EMS terms exhibited in Table 12.2. Here the factors are fixed and replications are considered as random.

Table 12.2 EMS Terms for Two-Factor Experiment in a Randomized Block

Source	df	3 F i	2 F j	4 R k	1 R m	EMS
R_k	3	3	2	1	1	$\sigma_\varepsilon^2 + 6\sigma_R^2$
A_i	2	0	2	4	1	$\sigma_\varepsilon^2 + 2\sigma_{RA}^2 + 8\phi_A$
RA_{ik}	6	0	2	1	1	$\sigma_\varepsilon^2 + 2\sigma_{RA}^2$
B_j	1	3	0	4	1	$\sigma_\varepsilon^2 + 3\sigma_{RB}^2 + 12\phi_B$
RB_{jk}	3	3	0	1	1	$\sigma_\varepsilon^2 + 3\sigma_{RB}^2$
AB_{ij}	2	0	0	4	1	$\sigma_\varepsilon^2 + \sigma_{RAB}^2 + 4\phi_{AB}$
RAB_{ijk}	6	0	0	1	1	$\sigma_\varepsilon^2 + \sigma_{RAB}^2$
$\varepsilon_{m(ijk)}$	0	1	1	1	1	σ_ε^2 (not retrievable)
Total	23					

Since the degrees of freedom are quite low for the tests indicated above, and since there is no separate estimate of error variance, it is customary to assume that $\sigma_{RA}^2 = \sigma_{RB}^2 = \sigma_{RAB}^2 = 0$ and to pool these three terms to serve as the error variance. The reduced model is then as shown in Table 12.3.

Table 12.3 EMS Terms for Two-Factor Experiment, Randomized Block Design, Reduced Model

Source	df	EMS
R_k	3	$\sigma_\varepsilon^2 + 6\sigma_R^2$
A_i	2	$\sigma_\varepsilon^2 + 8\phi_A$
B_j	1	$\sigma_\varepsilon^2 + 12\phi_B$
AB_{ij}	2	$\sigma_\varepsilon^2 + 4\phi_{AB}$
ε_{ijk}	15	σ_ε^2
Total	23	

Here all effects are tested against the 15-df error term. Although we usually pool the interactions of replications with treatment effects for the error term, this is not always necessary. It will depend upon the degrees of freedom available for testing various hypotheses in Table 12.2, and also whether or not any repeat measurements can be taken within a replication. If the latter is possible, a separate estimate of σ_ε^2 is available and all tests in Table 12.2 may be made.

Example 12.1 At Purdue, a researcher was interested in determining the effect of both nozzle types and operators on the rate of fluid flow in cubic centimeters through these nozzles. He considered three fixed nozzle types and five randomly chosen operators. Each of these 15 combinations was to be run in a random order on each of three days. These three days will be considered as three replications of the complete 3×5 factorial experiment. After subtracting 96.0 cm^3 from each reading and multiplying all readings by 10, the data layout and observed readings (in cubic centimeters) are as recorded in Table 12.4.

Table 12.4 Nozzle Example Data

Operator	1			2			3			4			5		
Nozzle	*A*	*B*	*C*	*A*	*B*	*C*	*A*	*B*	*C*	*A*	*B*	*C*	*A*	*B*	*C*
Replication															
I	6	13	10	26	4	−35	11	17	11	21	−5	12	25	15	−4
II	6	6	10	12	4	0	4	10	−10	14	2	−2	18	8	10
III	−15	13	−11	5	11	−14	4	17	−17	7	−5	−16	25	1	24

The model for this example is

$$Y_{ijk} = \mu + R_k + N_i + O_j + NO_{ij} + \varepsilon_{ijk}$$

with

$$k = 1, 2, 3 \qquad i = 1, 2, 3 \qquad j = 1, 2, \cdots, 5$$

The R_k levels are random, O_j levels are random, and N_i levels are fixed. If m repeat measurements are considered in the model, but $m = 1$, the EMS values are those appearing in Table 12.5.

Table 12.5 EMS for Nozzle Example

		3	5	3	1	
		F	*R*	*R*	*R*	
Source	df	*i*	*j*	*k*	*m*	EMS
R_k	2	3	5	1	1	$\sigma_\varepsilon^2 + 15\sigma_R^2$
N_i	2	0	5	3	1	$\sigma_\varepsilon^2 + 3\sigma_{NO}^2 + 15\phi_N$
O_j	4	3	1	3	1	$\sigma_\varepsilon^2 + 9\sigma_O^2$
NO_{ij}	8	0	1	3	1	$\sigma_\varepsilon^2 + 3\sigma_{NO}^2$
$\varepsilon_{m(ijk)}$	28	1	1	1	1	σ_ε^2
Total	44					

These EMS values indicate that all main effects and the interaction can be tested with reasonable precision (df). The sums of squares are computed in the usual manner from the subtotals in Table 12.6.

Table 12.6 Subtotals in Nozzle Example

			Operator					
Replication	Total	Nozzle	1	2	3	4	5	Nozzle Total
I	127	*A*	−3	43	19	42	68	169
II	92	*B*	32	19	44	−8	24	111
III	29	*C*	9	−49	−16	−6	30	−32
Totals	248		38	13	47	28	122	248

Using these subtotals, the usual sums of squares can easily be computed and the results displayed as in Table 12.7.

Table 12.7 ANOVA for Nozzle Example

Source	df	SS	MS	EMS	*F*	$F_{0.05}$
R_k	2	328.84	164.42	$\sigma_\varepsilon^2 + 15\sigma_R^2$	1.70	3.34
N_i	2	1426.97	713.48	$\sigma_\varepsilon^2 + 3\sigma_{NO}^2 + 15\phi_N$	3.13	4.46
O_j	4	798.79	199.70	$\sigma_\varepsilon^2 + 9\sigma_O^2$	2.06	2.71
NO_{ij}	8	1821.48	227.68	$\sigma_\varepsilon^2 + 3\sigma_{NO}^2$	2.35*	2.29
ε_{ijk}	28	2709.16	96.76	σ_ε^2		
Total	44	7085.24				

The results of tests of hypotheses on this model show only a significant nozzle × operator interaction at the 5-percent level of significance. The plot

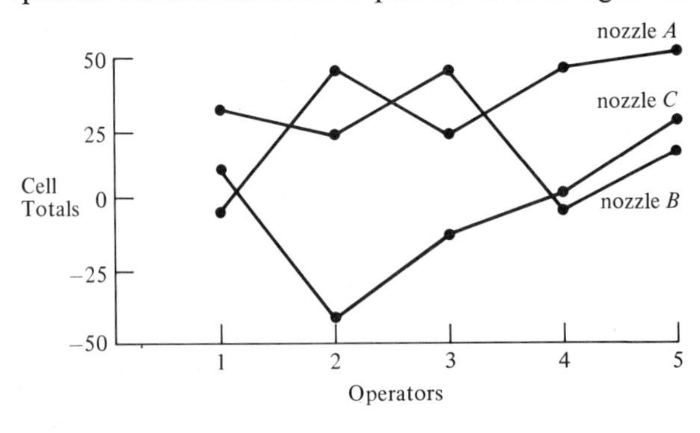

Figure 12.1 Interaction plot for nozzle × operator interaction in nozzle example.

in Figure 12.1 of all totals for the right-hand data of Table 12.6 shows this interaction graphically.

If desirable, the model of this example could be expanded to include an estimate of replication by operator and replication by nozzle interactions, using the three-way interaction ($N \times O \times$ Replication) as the error. The results of such a breakdown are recorded in Table 12.8.

Table 12.8 ANOVA for Nozzle Example with Replication Interactions

Source	df	SS	MS	EMS	F
R_k	2	328.84	164.42	$\sigma_\varepsilon^2 + 3\sigma_{RO}^2 + 15\sigma_R^2$	1.86
N_i	2	1426.97	713.48	$\sigma_\varepsilon^2 + 3\sigma_{NO}^2 + 5\sigma_{NR}^2 + 15\phi_N$	*
NR_{ik}	4	272.23	68.06	$\sigma_\varepsilon^2 + 5\sigma_{NR}^2$	<1
O_j	4	798.79	199.70	$\sigma_\varepsilon^2 + 3\sigma_{RO}^2 + 9\sigma_O^2$	2.26
RO_{jk}	8	705.60	88.20	$\sigma_\varepsilon^2 + 3\sigma_{RO}^2$	<1
NO_{ij}	8	1821.48	227.68	$\sigma_\varepsilon^2 + 3\sigma_{NO}^2$	2.10
ε_{ijk} or NRO_{ijk}	16	1731.33	108.21	σ_ε^2	
Totals	44	7085.24			

	Operator						Nozzle			
Replication	1	2	3	4	5	Total	A	B	C	Total
I	29	−5	39	28	36	127	89	44	−6	127
II	22	16	4	14	36	92	54	30	8	92
III	−13	2	4	−14	50	29	26	37	−34	29
Totals	38	13	47	28	122	248	169	111	−32	248

The F test values are shown in Table 12.8 except for the test on nozzles labeled *. Since there is no direct test, an F' test is used as described in Chapter 10. Here

$$\text{MS} = \text{MS}_{NR} + \text{MS}_{NO} - \text{MS}_{NRO}$$

Here the coefficients are 1, 1, and −1, and the corresponding mean squares and degrees of freedom are

$$\text{MS}_{NR} = 68.06 \quad \text{with } \nu_1 = 4 \text{ df}$$

$$\text{MS}_{NO} = 227.68 \quad \text{with } \nu_2 = 8 \text{ df}$$

$$\text{MS}_{NRO} = 108.21 \quad \text{with } \nu_3 = 16 \text{ df}$$

The F' test is then

$$F' = \frac{\text{MS}_N}{\text{MS}_{NR} + \text{MS}_{NO} - \text{MS}_{NRO}}$$

$$= \frac{713.48}{68.06 + 227.68 - 108.21}$$

$$= \frac{713.48}{187.53} = 3.80$$

and

$$v = \frac{(187.53)^2}{(1)^2[(68.06)^2/4] + (1)^2[(227.68)^2/8] + (-1)^2[(108.21)^2/16]} = 4.2$$

With 2 and 4 df the 5-percent F value is 6.94, and with 2 and 5 df it is 5.79; hence, the $F' = 3.80$ is not significant at the 5-percent level. Hence all F tests indicate no significant effects. The $N \times O$ interaction does not show significance as it did in Table 12.7, because the test is less sensitive due to the reduction in degrees of freedom from 28 to 16 in the error mean square. Since both the $N \times R$ and $R \times O$ interaction F tests are less than one, we are probably justified in assuming that σ^2_{NR} and σ^2_{RO} are zero and in pooling these with σ^2_{NRO} for a 28-df error mean square as in Table 12.7.

12.3 FACTORIAL EXPERIMENT IN A LATIN SQUARE DESIGN

If there are two restrictions on the randomization, and an equal number of restriction levels and treatment levels are used, Chapter 5 has shown that a Latin square design may be used. The model for this design is

$$Y_{ijk} = \mu + R_i + \tau_j + \gamma_k + \varepsilon_{ijk}$$

where τ_j represents the treatments, and both R_i and γ_k represent restrictions on the randomization. If a 4×4 Latin square is used for a given experiment, such as the one in Section 5.2, the four treatments A, B, C, and D could represent the four treatment combinations of a 2×2 factorial (Table 12.9). Here

$$\tau_j = A_m + B_q + AB_{mq}$$

Table 12.9 4×4 Latin Square (Same as Table 5.1)

	Car			
Position	I	II	III	IV
1	*C*	*D*	*A*	*B*
2	*B*	*C*	*D*	*A*
3	*A*	*B*	*C*	*D*
4	*D*	*A*	*B*	*C*

with its 3 df. Thus, a factorial experiment ($A \times B$) is run in a Latin square design. For the problem of Section 5.2, instead of four tire brands, the four treatments could consist of two brands (A_1 and A_2) and two types (B_1 and B_2). The types might be black- and white-walled tires. The complete model would then be

$$Y_{ikmq} = \mu + R_i + \gamma_k + A_m + B_q + AB_{mq} + \varepsilon_{ikmq}$$

and the degree-of-freedom breakdown would be that shown in Table 12.10.

Table 12.10 Degree-of-Freedom Breakdown for Table 12.9

Source	df
R_i	3
γ_k	3
A_m	1 }
B_q	1 } 3 df for τ_j
AB_{mq}	1 }
ε_{ikmq}	6
Total	15

This idea could be extended to running a factorial in a Graeco-Latin square, and so on.

12.4 REMARKS

In the preceding sections factorial experiments have been considered where the design of the experiment was a randomized block or a Latin square. It is also possible to run a nested experiment or a nested factorial experiment in a randomized block or Latin square design. In the case of a nested experiment, the treatment effect might be broken down as

$$\tau_m = A_i + B_{j(i)}$$

and when this experiment has one restriction on the randomization, the model is

$$Y_{ijk} = \mu + R_k + \underbrace{A_i + B_{j(i)}}_{\tau_m} + \varepsilon_{ijk}$$

If a nested factorial is repeated on several different days, the model would be

$$Y_{ijkm} = \mu + R_k + \underbrace{A_i + B_{j(i)} + C_m + AC_{im} + BC_{mj(i)}}_{\tau_m} + \varepsilon_{ijkm}$$

where the whole nested factorial is run in a randomized block design. The analysis of such designs follows from the methods of Chapter 11.

12.5 SUMMARY

The summary at the end of Chapter 11 may now be extended.

Experiment	*Design*	*Analysis*
II. Two or more factors		
A. Factorial (crossed)		
	1. Completely randomized $Y_{ijk} = \mu + A_i + B_j + AB_{ij} + \varepsilon_{k(ij)}, \cdots$ for more factors	1.
	a. General case	*a*. ANOVA with interactions
	b. 2^n case	*b*. Yates method or general ANOVA
	c. 3^n case	*c*. General ANOVA
	2. Randomized block	2.
	a. Complete $Y_{ijk} = \mu + R_k + A_i + B_j + AB_{ij} + \varepsilon_{ijk}$	*a*. Factorial ANOVA with replications R_k
	3. Latin square	3.
	a. Complete $Y_{ijkm} = \mu + R_k + \gamma_m + A_i + B_j + AB_{ij} + \varepsilon_{ijkm}$	*a*. Factorial ANOVA with replications and positions
B. Nested (hierarchical)		
	1. Completely randomized $Y_{ijk} = \mu + A_i + B_{j(i)} + \varepsilon_{k(ij)}$	1. Nested ANOVA
	2. Randomized block	2. Nested ANOVA with blocks R_k
	a. Complete $Y_{ijk} = \mu + R_k + A_i + B_{j(i)} + \varepsilon_{ijk}$	*a*. Nested ANOVA with blocks R_k
	3. Latin square	3.
	a. Complete $Y_{ijkm} = \mu + R_k + \gamma_m + A_i + B_{j(i)} + \varepsilon_{ijkm}$	*a*. Nested ANOVA with blocks and positions

Experiment	Design	Analysis
C. Nested factorial		
	1. Completely randomized $Y_{ijkm} = \mu + A_i + B_{j(i)} + C_k + AC_{ik} + BC_{kj(i)} + \varepsilon_{m(ijk)}$	1. Nested factorial ANOVA
	2. Randomized block	2.
	a. Complete $Y_{ijkm} = \mu + R_k + A_i + B_{j(i)} + C_m + AC_{im} + BC_{mj(i)} + \varepsilon_{ijkm}$	*a*. ANOVA with blocks R_k
	3. Latin square	3.
	a. Complete $Y_{ijkmq} = \mu + R_k + \gamma_m + A_i + B_{j(i)} + C_q + AC_{iq} + BC_{qj(i)} + \varepsilon_{ijkmq}$	*a*. ANOVA with blocks and positions

PROBLEMS

12.1 Data on the glass rating of tubes taken from two fixed stations and three shifts are recorded each week for three weeks. Considering the weeks as blocks, the six treatment combinations (two stations by three shifts) were tested in a random order each week, with results as follows:

	Station 1			Station 2		
Shift	1	2	3	1	2	3
Week 1	3	3	3	6	3	6
	6	4	6	8	9	8
	6	7	7	11	11	13
Week 2	14	8	11	4	15	4
	16	8	12	6	15	7
	19	9	17	7	17	10
Week 3	2	2	2	2	2	10
	3	3	4	5	4	12
	6	4	6	7	6	13

Analyze these data as a factorial run in a randomized block design.

12.2 In the example of Section 5.2 (Table 5.2) consider tire brands A, B, C, D as four combinations of two factors, ply of tires and type of tread, where

$$A = P_1T_1 \qquad B = P_1T_2 \qquad C = P_2T_1 \qquad D = P_2T_2$$

representing the four combinations of two ply values and two tread types. Analyze the data for this revised design—a 2^2 factorial run in a Latin square design.

12.3 If in Problem 11.5 the stock factor is replaced by "days," considered random, and the rest of the experiment is performed in a random manner on each of the two days, use the same numerical results and reanalyze with this restriction.

12.4 Again, in Problem 11.5 assume that the whole experiment (24 readings) was repeated on five randomly selected days. Set up an outline of this experiment, including the EMS column, df, and so on.

12.5 Discuss the similarities and differences between a nested experiment with two factors and a randomized block design with one factor.

12.6 If the nested experiment involving machines and heads in Section 11.2 were completely repeated on three subsequent randomly selected days, outline its mathematical model and ANOVA table complete with the EMS column.

12.7 For Problem 12.6 show a complete model with *all* interactions and the reduced model. Make suggestions regarding how this experiment might be made less costly and still give about the same information.

12.8 For the nested factorial example in Section 11.4 on gun loading, set up the model if the whole experiment is repeated on four more randomly selected days.

12.9 Examine both the full and reduced models for Problem 12.8 and comment.

13

Factorial Experiment—Split-Plot Design

13.1 INTRODUCTION

In many experiments where a factorial arrangement is desired, it may not be possible to completely randomize the order of experimentation. In the last chapter, restrictions on randomization were considered, and both randomized block and Latin square designs were discussed. There are still many practical situations in which it is not at all feasible to even randomize within a block. Under certain conditions, these restrictions will lead to a split-plot design. An example will show why such designs are quite common.

Example 13.1 The data in Table 13.1 were compiled on the effect of oven temperature T and baking time B on the life Y of an electrical component.[1]

Table 13.1 Electrical Component Life-Test Data

Baking Time B (min)	Oven Temperature T (°F)			
	580	600	620	640
5	217	158	229	223
	188	126	160	201
	162	122	167	182
10	233	138	186	227
	201	130	170	181
	170	185	181	201
15	175	152	155	156
	195	147	161	172
	213	180	182	199

[1] Taken from a problem of Wortham and Smith [19, p. 112].

Looking only at this table of data, we might think of this as a 4 × 3 factorial with three replications per cell and proceed to the analysis in Table 13.2. Here only the temperature has a significant effect on the life of the component at the 1-percent significance level.

Table 13.2 ANOVA for Electrical Component Life-Test Data as a Factorial

Source	df	SS	MS	EMS
T_i	3	12,494	4165	$\sigma_\varepsilon^2 + 9\phi_T$
B_j	2	566	283	$\sigma_\varepsilon^2 + 12\phi_B$
BT_{ij}	6	2601	434	$\sigma_\varepsilon^2 + 3\phi_{TB}$
$\varepsilon_{k(ij)}$	24	13,670	570	σ_ε^2
Totals	35	29,331		

We might go on now and seek linear, quadratic, and cubic effects of temperature; but before proceeding, a few questions on design should be raised. What was the order of experimentation? The data analysis above assumes a completely randomized design. This means that one of the four temperatures was chosen at random and the oven was heated to this temperature, then a baking time was chosen at random and an electrical component was inserted in the oven and baked for the time selected. After this run, the whole procedure was repeated until the data were compiled. Now, was the experiment really conducted in this manner? Of course the answer is "no." Once an oven is heated to temperature, all nine components are inserted; three are baked for five minutes, three for 10 minutes, and three for 15 minutes. We would argue that this is the only practical way to run the experiment. Complete randomization is too impractical as well as too expensive. Fortunately, this restriction on the complete randomization can be handled in a design called a *split-plot* design.

The four temperature levels are called *plots*. They could be called blocks, but in the previous chapter, blocks and replications have been used for a complete rerun of the whole experiment. (The word "plots" has been inherited from agricultural applications.) In such a setup, temperature—a main effect—is confounded with these plots. If conditions change from one oven temperature to another, these changes will show up as temperature differences. Thus, in such a design, a main effect is confounded with plots. This is necessary because it is often the most practical way to order the experiment. Now once a temperature has been set up by choosing one of the four temperatures at random, three components can be placed in the oven and one component can be baked for 5 minutes, another for 10 minutes, and the third one for 15 minutes. The specific component which is to be baked for 5, 10, and 15 minutes is again decided by a random selection. These three baking-time levels may be thought of as a splitting of the plot into three parts, one part for each baking time. This defines the three parts of a main plot that are called split plots. Note here that only three components were placed in the oven, not nine. The temperature is then changed to another level and three more components are placed in the oven for 5-, 10-, and 15-minutes baking time. This same procedure is followed for all four temperatures; after this the whole experiment may be replicated. These replications may be run several days after the initial experiment; in fact, it is often desirable to collect data from two or three replications and then decide if more replications are necessary.

The experiment conducted above was actually run as a split-plot experiment and laid out as shown in Table 13.3.

Table 13.3 Split-Plot Layout for Electrical Component Life-Test Data

Replication *R*	Baking Time *B* (min)	Oven Temperature *T* (°F)			
		580	600	620	640
I	5	217	158	229	223
	10	233	138	186	227
	15	175	152	155	156
II	5	188	126	160	201
	10	201	130	170	181
	15	195	147	161	172
III	5	162	122	167	182
	10	170	185	181	201
	15	213	180	182	199

Here temperatures are confounded with plots and the $R \times T$ cells are called whole plots. Inside a whole plot, the baking times are applied to one third of the material. These plots associated with the three baking times are called split plots. Since one main effect is confounded with plots and the other main effect is not, it is usually desirable to place in the split the main effect we are most concerned about testing, as this factor is not confounded.

We might think that factor *B* (baking time) is nested with the main plots but this is not the case, since the same levels of *B* are used in all plots. A model for this experiment would be

$$Y_{ijk} = \mu + \underbrace{R_i + T_j + RT_{ij}}_{\text{whole plot}} + \underbrace{B_k + RB_{ik} + TB_{jk} + RTB_{ijk}}_{\text{split plot}}$$

The first three variable terms in this model represent the whole plot, and the RT interaction is often referred to as the *whole plot error*. The usual assumption is that this interaction does not exist, that this term is really an estimate of the error within the main plot. The last four terms represent the split plot, and the RTB interaction is referred to as the *split-plot error*. Sometimes the RB term is also considered nonexistent and is combined with RTB as an error term. A separate error term might be obtained if it were feasible to repeat some observations within the split plot. The proper EMS values can be found by considering m repeat measurements where $m = 1$ (Table 13.4).

Table 13.4 EMS for Split-Plot Electrical Component Life-Test Data

			3	4	3	1	
			R	F	F	R	
	Source	df	i	j	k	m	EMS
Whole plot	R_i	2	1	4	3	1	$\sigma_\varepsilon^2 + 12\sigma_R^2$
	T_j	3	3	0	3	1	$\sigma_\varepsilon^2 + 3\sigma_{RT}^2 + 9\phi_T$
	RT_{ij}	6	1	0	3	1	$\sigma_\varepsilon^2 + 3\sigma_{RT}^2$
Split plot	B_k	2	3	4	0	1	$\sigma_\varepsilon^2 + 4\sigma_{RB}^2 + 12\phi_B$
	RB_{ik}	4	1	4	0	1	$\sigma_\varepsilon^2 + 4\sigma_{RB}^2$
	TB_{jk}	6	3	0	0	1	$\sigma_\varepsilon^2 + \sigma_{RTB}^2 + 3\phi_{TB}$
	RTB_{ijk}	12	1	0	0	1	$\sigma_\varepsilon^2 + \sigma_{RTB}^2$
	$\varepsilon_{m(ijk)}$	—	1	1	1	1	σ_ε^2 (not retrievable)
	Total	35					

Since the error mean square cannot be isolated in this experiment, $\sigma_\varepsilon^2 + \sigma_{RTB}^2$ is taken as the split-plot error, and $\sigma_\varepsilon^2 + 3\sigma_{RT}^2$ is taken as the whole-plot error. The main effects and interaction of interest TB can be tested, as seen from the EMS column, although no exact tests exist for the replication effect nor for interactions involving the replications. This is not a serious disadvantage for this design, since tests on replication effects are not of interest but are isolated only to reduce the error variance. The analysis of the data from Table 13.3 follows the methods given in Chapter 6. The results are shown in Table 13.5.

Testing the hypothesis of no temperature effect gives

$$F_{3,6} = \frac{4165}{296} = 14.1 \text{ (significant at the 1-percent level)}$$

Testing for baking time gives

$$F_{2,4} = \frac{283}{1755} < 1 \text{ (not significant)}$$

Testing for TB interaction gives

$$F_{6,12} = \frac{434}{243} = 1.79 \text{ (not significant at the 5-percent level)}$$

Table 13.5 ANOVA for Split-Plot Electrical Component Life-Test Data

Source	df	SS	MS	EMS
R_i	2	1963	982	$\sigma_\varepsilon^2 + 12\sigma_R^2$
T_j	3	12,494	4165	$\sigma_\varepsilon^2 + 3\sigma_{RT}^2 + 9\phi_T$
RT_{ij}	6	1774	296	$\sigma_\varepsilon^2 + 3\sigma_{RT}^2$
B_k	2	566	283	$\sigma_\varepsilon^2 + 4\sigma_{RB}^2 + 12\phi_B$
RB_{ik}	4	7021	1755	$\sigma_\varepsilon^2 + 4\sigma_{RB}^2$
TB_{jk}	6	2601	434	$\sigma_\varepsilon^2 + \sigma_{RTB}^2 + 3\phi_{TB}$
RTB_{ijk}	12	2912	243	$\sigma_\varepsilon^2 + \sigma_{RTB}^2$
Totals	35	29,331		

No exact tests are available for testing the replication effect or replication interaction with other factors, but these effects are present only to reduce the experimental error in this split-plot design.

The results of this split-plot analysis are not too different from the results using the incorrect method of Table 13.2, but this split-plot design shows the need for careful consideration of the method of randomization before starting an experiment. This split-plot design represents a restriction on the randomization over a complete randomization in a factorial experiment.

Since this type of design is encountered often in industrial experiments, another example will be considered.

Example 13.2 A defense-related organization was to study the pulloff force necessary to separate boxes of chaff from a tape on which they are affixed. These boxes are made of a cardboard material, approximately 3 by 3 by 1 inches, and are mounted on a strip of cloth tape that has an adhesive backing. The tape is 2 inches wide, and the boxes are placed 7 inches center to center on the strip. There are 75 boxes mounted on each strip.

The tape is pulled from the box at a 90° angle as it is wound onto a drum. During this separation process, the portion of the tape still adhering to the box carries the box onto a platform. The box trips a microswitch, which energizes a plunger. The plunger then kicks the box out of the machine.

After this problem was discussed with the plant engineers, several factors were listed which might affect this pulloff force. The most important factors were temperature and humidity. It was agreed to use three fixed temperature levels, −55°C, 25°C, and 55°C, and three fixed humidity levels, 50 percent, 70 percent, and 90 percent. These gave nine basic treatment combinations. Since there might be differences in pulloff force as a result of the strip selected, it was decided to choose five different strips at random for use in this experiment. There might also be differences within a strip, so two boxes were chosen at random and cut from each strip for the test.

The test in the laboratory was accomplished by hand-holding the package, attaching a spring scale to the strip by means of a hole previously punched in the strip, and pulling the tape from the package in a direction perpendicular to the package.

In discussing the design of this experiment, it seemed best to set the climatic condition (a combination of temperature and humidity) at random from one of the $3 \times 3 = 9$ conditions, and then test two boxes from each of the five strips while these conditions were maintained. Then another of the nine conditions is set, and the results again determined on two boxes from each of the five strips. This is then a restriction on the randomization, and the resulting design is a split-plot design. It was agreed to replicate the complete experiment four times. A layout for this experiment is shown in Table 13.6.

In Table 13.6, each replication is a repeat of Replication I, but a new order of randomizing the nine atmospheric conditions is taken in each

Table 13.6 Split-Plot Design of Chaff Experiment

			Temperature T (°C)								
			−55			25			55		
			Humidity H (percent)								
	Strip S	Box	50	70	90	50	70	90	50	70	90
Replication I	1	1									
		2									
R	2	3									
		4									
	3	5									
		6									
	4	7									
		8									
	5	9									
		10									
Replication II (Repeat as above)											
Replication III (Repeat as above)											
Replication IV (Repeat as above)											

replication. In this design, atmospheric conditions and replications form the whole plot and strips are in the split-plot.

The model for this design and its associated EMS relations are set up in Table 13.7.

Table 13.7 EMS for Chaff Experiment

	Source	df	4 R i	3 F j	3 F k	5 R m	2 R q	EMS
	R_i	3	1	3	3	5	2	$\sigma_\varepsilon^2 + 18\sigma_{RS}^2 + 90\sigma_R^2$
	T_j	2	4	0	3	5	2	$\sigma_\varepsilon^2 + 6\sigma_{RTS}^2 + 24\sigma_{TS}^2 + 30\sigma_{RT}^2 + 120\phi_T$
Whole	RT_{ij}	6	1	0	3	5	2	$\sigma_\varepsilon^2 + 6\sigma_{RTS}^2 + 30\sigma_{RT}^2$
plot	H_k	2	4	3	0	5	2	$\sigma_\varepsilon^2 + 6\sigma_{RHS}^2 + 24\sigma_{HS}^2 + 30\sigma_{RH}^2 + 120\phi_H$
	RH_{ik}	6	1	3	0	5	2	$\sigma_\varepsilon^2 + 6\sigma_{RHS}^2 + 30\sigma_{RH}^2$
	TH_{jk}	4	4	0	0	5	2	$\sigma_\varepsilon^2 + 2\sigma_{RTHS}^2 + 8\sigma_{THS}^2 + 10\sigma_{RTH}^2 + 40\phi_{TH}$
	RTH_{ijk}	12	1	0	0	5	2	$\sigma_\varepsilon^2 + 2\sigma_{RTHS}^2 + 10\sigma_{RTH}^2$
Split	S_m	4	4	3	3	1	2	$\sigma_\varepsilon^2 + 18\sigma_{RS}^2 + 72\sigma_S^2$
plot	RS_{im}	12	1	3	3	1	2	$\sigma_\varepsilon^2 + 18\sigma_{RS}^2$
	TS_{jm}	8	4	0	3	1	2	$\sigma_\varepsilon^2 + 6\sigma_{RTS}^2 + 24\sigma_{TS}^2$
	RTS_{ijm}	24	1	0	3	1	2	$\sigma_\varepsilon^2 + 6\sigma_{RTS}^2$
	HS_{km}	8	4	3	0	1	2	$\sigma_\varepsilon^2 + 6\sigma_{RHS}^2 + 24\sigma_{HS}^2$
	RHS_{ikm}	24	1	3	0	1	2	$\sigma_\varepsilon^2 + 6\sigma_{RHS}^2$
	THS_{jkm}	16	4	0	0	1	2	$\sigma_\varepsilon^2 + 2\sigma_{RTHS}^2 + 8\sigma_{THS}^2$
	$RTHS_{ijkm}$	48	1	0	0	1	2	$\sigma_\varepsilon^2 + 2\sigma_{RTHS}^2$
	$\varepsilon_{q(ijkm)}$	180	1	1	1	1	1	σ_ε^2
	Total	359						

From the EMS column tests can be made on the effects of replications, replications by temperature interaction, replications by humidity interaction, replications by temperature by humidity interaction, strips, and strips by all other factor interactions. Unfortunately no exact tests are available for the factors of chief importance: temperature, humidity, and temperature by humidity interaction. Of course, we could first test the hypotheses that can be tested, and if some of these are not significant at a reasonably high level (say 25 percent), assume they are nonexistent and then remove these terms from the EMS column. For example, if the *RT* interaction can be assumed as zero ($\sigma_{RT}^2 = 0$), the temperature mean square can be tested against the *TS* interaction mean square with 2 and 8 df. Or, if the *TS* interaction can be assumed as zero ($\sigma_{TS}^2 = 0$), the temperature mean square may be tested against the RT interaction with 2 and 6 df. If neither of these assumptions is reasonable, a pseudo-F (F') test can be used such as discussed in Chapter 10.

Unfortunately, data were available only for the first replication of this experiment, so its complete analysis cannot be given. The method of analysis is the same as given in Chapter 6, even though this experiment is rather complicated. It is presented here only to show another actual example of a split-plot design.

13.2 A SPLIT-SPLIT-PLOT DESIGN

In a study of the cure rate index on some samples of rubber, three laboratories, three temperatures, and three types of mix, were involved. Material for the three mixes was sent to one of the three laboratories where they ran the experiment on the three mixes at the three temperatures (145°C, 155°C, and 165°C). However, once a temperature was set, all three mixes were subjected to that temperature and then another temperature was set and again all three mixes were involved, and finally the third temperature was set. Material was also sent to the second and third laboratories and similar experimental procedures were performed. There are therefore two restrictions on randomization since the laboratory is chosen, then the temperature is chosen, and then mixes can be randomized at that particular temperature and laboratory.

By complete replication of the whole experiment to achieve four replications, the three laboratories and four replications form the whole plots. Then the temperature levels may be randomized at each laboratory and in each replication to form a split-plot. Then at each temperature–laboratory–replication combination, the three mixes are randomly applied forming what is called a *split-split-plot* indicating that two main effects (laboratory and temperature) are confounded with blocks.

Table 13.8 shows one possible order of the first nine experiments in one replication of (a) a completely randomized, (b) a split-plot design, and (c) a split-split-plot design. The resultant data will not reveal how the experiment was ordered so this must be introduced into the design phase as it will change the basic design and how one interprets the results. Table 13.9 gives the data for the problem on cure rate index.

The model and its EMS values are given in Table 13.10.

The EMS column indicates that F tests can be made on all three fixed effects and their interactions without a separate error term. Since the effect of replication is often considered of no great interest and the interaction of replication with the other factors is often assumed to be nonexistent, this appears to be a feasible experiment. If examination of the EMS column should indicate that some F tests have too few degrees of freedom in their

Table 13.8 Order of First Nine Rubber Cure Rate Index Experiments

	Temperature (°C)									
	145			155			165			
	Mix			Mix			Mix			
Laboratory	*A*	*B*	*C*	*A*	*B*	*C*	*A*	*B*	*C*	
1			3				4			(a)
2	8	1			5		7		9	Completely randomized
3			2						6	
1										(b)
2	4	6	1	7	9	3	2	8	5	Split-plot
3										
1										(c)
2	8	9	7	3	1	2	4	6	5	Split-split-plot
3										

Table 13.9 Rubber Cure Rate Index Data

		Temperature (°C)								
		145°			155			165		
		Mix			Mix			Mix		
Replication	Laboratory	*A*	*B*	*C*	*A*	*B*	*C*	*A*	*B*	*C*
I	1	18.6	14.5	21.1	9.5	7.8	11.2	5.4	5.2	6.3
	2	20.0	18.4	22.5	11.4	10.8	13.3	6.8	6.0	7.7
	3	19.7	16.3	22.7	9.3	9.1	11.3	6.7	5.7	6.6
II	1	17.0	15.8	20.8	9.4	8.3	10.0	5.3	4.9	6.4
	2	20.1	18.1	22.7	11.5	11.1	14.0	6.9	6.1	8.0
	3	18.3	16.7	21.9	10.2	9.2	11.0	6.0	5.5	6.5
III	1	18.7	16.5	21.8	9.5	8.9	11.5	5.7	4.3	5.8
	2	19.4	16.5	21.5	11.4	9.5	12.0	6.0	5.0	6.6
	3	16.8	14.4	19.3	9.8	8.0	10.9	5.0	4.6	5.9
IV	1	18.7	17.6	21.0	10.0	9.1	11.1	5.3	5.2	5.6
	2	20.0	16.7	21.3	11.5	9.7	11.5	5.7	5.2	6.3
	3	17.1	15.2	19.3	9.5	9.0	11.4	4.8	5.4	5.8

Table 13.10 EMS for Rubber Cure Rate Index Experiment

	Source	df	4 R i	3 F j	3 F k	3 F m	1 R q	EMS
Whole plot	R_i	3	1	3	3	3	1	$\sigma_\varepsilon^2 + 27\sigma_R^2$
	L_j	2	4	0	3	3	1	$\sigma_\varepsilon^2 + 9\sigma_{RL}^2 + 36\phi_L$
	RL_{ij}	6	1	0	3	3	1	$\sigma_\varepsilon^2 + 9\sigma_{RL}^2$
Split-plot	T_k	2	4	3	0	3	1	$\sigma_\varepsilon^2 + 9\sigma_{RT}^2 + 36\phi_T$
	RT_{ik}	6	1	3	0	3	1	$\sigma_\varepsilon^2 + 9\sigma_{RT}^2$
	LT_{jk}	4	4	0	0	3	1	$\sigma_\varepsilon^2 + 3\sigma_{RLT}^2 + 12\phi_{LT}$
	RLT_{ijk}	12	1	0	0	3	1	$\sigma_\varepsilon^2 + 3\sigma_{RLT}^2$
Split-split-plot	M_m	2	4	3	3	0	1	$\sigma_\varepsilon^2 + 9\sigma_{RM}^2 + 36\phi_M$
	RM_{im}	6	1	3	3	0	1	$\sigma_\varepsilon^2 + 9\sigma_{RM}^2$
	LM_{jm}	4	4	0	3	0	1	$\sigma_\varepsilon^2 + 3\sigma_{RLM}^2 + 12\phi_{LM}$
	RLM_{ijm}	12	1	0	3	0	1	$\sigma_\varepsilon^2 + 3\sigma_{RLM}^2$
	TM_{km}	4	4	3	0	0	1	$\sigma_\varepsilon^2 + 3\sigma_{RTM}^2 + 12\phi_{TM}$
	RTM_{ikm}	12	1	3	0	0	1	$\sigma_\varepsilon^2 + 3\sigma_{RTM}^2$
	LTM_{jkm}	8	4	0	0	0	1	$\sigma_\varepsilon^2 + \sigma_{RLTM}^2 + 4\phi_{LTM}$
	$RLTM_{ijkm}$	24	1	0	0	0	1	$\sigma_\varepsilon^2 + \sigma_{RLTM}^2$
	$\varepsilon_{q(ijkm)}$	0	1	1	1	1	1	σ_ε^2 (not retrievable)
	Total	107						

denominator (some statisticians say less than 6), the experimenter might consider increasing the number of replications to increase the precision of the test. In the above experiment, if five replications were used, the whole plot error *RL* would have 8 df instead of 6, the split-plot error *RLT* would have 16 instead of 12, and so on.

Another technique that is sometimes used to increase the degrees of freedom is to pool all interactions with replications into the error term for that section of the table. Here one might pool *RT* with *RLT* and *RM*, *RLM*, and *RTM* with *RLTM* if additional degrees of freedom are believed necessary. One way to handle experiments with several replications is to stop after two or three replications, compute the results, and see if significance has been achieved or that the *F*'s are large even if not significant. Then add another replication, and so on, increasing the precision (and the cost) of the experiment in the hope of detecting significant effects if they are there.

The numerical results of the rubber cure rate index are given in Table 13.11.

Here laboratory effect is significant at the 5-percent level and temperature, mixes, and temperature–mix interaction are all highly significant (<0.001).

Table 13.11 ANOVA for Rubber Cure Rate Index Experiment

Source	df	SS	MS
R	3	9.41	3.14
L	2	40.66	20.33*
RL	6	16.11	2.68
T	2	3119.51	1559.76***
RT	6	2.07	0.34
LT	4	4.94	1.24
RLT	12	7.81	0.65
M	2	145.71	72.86***
RM	6	3.29	0.55
LM	4	0.35	0.09
RLM	12	2.93	0.24
TM	4	43.69	10.92***
RTM	12	2.22	0.18
LTM	8	1.07	0.14
$RLTM$	24	4.97	0.21
Totals	107	3404.74	

* One asterisk indicates significance at the 5-percent level; three, the 0.1-percent level.

13.3 SUMMARY

The summary of Chapter 12 may now be extended.

Experiment	*Design*	*Analysis*
II. Two or more factors		
A. Factorial (crossed)		
	1. Completely randomized	1.
	2. Randomized block	2.
	a. Complete $Y_{ijk} = \mu + R_k + A_i + B_j + AB_{ij} + \varepsilon_{ijk}$	*a*. Factorial ANOVA
	b. Incomplete, confounding:	*b*.
	i. Main effect—split-plot $Y_{ijk} = \mu + \underbrace{R_i + A_j + RA_{ij}}_{\text{whole plot}} + \underbrace{B_k + RB_{ik} + AB_{jk} + RAB_{ijk}}_{\text{split-plot}}$	i. Split-plot ANOVA

PROBLEMS

13.1 As part of the experiment discussed in Example 13.2, a chamber was set at 50-percent relative humidity and two boxes from each of three strips were inserted and the pull-off force determined. This was repeated for 70 percent and 90 percent, the order of humidities being randomized. Two replications of the whole experiment were made and the results follow:

		Humidity (percent)					
Replication	Strip	50		70		90	
I	1	1.12	1.75	3.50	0.50	1.00	0.75
	2	1.13	3.50	0.75	1.00	0.50	0.50
	3	2.25	3.25	1.75	1.88	1.50	0.00
II	1	1.75	1.88	1.75	0.75	1.50	0.75
	2	5.25	5.25	0.75	2.25	1.50	1.50
	3	1.50	3.50	1.62	2.50	0.75	0.50

Set up a split-plot model for this experiment and outline the ANOVA table, including EMS.

13.2 Analyze Problem 13.1.

13.3 In Problem 12.3, perform a pseudo-F test on the type effect.

13.4 Set up F' tests for the effects in Problem 12.4 that cannot be tested directly.

13.5 In a time study two types of stimuli are used—two- and three-dimensional film—and combinations of stimuli and job are repeated once to give eight strips of film. The order of these eight is completely randomized on one long film and presented to five analysts for rating at one sitting. Since this latter represents a restriction on the randomization, the whole experiment is repeated at three sittings (four in all) and the experiment is considered as a split-plot design. Set up a model for this experiment and indicate the tests to be made. (Consider analysts as random.)

13.6 Determine an F' test for any effects in Problem 13.5 that cannot be run directly.

13.7 Three replications are made of an experiment to determine the effect of days of the week and operators on screen-color difference of a TV tube in °K. On a given day, each of four operators measured the screen-color difference on a given tube. The order of measuring by the four operators is randomized each day for five days of the week. The results are

Replication	Operator	Monday	Tuesday	Wednesday	Thursday	Friday
I	*A*	800	950	900	740	880
	B	760	900	820	740	960
	C	920	920	840	900	1020
	D	860	940	820	940	980
II	*A*	780	810	880	960	920
	B	820	940	900	780	820
	C	740	900	880	840	880
	D	800	900	800	820	900
III	*A*	800	1030	800	840	800
	B	900	920	880	920	1000
	C	800	1000	920	760	920
	D	900	980	900	780	880

Set up the model and outline the ANOVA table for this problem.

13.8 Considering days as fixed, operators as random, and replications as random, analyze Problem 13.7.

13.9 Explain how a split-plot design differs from a nested design, since both have a factor within levels of another factor.

13.10 From the data of Table 13.9 on the rubber cure rate index experiment, verify the results of Table 13.11.

13.11 Since two qualitative factors in Problem 13.10 show significance, compare their means by a Newman–Keuls test.

13.12 Since temperature in Problem 13.10 is a highly significant quantitative effect, fit a proper polynomial (or polynomials) to the data.

13.13 Re-do Table 13.10, assuming that the three laboratories are chosen at random from a large number of possible laboratories.

13.14 Analyze the results of Table 13.9, based on the EMS values of Problem 13.13.

14
Factorial Experiment—Confounding in Blocks

14.1 INTRODUCTION

As seen in Chapter 13, there are many situations in which randomization is restricted. The split-plot design is an example of a restriction on randomization where a main effect is confounded with blocks. Sometimes these restrictions are necessary because the complete factorial cannot be run in one day, or in one environmental chamber. When such restrictions are imposed on the experiment, a decision must be made as to what information may be sacrificed and thus on what is to be confounded. A simple example may help to illustrate this point.

A chemist was studying the disintegration of a chemical in solution. He had a beaker of the mixture and wished to determine whether or not the depth in the beaker would affect the amount of disintegration and whether material from the center of the beaker would be different than material from near the edge. He decided on a 2^2 factorial experiment—one factor being depth, the other radius. A cross section of the beaker (Figure 14.1) shows the four treatment combinations to be considered.

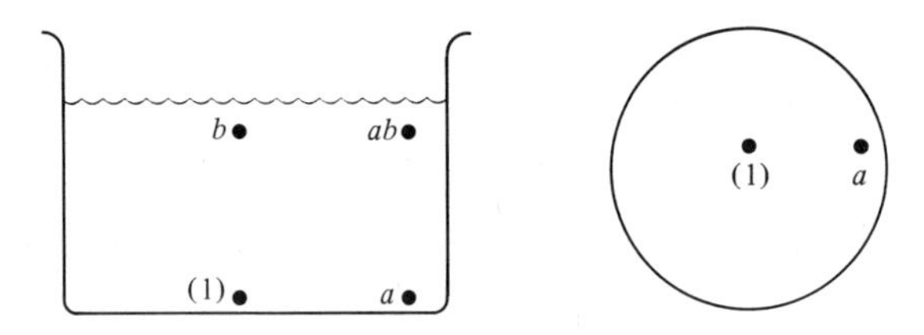

Figure 14.1

In Figure 14.1, (1) represents radius zero, depth at lower level (near bottom of beaker); *a* represents the radius at near maximum (near the edge of the beaker) and a low depth level; *b* is radius zero, depth at a higher level;

and *ab* is both radius and depth at higher levels. In order to get a reading on the amount of disintegration of material, samples of solution must be drawn from these four positions. This was accomplished by dipping into the beaker with an instrument designed to "capture" a small amount of solution. The only problem was that the experimenter had only two hands and could get only two samples at the same time. By the time he returned for the other two samples, conditions might have changed and the amount of disintegration would probably be greater. Thus there is a physical restriction on this 2^2 factorial experiment; only two observations can be made at one time. The question is, which two treatment combinations should be taken on the first dip? Considering the two dips as blocks, we are forced to use an incomplete block design, as the whole factorial cannot be performed in one dip.

Three possible block patterns are shown in Table 14.1.

Table 14.1 Three Possible Blockings of a 2^2 Factorial

	Plan					
Dip	I		II		III	
1	(1)	*b*	(1)	*a*	(1)	*ab*
2	*a*	*ab*	*b*	*ab*	*a*	*b*

If plan I of Table 14.1 is used, the blocks (dips) are confounded with the radius effect, since all zero-radius readings are in dip 1 and all maximum radius readings are in dip 2. In plan II, the depth effect is confounded with the dips, as low-level depth readings are both in dip 1 and high-level depth readings are in dip 2. In plan III, neither main effect is confounded, but the interaction is confounded with the dips. To see that the interaction is confounded, recall the expression for the *AB* interaction in a 2^2 factorial found in Chapter 7.

$$AB = \tfrac{1}{2}[(1) - a - b + ab]$$

Note that the two treatment effects with the plus sign, (1) and *ab*, are both in dip 1, and the two with a minus sign, *a* and *b*, are in dip 2. Hence we cannot distinguish between a block effect (dips) and the interaction effect (*AB*). In most cases it is better to confound an interaction than to confound a main effect. The hope is, of course, that there is no interaction, and that information on the main effects can still be found from such an experiment.

14.2 CONFOUNDING SYSTEMS

In order to design an experiment in which the number of treatments which can be run in a block is less than the total number of treatment combinations, the experimenter must first decide on what effects he is willing to confound.

If, as in the example above, he has only one interaction in the experiment and decides that he can confound this interaction with blocks, the problem is simply which treatment combinations to place in each block. One way to accomplish this is to place in one block those treatment combinations with a plus sign in the effect to be confounded, and those with a minus sign in the other block. A more general method is necessary when the number of blocks and the number of treatments increases.

First, a *defining contrast* is set up. This is merely an expression stating which effects are to be confounded with blocks. In the simple example above, to confound AB, write AB as the defining contrast. Once the defining contrast has been set up, several methods are available for determining which treatment combinations will be placed in each block. One method was shown above—place in one block the treatment combinations which have a plus sign in AB and those with a minus sign in the other block. Another method is to consider each treatment combination. Those which have an even number of letters in common with the effect letters in the defining contrast go in one block. Those which have an odd number of letters in common with the defining contrast go into the other block. Here (1) has no letters in common with AB or an even number. Type a has one letter in common with AB (a and A) or an odd number. b also has one letter in common with AB. ab has two letters in common with AB. Thus the block contents are those in Figure 14.2.

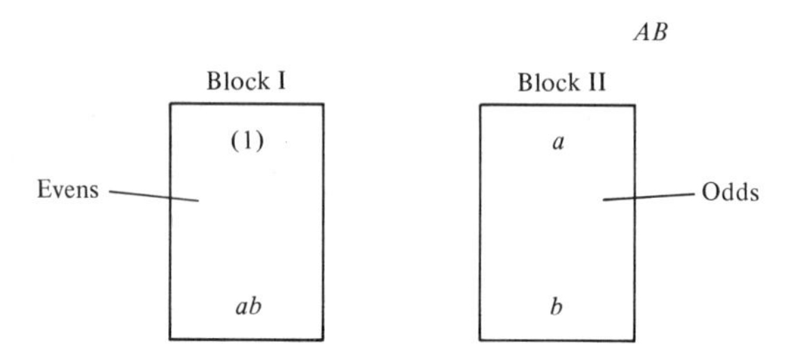

Figure 14.2

One disadvantage of the two methods given above is that they are only good on 2^n factorials. A more general method is attributed to Kempthorne [11]. Consider the linear expression

$$A_1X_1 + A_2X_2 + \cdots + A_nX_n = L$$

Here A_i is the exponent on the ith factor appearing in each independent defining contrast, and X_i is the level of the ith factor appearing in a given treatment combination. Every treatment combination with the same L value will be placed in the same block. For the simple example above where the defining contrast is AB, $A_1 = 1$, $A_2 = 1$, and all other A_i's $= 0$. Hence

$$L = 1 \cdot X_1 + 1 \cdot X_2$$

For each treatment combination the L values are

$$(1)\colon L = 1 \cdot 0 + 1 \cdot 0 = 0$$

$$a\colon L = 1 \cdot 1 + 1 \cdot 0 = 1$$

$$b\colon L = 1 \cdot 0 + 1 \cdot 1 = 1$$

$$ab\colon L = 1 \cdot 1 + 1 \cdot 1 = 2$$

Hence 0 and 2 are both 0, since a 2^n factorial is of modulus 2 (see Section 9.2). The assignment of treatment combinations to blocks is then

$L = 0$	block 1	(1) *ab*
$L = 1$	block 2	*a* *b*

For a more complex defining contrast such as ABC^2, the expression is

$$L = X_1 + X_2 + 2X_3$$

and this can be used to decide which treatment combinations in a 3^3 experiment go into a common block.

The block containing the treatment combination (1) is called the *principal block*. The treatment combinations in this block are elements of a group where the operation on the group is multiplication, modulus 2. The elements of the other block or blocks may be generated by multiplying one element in the new block by each element in the principal block. Multiplying elements together within the principal block will generate more elements within the principal block. This is best seen with a slightly more complex factorial, say 2^3. If the highest order interaction is to be confounded and the eight treatment combinations are to be placed into two blocks of four treatments each, then confounding ABC gives

$$L = X_1 + X_2 + X_3$$

Testing each of the eight treatment combinations gives

$$(1)\colon L = 0 + 0 + 0 = 0$$

$$a\colon L = 1 + 0 + 0 = 1$$

$$b\colon L = 0 + 1 + 0 = 1$$

$$ab\colon L = 1 + 1 + 0 = 2 = 0$$

$$c\colon L = 0 + 0 + 1 = 1$$

$$ac\colon L = 1 + 0 + 1 = 2 = 0$$

$$bc\colon L = 0 + 1 + 1 = 2 = 0$$

$$abc\colon L = 1 + 1 + 1 = 3 = 1$$

The blocks are then

block 1	block 2
(1)	a
ab	b
ac	c
bc	abc
$L = 0$	$L = 1$

Group theory can simplify this procedure. Determine two elements in the principal block, for example, ab and bc. Multiply these two to get $ab^2c = ac$, the fourth element. Generate block 2 by finding one element, say a. Multiply this element by each element in the principal block, thus generating all the elements of the second block as follows:

$$a \cdot (1) = a$$

$$a \cdot ab = a^2b = b$$

$$a \cdot ac = a^2c = c$$

$$a \cdot bc = abc$$

These concepts may be extended to more complex confounding examples, as will be shown in later sections.

14.3 BLOCK CONFOUNDING WITH REPLICATION

Whenever an experiment is restricted so that all treatments cannot appear in one block, some interaction is usually confounded with blocks. If several replications of the whole experiment (all blocks) are possible, as in the case of the split-plot, the same interaction may be confounded in all replications. In this case, the design is said to be *completely confounded*. If, on the other hand, one interaction is confounded in the first replication, a different interaction is confounded in the second replication, and so on, the design is said to be *partially confounded.*

Complete Confounding

Considering a 2^3 factorial experiment in which only four treatment combinations can be finished in one day yields a 2^3 factorial run in two incomplete blocks of four treatment combinations each. We might confound the highest order interaction ABC as follows:

$$ABC$$

As seen in Section 14.2, the blocks would be

block 1	block 2
(1)	*a*
ab	*b*
ac	*c*
bc	*abc*

If this whole experiment (2^3 in two blocks of four each) can be replicated, say three times, the layout might be

replication I		replication II		replication III	
block 1	block 2	block 1	block 2	block 1	block 2
ac	*a*	*c*	(1)	*ab*	*c*
(1)	*c*	*abc*	*ac*	(1)	*b*
ab	*abc*	*b*	*bc*	*ac*	*abc*
bc	*b*	*a*	*ab*	*bc*	*a*

The confounding scheme in the above (ABC) is the same for all three replications, but the order of experimentation has been randomized within each block. Also the decision as to which block is to be run first in each replication is made at random. An analysis layout for this experiment appears in Table 14.2.

Here the replication interaction with all three main effects and their interactions are usually taken as the error term for testing the important effects. The replication effect and the block (or ABC) effect could be tested

Table 14.2 Analysis Layout for Completely Confounded 2^3 Factorial

Source	df	df
Replications	2	
Blocks or ABC	1	
Replications × block interactions	2	5 between plots
A	1	
B	1	
AB	1	
C	1	
AC	1	
BC	1	
Replications × all others	12	18 within plots
Totals	23	23

against the replication × block interaction, but the degrees of freedom are low and the power of such a test is poor. Such a design is quite powerful, however, in testing the main effects *A*, *B*, and *C* and their first-order interactions. However, it will give no clean information on the *ABC* interaction.

Partial Confounding

In the example just considered, we might be concerned with some test on the *ABC* interaction. This could be found by confounding some interactions other than *ABC* in some of the replications. We might use four replications and confound *ABC* in the first one, *AB* in the second, *AC* in the third, and *BC* in the fourth. Thus the four replications will yield full information on *A*, *B*, and *C* but three-fourths information on *AB*, *AC*, *BC*, and *ABC*, since we can compute an unconfounded interaction such as *AB* in three out of four of the replications. The following layout shows the proper block entries for each replication:

	replication I		replication II		replication III		replication IV	
confound	*ABC*		*AB*		*AC*		*BC*	
	(1)	*a*	(1)	*a*	(1)	*a*	(1)	*b*
	ab	*b*	*c*	*b*	*ac*	*c*	*bc*	*c*
	ac	*c*	*ab*	*ac*	*abc*	*bc*	*a*	*ab*
	bc	*abc*	*abc*	*bc*	*b*	*ab*	*abc*	*ac*
	$L = X_1 + X_2 + X_3$		$L = X_1 + X_2$		$L = X_1 + X_3$		$L = X_2 + X_3$	

The analysis layout would be that shown in Table 14.3.

Table 14.3 ANOVA for Partially Confounded 2^3 Factorial

Source	df			df
Replications	3			
Blocks within replications	4			7 between plots
ABC (Replication I)		1		
AB (Replication II)		1		
AC (Replication III)		1		
BC (Replication IV)		1		
A	1			
B	1			
C	1			
AB	1		only from	
AC	1		replications	
BC	1		where not	
ABC	1		confounded	
Replications × all effects	17			24 within plots
Totals	31			31

The residual term with its 17 df can be explained as follows. In all four replications, the main effects A, B, and C can be isolated. This gives 3×1 df for each replication by main-effect interaction, or a total of 9 df. The AB, AC, BC, and ABC interactions can be isolated in only three out of the four replications. Hence the replications by AC have $2 \times 1 = 2$ df, or a total of 8 df for all such interactions. This gives $9 + 8 = 17$ df for effects by replications, which is usually taken as the error estimate. The blocks and replications are separated out in the top of Table 14.3 only to reduce the experimental error. The numerical analysis of problems like this is carried out in the same manner as given in earlier chapters.

Consider the ceramic tool data of Chapters 1, 6, and 7 when there is a restriction that only four experiments can be run in one day. One might then block each of the four replications into two blocks of four and run one block each day. This would require eight days to complete the 32 observations. If a different interaction is confounded for each replication, a partially confounded design such as given above might yield

	replication I		replication II	
confound	ABC		AB	
	$(1) = 2$	$a = 0$	$(1) = -3$	$a = 1$
	$ab = 3$	$b = 1$	$c = -6$	$b = 1$
	$ac = -7$	$c = 0$	$ab = 8$	$ac = -6$
	$bc = -2$	$abc = -1$	$abc = 0$	$bc = 2$

	replication III		replication IV	
confound	AC		BC	
	$(1) = 5$	$a = 0$	$(1) = -2$	$b = 9$
	$ac = 0$	$c = -3$	$bc = -1$	$c = -3$
	$abc = -2$	$bc = -1$	$a = -6$	$ab = 0$
	$b = 4$	$ab = 2$	$abc = -4$	$ac = -4$

Its analysis is given in Table 14.4. Using the 17-df term as error, the effect of B and C (bevel angle and type of cut) stand out.

14.4 BLOCK CONFOUNDING—NO REPLICATION

In many cases an experimenter cannot afford several replications of an experiment, and he is further restricted in that he cannot run the complete factorial in one block or at one time. Again, the experiment may be blocked and information recovered on all but some high-order interactions, which may be confounded. Designs for such experiments will be considered for the special cases of 2^n and 3^n factorials because they appear often in practice.

Table 14.4 ANOVA for Partially Confounded Ceramic Tool Experiment

Source	df	SS	MS
Replications	3	16.10	5.37
Blocks within replications	4	40.36	10.09
A (tool types)	1	11.28	11.28
B (bevel angle)	1	81.28	81.28**
C (type of cut)	1	124.03	124.03**
AB	1	0.67	0.67
AC	1	4.17	4.17
BC	1	0.67	0.67
ABC	1	3.38	3.38
Replications × all effects	17	153.78	9.05
Totals	31	435.72	

** Two asterisks indicate significance at the 1 percent level.

Confounding in 2^n Factorials

The methods given earlier in this chapter can be used to determine the block composition for a specific confounding scheme. If only one replication is possible, one is run, and some of the higher order interaction terms must be used as experimental error unless some independent measure of error is available from previous data. This type of design is used mostly when there are several factors involved (say four or more), some high-order interactions may be confounded with blocks, and some others are yet available for an error estimate. For example, consider a 2^4 factorial where only eight treatment combinations can be run at one time. One possible confounding scheme would be

confound $ABCD$ and $L = X_1 + X_2 + X_3 + X_4$

block 1	(1) *ab* *bc* *ac* *abcd* *cd* *ad* *bd*	$L = 0$
block 2	*a* *b* *abc* *c* *bcd* *acd* *d* *abd*	$L = 1$

and the analysis would be as shown in Table 14.5. The three-way interactions may be pooled here to give 4 df for error, assuming that these interactions are actually nonexistent. All main effects and first-order interactions could then be tested with 1 and 4 df. After examining the results, it might be possible to pool some of the two-way interactions that have small mean squares with the

Table 14.5 2^4 Factorial in Two Blocks

Source	df
A	1
B	1
C	1
D	1
AB	1
AC	1
AD	1
BC	1
BD	1
CD	1
ABC	1 (4 df for error)
ABD	1 (4 df for error)
ACD	1 (4 df for error)
BCD	1 (4 df for error)
Blocks or $ABCD$	1
Total	15 df

three-way interactions if more degrees of freedom are desired in the error term.

If only four treatment combinations can be run in a block, one may confound the 2^4 factorial in four blocks of four treatment combinations each. For four blocks, 3 df must be confounded with blocks. If two interactions are confounded, the product (modulus 2) of these two is also confounded, since the product of the signs in two effects gives the signs for the product. Thus, if AB and CD are confounded, so also will $ABCD$ be confounded. This scheme would confound two first-order interactions and one third order. It might be better to confound two second-order interactions and one first order:

$$ABC \qquad BCD \qquad AD$$

Note here that $(ABC)(BCD) = AB^2C^2D = AD$ (modulus 2). Hence 3 df are confounded with blocks. Before proceeding to the block compositions, consider a problem where four factors (A, B, C, and D) are involved, each at two levels, and only four treatment combinations can be run in one day. The measured variable is yield. The confounding scheme given above gives two independent equations

$$L_1 = X_1 + X_2 + X_3$$

$$L_2 = X_2 + X_3 + X_4$$

Each treatment combination will give one of the following four pairs when substituted into both L_1 and L_2 = 00, 01, 10, 11. For example,

$$(1)\colon L_1 = 0 \qquad L_2 = 0$$

$$a\colon L_1 = 1 \qquad L_2 = 0$$

$$b\colon L_1 = 1 \qquad L_2 = 1$$

$$ab\colon L_1 = 2 = 0 \qquad L_2 = 1$$

and so on. All treatment combinations with the same pair of L values will be placed together in one block. Thus the 16 treatment combinations will be assigned to four blocks:

block 1	block 2	block 3	block 4
$(1) = 82$	$a = 76$	$b = 79$	$ab = 85$
$bc = 55$	$abc = 74$	$c = 71$	$ac = 84$
$acd = 81$	$cd = 72$	$abcd = 89$	$bcd = 84$
$abd = 88$	$bd = 73$	$ad = 79$	$d = 80$
$L_1 = 0$	$L_1 = 1$	$L_1 = 1$	$L_1 = 0$
$L_2 = 0$	$L_2 = 0$	$L_2 = 1$	$L_2 = 1$

Here the order of experimentation within each block is randomized and the resulting responses are given for each treatment combination. The proper treatment combinations for each block can be generated from the principal block. If bc and acd are in the principal block, then their product

$$(bc)(acd) = abc^2d = abd$$

is also in the principal block. For another block, consider a and multiply a by all the combinations in the principal block, giving

$$a(1) = a, \quad a(bc) = abc, \quad a(acd) = cd, \quad a(abd) = bd$$

the entries in the second block. In the same manner, the other two blocks may be generated. This means that only a few readings in the principal block need to be determined by the L_1, L_2 values, and all the rest of the design can be generated from these few treatment combinations. Tables are also available for designs when certain effects are confounded [7].

This experiment is then a 2^4 factorial experiment. The design is a randomized incomplete block design where blocks are confounded with the interactions ABC, BCD, and AD. For the analysis, we might use the Yates method of Chapter 7 (Table 14.6).

Table 14.6 Yates Method Analysis for 2^4 Example

Treatment Combination	Response	(1)	(2)	(3)	(4)	SS
(1)	82	158	322	606	1252	—
a	76	164	284	646	+60	225.00
b	79	155	320	+32	2	0.25
ab	85	129	326	28	30	56.25
c	71	159	0	−20	−32	64.00
ac	84	161	+32	22	+32	64.00
bc	55	153	14	+18	−14	12.25
abc	74	173	14	12	−26	42.25*
d	80	−6	6	−38	40	100.00
ad	79	+6	−26	6	−4	1.00*
bd	73	+13	2	+32	42	110.25
abd	88	+19	20	0	−6	2.25
cd	72	−1	12	−32	44	121.00
acd	81	15	+6	18	−32	64.00
bcd	84	9	16	−6	50	156.25*
abcd	89	5	−4	−20	−14	12.25
Total						1031.00

* Confounded with blocks.

In Table 14.6 the sums of squares in the last column correspond to the effect for the treatment combination on the left. That is,

$$SS_A = 225.00 \qquad SS_B = 0.25 \qquad \text{and so on}$$

Note that the effects *ABC*, *BCD*, and *AD* are confounded with blocks. Their total sum of squares is 42.25 + 156.25 + 1.00 = 199.50. If the above block totals are computed, they yield

block 1 — 306 block 3 — 318

block 2 — 295 block 4 — 333

and the sum of squares between these blocks is

$$\frac{(306)^2 + (295)^2 + (318)^2 + (333)^2}{4} - \frac{(1252)^2}{16}$$

$$SS_{\text{block}} = 98{,}168.50 - 97{,}969.00 = 199.50$$

which is identical with the sum of the three interactions with which blocks are confounded. If all three-way and four-way interactions that are not confounded are pooled for an error term, the resulting analysis is that shown in Table 14.7.

Table 14.7 ANOVA for 2^4 Example in Four Blocks

Source	df	SS	MS
A	1	225.00	225.00
B	1	0.25	0.25
C	1	64.00	64.00
D	1	100.00	100.00
AB	1	56.25	56.25
AC	1	64.00	64.00
BC	1	12.25	12.25
BD	1	110.25	110.25
CD	1	121.00	121.00
Error or ABD, ACD, $ABCD$	3	78.50	26.17
Blocks or ABC, BCD, AD	3	199.50	66.50
Totals	15	1031.00	

With only 1 and 3 df, none of the four main effects nor the five two-way interactions can be declared significant at the 5-percent significance level. Since the B effect and the BC interaction are not significant even at the 25-percent level, we might wish to pool these with the error to increase the error degrees of freedom. The resulting error sum of squares would then be

$$0.25 + 12.25 + 78.50 = 91.00$$

with 5 df. The error mean square then $= 91.00/5 = 18.20$. Using this error mean square, A and CD are now significant at the 5-percent level. We might conclude that the effect of factor A is important, that B is not present, and that there may be a CD interaction. As 2^n experiments are often run to get an overall picture of the important factors, each set at the extremes of its range, another experiment might now be planned without factor B, since its effect is negligible over the range considered here. The next experiment might include factors A, C, and D at, perhaps, three levels each, or a 3^3 experiment. This experiment might also require blocking.

Confounding in 3^n Factorials

When a 3^n-factorial experiment cannot be completely randomized, it is usually blocked in blocks which are multiples of 3. The use of the I and J components of interaction introduced in Section 9.2 is helpful in confounding only part of an interaction with blocks. Kempthorne's rule also applies, using such interactions as AB, AB^2, $ABC^2, \cdots$, and treatment combinations in the form 00, 10, 01, 11 instead of (1), a, b, ab as in 2^n. Here treatment combinations can be multiplied which will actually add these exponents (modulus 3).

If a 3^2 factorial is restricted so that only three of the nine treatment combinations can be run in one block, we usually confound AB or AB^2, as each carries 2 df. The defining contrast might be AB^2 which gives $L = X_1 + 2X_2$.

Each of the nine treatment combinations then yields

00, $L = 0$	01, $L = 2$	02, $L = 4 = 1$
10, $L = 1$	11, $L = 3 = 0$	12, $L = 5 = 2$
20, $L = 2$	21, $L = 4 = 1$	22, $L = 6 = 0$

Placing treatment combinations with the same L value in common blocks gives

block 1	block 2	block 3
$L = 0$	$L = 1$	$L = 2$
00	10	20
11	21	01
22	02	12

Block 1 with 00 in it is the principal block. Block 2 is generated by *adding* one treatment combination in block 2 to all those in block 1, giving

$$00 + 10 = 10 \qquad 11 + 10 = 21 \qquad 22 + 10 = 32 = 02$$

Similarly for block 3,

$$00 + 20 = 20 \qquad 11 + 20 = 31 = 01 \qquad 22 + 20 = 42 = 12$$

Since there are three blocks, 2 df must be confounded. Here AB^2 is confounded with blocks. If AB with its 2 df were confounded, $L = X_1 + X_2$, and the blocking would be

block 1	block 2	block 3
$L = 0$	$L = 1$	$L = 2$
00	10	20
12	22	02
21	01	11

If the data of Table 9.1 are used as an example, and AB^2 is confounded, the responses are

block 1	block 2	block 3
$00 = 1$	$10 = -2$	$20 = 3$
$11 = 4$	$21 = 1$	$01 = 0$
$22 = 2$	$02 = 2$	$12 = -1$

The block sums of squares are determined from the three block totals of 7, 1, and 2:

$$SS_{\text{block}} = \frac{7^2 + 1^2 + 2^2}{3} - \frac{(10)^2}{9} = 6.89$$

which is identical with the $I(AB)$ interaction component computed in Section 9.2 (Table 9.6) as it should be. The remainder of the analysis proceeds the same as in Section 9.2, and the resulting ANOVA table is that exhibited in Table 14.8.

Table 14.8 ANOVA for 3^2 Example

Source	df	SS	MS
A	2	4.22	2.11
B	2	1.56	0.78
AB	2	16.22	8.11
Blocks or AB^2	2	6.89	3.44
Totals	8	28.89	

If these were real data, the AB part of the interaction might be considered as error, but then neither main effect is significant. The purpose here is only to illustrate the blocking of a 3^2 experiment.

Consider now a 3^3 experiment in which the 27 treatment combinations (such as given in Table 9.7) cannot all be completely randomized. If nine can be randomized and run on one day, nine on another day, and so on, we might use a 3^3 factorial in three blocks of nine treatment combinations each. This requires 2 df to be confounded with blocks. Since the ABC interaction with 8 df can be partitioned into ABC, ABC^2, AB^2C, and AB^2C^2, each with 2 df, one of these four could be confounded with blocks. If AB^2C is confounded,

$$L = X_1 + 2X_2 + X_3$$

and the three blocks are

$L = 0$	$L = 1$	$L = 2$
000	100	200
011	111	211
110	210	010
121	221	021
102	202	002
212	012	112
220	020	120
022	122	222
201	001	101

The analysis of data for this design would yield Table 14.9.

Table 14.9 3^3 Factorial in Three Blocks

Source	df
A	2
B	2
AB	4
C	2
AC	4
BC	4
Error or ABC, ABC^2, AB^2C^2	6
Blocks or AB^2C	2
Total	26

This is a useful design, since we can retrieve all main effects (A, B, C) and all two-way interactions, if we are willing to pool the 6 df in ABC as error. The determination of sums of squares, and so forth, is the same as for a complete 3^3 given in Example 9.1. Confounding other parts of the ABC interaction would, of course, yield different blocking arrangements, although the outline of the analysis would be the same. In practice, different blocking arrangements might yield different responses.

More complex confounding schemes can be found in tables such as in Chapter 9 of Davies [7].

For practice, consider confounding a 3^3 in nine blocks of three treatment combinations each. Here 8 df must be confounded with blocks. One confounding scheme might be

$$AB^2C^2 \qquad AB \qquad BC^2 \qquad AC$$

Note that

$$(AB^2C^2)(AB) = A^2C^2 = AC$$

and

$$(AB)(BC^2) = AB^2C^2$$

These are not all independent. In fact only two of the expressions are. For this reason, we need only two expressions for the L values:

$$L_1 = X_1 + 2X_2 + 2X_3$$

$$L_2 = X_1 + X_2$$

as these two will yield nine pairs of numbers—one pair for each block. First determine the principal block, where both L_1 and L_2 are zero. One treatment combination is obviously 000. Another is 211, as

$$L_1 = 1(2) + 2(1) + 2(1) = 6 = 0$$

$$L_2 = 1(2) + 1(1) = 3 = 0$$

A third is 122, as

$$L_1 = 1(1) + 2(2) + 2(2) = 9 = 0$$

$$L_2 = 1(1) + 1(2) = 3 = 0$$

The other eight blocks can now be generated from this principal block, giving

block	1	2	3	4	5	6	7	8	9
	000	001	002	010	020	100	200	110	101
	211	212	210	221	201	011	111	021	012
	122	120	121	102	112	222	022	202	220

An analysis layout would be that shown in Table 14.10.

Table 14.10 3^3 Factorial in Nine Blocks

Source	df	
A	2	
B	2	
C	2	
AB^2	2	
AC^2	2	
BC	2	
ABC	2	
AB^2C	2	6 df for error
ABC^2	2	
Blocks or AB, BC^2, AC, AB^2C^2	8	
Total	26	

This design might be reasonable if interest centered only in the main effects A, B, and C. Such a design might be necessary where three factors are involved, each at three levels, but only three treatment combinations can be run in one day. These three might involve an elaborate environmental test where at best only three sets of environmental conditions, temperature, pressure, and humidity, can be simulated in one day.

14.5 SUMMARY

The summary of Chapter 13 may now be extended.

Experiment	*Design*	*Analysis*
II. Two or more factors		
A. Factorial (crossed)	1. Completely randomized	
	2. Randomized block	2.
	a. Complete $Y_{ijk} = \mu + R_k + A_i + B_j + AB_{ij} + \varepsilon_{ijk}$	*a*. Factorial ANOVA
	b. Incomplete, confounding:	*b*.
	i. Main effect—split-plot $Y_{ijk} = \mu + \underbrace{R_i + A_j + RA_{ij}}_{\text{whole plot}} \underbrace{+ B_k + RB_{ik} + AB_{jk} + RAB_{ijk}}_{\text{split-plot}}$	i. Split-plot ANOVA
	ii. Interactions in 2^n and 3^n	ii.
	(1) Several replications $Y_{ijkq} = \mu + R_i + \beta_j + R\beta_{ij} + A_k + B_q + AB_{kq} + \varepsilon_{ikq}$	(1) Factorial ANOVA with replications R_i and blocks β_j or confounded interaction
	(2) One replication only $Y_{ijk} = \mu + \beta_i + A_j + B_k + AB_{jk}, \cdots$	(2) Factorial ANOVA with blocks β_j or confounded interaction

PROBLEMS

14.1 Assuming that Example 7.2 could not be run all in one day, set up a confounding scheme to confound the *AC* interaction using two blocks of four observations each. With the results of this problem (Table 7.9) show that you have indeed confounded *AC* with blocks.

14.2 For the data of Problem 7.6, consider running the 16 treatment combinations in four blocks of four (two replications per treatment). Confound *ACD*, *BCD*, and *AB* and determine the block compositions.

14.3 From the results of Problem 7.6, show that the blocks are confounded with the interactions in Problem 14.2.

14.4 For Problem 9.5, consider this experiment as run in three blocks of nine with two replications for each treatment combination. Confound *ABC* (2 df) where *A* is surface thickness, *B* is base thickness, and *C* is subbase thickness, and determine the design. Also use the numerical results of Problem 9.7 to show that *ABC* is indeed confounded in your design.

14.5 Set up a scheme for confounding a 2^4 factorial in two blocks of eight each, confounding *ABD* with blocks.

14.6 Repeat Problem 14.5 and confound *BCD*.

14.7 Work out a confounding scheme for a 2^5 factorial, confounding in four blocks of eight each. Show the outline of an ANOVA table.

14.8 Repeat Problem 14.7 in four replications confounding different interactions in each replication. Show the ANOVA table outline.

14.9 Work out the confounding of a 3^4 factorial in three blocks of 27 each.

14.10 Repeat Problem 14.9 in nine blocks of nine each.

14.11 Verify the results in Table 14.4 based on the data preceding the table. Check out all the components of the 17-df term rather than arriving at it by subtraction.

14.12 A systematic test was made to determine the effects on coil breakdown voltage of the following six factors, each at two levels as indicated:

1. Firing furnace	Number 1 or 3
2. Firing temperature	1650°C or 1700°C
3. Gas humidification	No or yes
4. Coil outer diameter	below 0.0300 inch or above 0.0305 inch
5. Artificial chipping	No or yes
6. Sleeve	Number 1 or 2

All 64 experiments could not be performed under the same conditions so the whole experiment was run in eight subgroups of eight experiments. Set up a reasonable confounding scheme for this problem and outline its ANOVA.

14.13 An experiment was run on the effects of several annealing variables on the magnetic characteristic of a metal. The following factors were to be considered:

1. Temperature	1375°, 1450°, 1525°, 1600°
2. Time in hours	2, 4
3. Exogas ratio	6.5 to 1, 8.5 to 1
4. Dew point	10°, 30°
5. Cooling rate in hours	4, 8
6. Core material	*A*, *B*.

Since all factors are at two levels or multiples of two levels, outline a scheme for this experiment in two blocks of 64 observations.

14.14 Outline a Yates method for handling the data of Problem 14.13 (see Problem 7.9).

14.15 If X is the surface rating of a steel sample and material is tested from eight heats, two chemical treatments, and three positions on the ingot, the resulting factorial is an $8 \times 2 \times 3$ or $2^3 \times 2 \times 3$ or $2^4 \times 3$ which is called a mixed factorial in the form $2^m \cdot 3^n$. Set up an ANOVA table for such a problem treating the eight heats as 2^3 pseudo factors.

14.16 If Problem 14.15 results must be fired in furnaces which can handle only 24 steel samples at a time, devise a confounding scheme for the experiment.

14.17 Complete the analysis of Problem 14.16 for the data below (coded):

	Position					
	1		2		3	
	Chemical Treatment		Chemical Treatment		Chemical Treatment	
Heat	1	2	1	2	1	2
1	1	0	0	0	0	2
2	−3	2	−1	−2	−2	−2
3	1	2	0	3	0	4
4	0	0	−2	1	0	1
5	−3	1	0	2	−2	0
6	−2	1	−3	2	0	0
7	3	3	3	3	1	−1
8	1	1	0	5	1	2

15
Fractional Replication

15.1 INTRODUCTION

As the number of factors to be considered in a factorial experiment increases, the number of treatment combinations increases very rapidly. This can be seen with a 2^n factorial where $n = 5$ requires 32 experiments for one replication, $n = 6$ requires 64, $n = 7$ requires 128, and so on. Along with this increase in the amount of experimentation comes an increase in the number of high-order interactions. Some of these high-order interactions may be used as error, as those above second order (three way) would be difficult to explain if found significant. Table 15.1 gives some idea of the number of main effects, first-order, second-order, $\cdots$, interactions that can be recovered if a complete 2^n factorial can be run.

Table 15.1 Buildup of 2^n-Factorial Effects

n	2^n	Main Effect	Order of Interaction 1st	2nd	3rd	4th	5th	6th	7th
5	32	5	10	10	5	1			
6	64	6	15	20	15	6	1		
7	128	7	21	35	35	21	7	1	
8	256	8	28	56	70	56	28	8	1

Considering $n = 7$, there will be 7 df for the seven main effects, 21 df for the 21 first-order interactions, 35 df for the 35 second-order interactions, leaving

$$35 + 21 + 7 + 1 = 64 \text{ df}$$

for an error estimate, assuming no blocking. Even if this experiment were confounded in blocks, there is still a large number of degrees of freedom for the error estimate. In such cases, it may not be economical to run a whole replicate of 128 observations. Nearly as much information can be gleaned from half as many observations. When only a fraction of a replicate is run, the design is called a *fractional replication* or a *fractional factorial.*

For running a fractional replication, the methods of Chapter 14 are used to determine a confounding scheme such that the number of treatment combinations in a block is within the economic range of the experimenter. If, for example, he can run 60 or 70 experiments and he is interested in seven factors, each at two levels, he can use a one-half replicate of a 2^7 factorial. The complete $2^7 = 128$ experiment is laid out in two blocks of 64 by confounding some high-order interaction. Then only one of these two blocks is run; the decision is made as to which one by the toss of a coin.

15.2 ALIASES

To see how a fractional replication is run, consider a simple case. Three factors are of interest, each at two levels, but the experimenter cannot afford $2^3 = 8$ experiments. He will, however, settle for four. This suggests a one-half replicate of a 2^3 factorial. Suppose ABC is confounded with blocks. The two blocks are then

$$I = ABC$$

block 1	(1)	ab	bc	ac
block 2	a	b	c	abc

A coin is flipped and the decision is made to run only block 2. What information can be gleaned from block 2 and what information is lost when only half of the experiment is run?

Referring to Table 7.7, which shows the proper coefficients ($+1$ or -1) for the treatment combinations in a 2^3 factorial to give the effects desired, box only those treatment combinations in block 2 which are to be run (Table 15.2).

Note, in the boxed area, that ABC has all plus signs in block 2, which is a result of confounding blocks with ABC. Note also that the effect of A is

$$A = +a - b - c + abc$$

and

$$BC = +a - b - c + abc$$

so that we cannot distinguish between A and BC in block 2. Two or more effects which have the same numerical value are called *aliases*. We cannot tell them apart. B and AC are likewise aliases, as are C and AB. Because of this confounding when only a fraction of the experiment is run, we must check the aliases and be reasonably sure they are not both present if such a design

Table 15.2 One-Half Replication of a 2^3 Factorial

Treatment Combination	Effect						
	A	B	AB	C	AC	BC	ABC
(1)	−	−	+	−	+	+	−
a	+	−	−	−	−	+	+
b	−	+	−	−	+	−	+
ab	+	+	+	−	−	−	−
c	−	−	+	+	−	−	+
ac	+	−	−	+	+	−	−
bc	−	+	−	+	−	+	−
abc	+	+	+	+	+	+	+

is to be of value. An analysis of this one-half replication of a 2^3 would be as shown in Table 15.3.

Table 15.3 One-Half Replication of a 2^3 Factorial

Source	df
A (or BC)	1
B (or AC)	1
C (or AB)	1
Total	3

This would hardly be a practical experiment unless the experimenter is sure that no first-order interactions exist and that he has some external source of error to use in testing A, B, and C. The real advantage of such fractionating will be seen on designs with larger n values.

If block 1 were run for the experiment instead of block 2, A has the value $-(1) + ab + ac - bc$, and BC has the value $+(1) - ab - ac + bc$, which gives the same total except for sign. The sum of squares due to A and BC are therefore the same; again they are said to be aliases. Here

$$A = -BC \qquad B = -AC \qquad C = -AB$$

The definition given above still holds where one effect is the alias of another if they have the same *numerical* value, or value regardless of sign.

A quick way to find the aliases of an effect in a fractional replication of a 2^n-factorial experiment is to multiply the effect by the terms in the defining contrast, modulus 2. The results will be aliases of the original effect. In the example above,

$$I = ABC$$

The alias of A is

$$A(ABC) = A^2BC = BC$$

The alias of B is

$$B(ABC) = AB^2C = AC$$

The alias of C is

$$C(ABC) = ABC^2 = AB$$

This simple rule works for any fractional replication for a 2^n factorial. It works also with a slight modification for a 3^n factorial run as a fractional replication.

15.3 FRACTIONAL REPLICATIONS

2^n Factorials

As an example, consider the problem suggested in the introduction. An experimenter wishes to study the effect of seven factors, each at two levels, but he cannot afford to run all 128 experiments. He will settle for 64, or a one-half replicate of a 2^7. Deciding to confound the highest order interaction with blocks, he has

$$I = ABCDEFG$$

The two blocks are found by placing (1) and all pairs, quadruples, and sextuples of the seven letters in one block and the single letters, triples, quintuples and one septuple of the seven letters in the other block. One of the two blocks is chosen at random and run. Before carrying out this experiment, he should check on the aliases. From the defining contrast

$$I = ABCDEFG$$

the alias of A is $A(ABCDEFG) = BCDEFG$, and all main effects are likewise aliased with fifth-order interactions.

AB is aliased with $AB(ABCDEFG) = CDEFG$, a fourth-order interaction. So all first-order interactions are aliased with fourth-order interactions. A second-order interaction such as ABC is aliased with a third-order interaction as

$$ABC(ABCDEFG) = DEFG$$

If the second- and third-order interactions are taken as error, the analysis would be as in Table 15.4.

This is a very practical design, as there are good tests on all main effects and first-order interactions, assuming all higher-order interactions are zero. The degrees of freedom for each test would be 1 and 35. If, for some reason known to the experimenter, he suspects some second-order interaction, he

Table 15.4 One-Half Replication of a 2^7 Factorial

Source	df
Main effects $A, B, \cdots, G$ (or fifth order)	1 each for 7
First-order interaction AB, AC (or fourth order)	1 each for 21
Second-order interaction ABC, ABD (or third order)	1 each for 35 } use as error
Total	63

could leave it out of the error estimate and still have sufficient degrees of freedom for the error estimate. The analysis of such an experiment follows the methods given in Chapter 7.

If this same experimenter is further restricted and can only afford to run 32 experiments, he might try a one-fourth replication of a 2^7. Here 3 df must be confounded with blocks. If two fourth-order interactions are confounded with blocks, one third-order is automatically confounded also, as

$$I = ABCDE = CDEFG = ABFG$$

which confounds 3 df with the four blocks. In this design, if only one of the four blocks of 32 observations is run, each effect has three aliases. These are

$$A = BCDE = ACDEFG = BFG$$
$$B = ACDE = BCDEFG = AFG$$
$$C = ABDE = DEFG = ABCFG$$
$$D = ABCE = CEFG = ABDFG$$
$$E = ABCD = CDFG = ABEFG$$
$$F = ABCDEF = CDEG = ABG$$
$$G = ABCDEG = CDEF = ABF$$
$$AB = CDE = ABCDEFG = FG$$
$$AC = BDE = ADEFG = BCFG$$
$$AD = BCE = ACEFG = BDFG$$
$$AE = BCD = ACDFG = BEFG$$
$$AF = BCDEF = ACDEG = BG$$
$$AG = BCDEG = ACDEF = BF$$
$$BC = ADE = BDEFG = ACFG$$
$$BD = ACE = BCEFG = ADFG$$
$$BE = ACD = BCDFG = AEFG$$
$$CD = ABE = EFG = ABCDFG$$
$$CE = ABD = DFG = ABCEFG$$
$$CF = ABDEF = DEG = ABCG$$
$$CG = ABDEG = DEF = ABCF$$
$$DE = ABC = CFG = ABDEFG$$
$$DF = ABCEF = CEG = ABDG$$
$$DG = ABCEG = CEF = ABDF$$
$$EF = ABCDF = CDG = ABEG$$
$$EG = ABCDG = CDF = ABEF$$
$$ACF = BDEF = ADEG = BCG$$
$$ACG = BDEG = ADEF = BCF$$
$$ADF = BCEF = ACEG = BDG$$
$$ADG = BCEG = ACEF = BDF$$
$$AEG = BCDG = ACDF = BEF$$
$$BEG = ACDG = BCDF = AEF$$

This is quite a formidable list of aliases; but when only one block of 32 is run, there are 31 df within the block. The above list accounts for these 31 df. If, in this design, all second-order (three-way) and higher interactions can be considered negligible, the main effects are all clear of first-order interactions. Three first-order interactions (AB, AF, and AG) are each aliased with another first-order interaction (FG, BG, and BF). If the choice of factors A, B, F, and G can be made with the assurance that the above interactions are either negligible or not of sufficient interest to be tested, these 3 df can be either pooled with error or left out of the analysis. The remaining 15 first-order interactions are all clear of other interactions except second order or higher. There are also 6 df left over for error involving only second-order interactions or higher. An analysis might be that shown in Table 15.5.

Table 15.5 One-Fourth Replication of a 2^7 Factorial

Source	df
Main effects $A, B, \cdots, G$	1 each for 7
First-order interaction $AC, AD, \cdots$	1 each for 15
AB (or FG), AF (or BG), AG (or BF) $\cdots$	1 each for 3
Second-order interaction or higher ($ACF, \cdots$)	1 each for 6
Total	31

Here tests can be made with 1 and 6 df or 1 and 9 df if the last two lines can be pooled for error. This is a fairly good design when it is necessary to run only a one-fourth replication of a 2^7 factorial.

3^n Factorials

The simplest 3^n factorial is a 3^2 requiring nine experiments for a complete factorial. If only a fraction of these can be run, we might consider a one-third replication of a 3^2 factorial. First we confound either AB or AB^2 with the three blocks. Confounding AB^2 gives

$$I = AB^2$$

$$L = X_1 + 2X_2$$

and the three blocks are

$L = 0$	$L = 1$	$L = 2$
00	10	20
11	21	01
22	02	12

If one of these three is run, the aliases are

$$A(AB^2) = A^2B^2 = A^4B^4 = AB$$

since the exponent on the first element is never left greater than one. A^2B^2 is squared to give $A^4B^4 = AB$ (modulus 3). However, the alias of B is

$$B(AB^2) = AB^3 = A$$

Hence all three effects A, B, and AB are aliases and mutually confounded. In order to get both aliases from one multiplication using the defining contrast, one multiplies the effect by the square of I as well as I. This gives

$$A[(AB^2)]^2 = A(A^2B^4) = A^3B = B$$

Hence, $A = AB = B$; this is a poor experiment, since, within the block that is run, both degrees of freedom confound A, B, and AB. In addition, no degrees of freedom are left for error. This example is cited only to show the slight modification necessary when determining the aliases in a 3^n factorial run as a fractional replication. The value of these fractional replications is seen when more factors are involved.

Consider a 3^3 factorial in three blocks of nine treatment combinations each. The effects can be broken down into 13 2-df effects: A, B, C, AB, AB^2, AC, AC^2, BC, BC^2, ABC, AB^2C, ABC^2, and AB^2C^2. Confounding ABC^2 with the three blocks gives

$$I = ABC^2 \qquad L = X_1 + X_2 + 2X_3$$

and the blocks are

$L = 0$	000	011	022	101	112	120	210	221	202
$L = 1$	100	111	122	201	212	220	010	021	002
$L = 2$	200	211	222	001	012	020	110	121	102

If only one of these blocks is now run, the aliases are

$$A = A(ABC^2) = A^2BC^2 = AB^2C$$

and

$$A = A(ABC^2)^2 = A^3B^2C^4 = B^2C = B^4C^2 = BC^2$$

$$B = B(ABC^2) = AB^2C^2$$

and

$$B = B(ABC^2)^2 = A^2B^3C^4 = A^4C^8 = AC^2$$

$$C = C(ABC^2) = ABC^3 = AB$$

and

$$C = C(ABC^2)^2 = A^2B^2C^5 = A^4B^4C^{10} = ABC$$

$$AB^2 = AB^2(ABC^2) = A^2B^3C^2 = A^4B^6C^4 = AC$$

and

$$AB^2 = AB^2(ABC^2)^2 = A^3B^4C^4 = BC$$

The analysis would be that in Table 15.6.

Table 15.6 One-Third Replication of a 3^3 Factorial

Source	df
A (or BC^2 or AB^2C)	2
B (or AB^2C^2 or AC^2)	2
C (or AB or ABC)	2
AC (or AB^2 or BC)	2
Total	8

This design would be somewhat practical if we could consider all interactions negligible and be content with 2 df for error.

These methods are easily extended to higher order 3^n experiments. Analysis proceeds as in Chapter 9.

It might be instructive to examine somewhat further this design of a one-third replication of a 3^3. A layout for 3^3 as a complete factorial was given in Table 9.7. If only one third of this were run, say block $L = 1$, suppose the other entries in Table 9.7 were crossed out, giving the results shown in Table 15.7.

Table 15.7 One-Third Replication of a 3^3 Factorial

			Factor A	
Factor B	Factor C	0	1	2
0	0	~~000~~	100	~~200~~
	1	~~001~~	~~101~~	201
	2	002	~~102~~	~~202~~
1	0	010	~~110~~	~~210~~
	1	~~011~~	111	~~211~~
	2	~~012~~	~~112~~	212
2	0	~~020~~	~~120~~	220
	1	021	~~121~~	~~221~~
	2	~~022~~	122	~~222~~

The remaining nine entries in Table 15.7 can now be summarized by levels of factor C only to give the result shown in Table 15.8.

A second glance at Table 15.8 reveals that this design is none other than a Latin square! In a sense our designs have now come full circle, from a Latin square as two restrictions on a single-factor experiment (Chapter 5) to a Latin square as a one-third replication of a 3^3 factorial. The designs come out the same, but they result from entirely different objectives. In a single-factor

Table 15.8 One-Third Replication of a 3^3 in Terms of Factor C

	Factor A		
Factor B	0	1	2
0	2	0	1
1	0	1	2
2	1	2	0

experiment, the general model

$$Y_{ij} = \mu + \tau_j + \varepsilon_{ij}$$

is partitioned by refining the error term due to restrictions on the randomization to give $Y_{ijk} = \mu + \tau_j + \beta_i + \gamma_k + \varepsilon'_{ij}$ where the block and position effects are taken from the original error term ε_{ij}.

In the one-third replication of a 3^3, there are three factors of interest which comprise the treatments, giving

$$Y_{ijk} = \mu + A_i + B_j + C_k + \varepsilon_{ijk}$$

where A, B, and C are taken from the treatment effect. Both models look alike but derive from different types of experiments. In both cases, the assumption is made that there is no interaction among the factors. This is more reasonable when the factors represent only randomization restrictions such as blocks and positions, rather than when each of the three factors are of vital concern to the experimenter.

By choosing each of the other two blocks in the confounding of ABC^2 ($L = 0$ or $L = 2$), two other Latin squares are obtained. If another confounding scheme is chosen, such as AB^2C, ABC, or AB^2C^2, more and different Latin squares can be generated. In fact, with four parts of the three-way interaction and three blocks per confounding scheme, 12 distinct Latin squares may be generated. This shows how it is possible to select a Latin square at random for a particular design.

It cannot be overstressed that we must be quite sure that no interaction exists before using these designs on a 3^3 factorial. It is not enough to say that there is no interest in the interactions, because, unfortunately, the interactions are badly confounded with main effects as aliases.

15.4 SUMMARY

Continuing the summary at the end of Chapter 14 under incomplete randomized blocks gives

Experiment	Design	Analysis
II. Two or more factors		
A. Factorial (crossed)		
	1. Completely randomized	
	2. Randomized block	
	a. Incomplete, confounding	
	i. Main effect—split-plot	
	ii. Interactions in 2^n and 3^n	
	1. Several replications $Y_{ijkq} = \mu + R_i + \beta_j + R\beta_{ij} + A_k + B_q + AB_{kq} + \varepsilon_{ikq}$	1. Factorial ANOVA with replications and blocks
	2. One replication only $Y_{ijk} = \mu + \beta_i + A_j + B_k + AB_{jk}, \cdots$	2. Factorial ANOVA with blocks or confounded interaction
	3. Fractional replication—aliases *a*. in 2^n *b*. in 3^n; $n = 3$: Latin square $Y_{ijk} = \mu + A_i + B_j + AB_{ij}, \cdots$	3. Factorial ANOVA with aliases or aliased interactions

PROBLEMS

15.1 If not all of Problem 14.1 can be run, determine the aliases if a one-half replication is run. Analyze the data for the principal block only.

15.2 If only a one-fourth replication can be run in Problem 14.2, determine the aliases.

15.3 For Problem 14.4, use a one-third replication and determine the aliases.

15.4 Run an analysis on the block in Problem 15.3 that contains treatment combination 022.

15.5 What would be the aliases in Problem 14.5 if a one-half replication is run?

15.6 Determine the aliases in a one-fourth replication of Problem 14.7.

15.7 Determine the aliases in Problem 14.9 if a one-third replication is run.

15.8 If a one-eighth replication of Problem 14.12 is run, what are the aliases? Comment on its ANOVA table.

15.9 What are the aliases and the ANOVA outline if only one block of Problem 14.13 is run?

15.10 If data are taken from one furnace only in Problem 14.16, what are the aliases and the ANOVA outline?

15.11 What are the numerical results and conclusions from just one furnace of Problem 14.17?

16
Miscellaneous Topics

16.1 INTRODUCTION

There are several techniques that have been developed, based on the general design principles presented in the first 15 chapters of this book and on other well-known statistical methods. No attempt will be made to discuss these in detail, as they are presented very well in the references. It is hoped that a background in experimental design as given in the first 15 chapters will make it possible for the experimenter to read and understand the references.

The methods to be discussed are covariance analysis, response surface experimentation, evolutionary operation, and analysis of attribute data.

16.2 COVARIANCE ANALYSIS

Philosophy

Occasionally when a study is being made of the effect of one or more factors on some response variable, say Y, there is another variable, or variables, which vary along with Y. It is often not possible to control this other variable (or variables) at some constant level throughout the experiment, but the variable can be measured along with the response variable. This variable is referred to as a *concomitant variable* X as it "runs along with" the response variable Y. In order, then, to assess the effect of the treatments on Y, one should first attempt to remove the effects of this concomitant variable X. This technique of removing the effect of X (or several X's) on Y and then analyzing the residuals for the effect of the treatments on Y is called *covariance analysis*.

Several examples of the use of this technique might be cited. If one wishes to study the effect on student achievement of several teaching methods, it is customary to use the difference between a pretest and a posttest score as the response variable. It may be, however, that the gain Y is also affected by the student's pretest score X as gain, and pretest scores may be correlated.

Covariance analysis will provide for an adjustment in the gains due to differing pretest scores assuming some type of regression of gain on pretest scores. The advantage of using this technique is that one is not limited to running an experiment on only those pupils who have approximately the same pretest scores or matching pupils with the same pretest scores and randomly assigning them to control and experimental groups. Covariance analysis has the effect of providing "handicaps" as if each student had the same pretest scores while actually letting the pretest scores vary along with the gains.

This technique has often been used to adjust weight gains in animals by their original weights in order to assess the effect of certain feed treatments on these gains. Many industrial examples might be cited such as the one given below where original weight of a bracket may affect the weight of plating applied to the bracket by different methods.

Snedecor [18] gives several examples of the use of covariance analysis. Ostle [15] is an excellent reference on the extension of covariance analysis to many different experimental designs. The *Biometrics* article of 1957 [4] devotes most of its pages to a discussion of many of the fine points of covariance analysis. The problem discussed below will illustrate the technique for the case of a single-factor experiment run in a completely randomized design. Since the covariance technique represents a marriage between regression and the analysis of variance, the methods of Chapter 8 and Chapter 3 will be used.

Example 16.1 Several steel brackets were sent to three different vendors to be zinc plated. The chief concern in this process is the thickness of the zinc plating and whether or not there was any difference in this thickness between the three vendors. Data on this thickness in hundred thousandths of an inch (10^{-5}) for four brackets plated by the three vendors are given in Table 16.1.

Table 16.1 Plating Thickness in 10^{-5} Inches from Three Vendors

	Vendor	
A	*B*	*C*
40	25	27
38	32	24
30	13	20
47	35	13

Assuming a single-factor experiment and a completely randomized design, an analysis of variance model for this experiment would be:

$$Y_{ij} = \mu + \tau_j + \varepsilon_{ij} \tag{16.1}$$

where Y_{ij} is the observed thickness of the ith bracket from the jth vendor, μ is a common effect, τ_j represents the vendor effect, and ε_{ij} represents the random error. Considering vendors as a fixed factor and random errors as normally and independently distributed with mean zero and common variance σ_ε^2, an ANOVA table was compiled (Table 16.2).

Table 16.2 ANOVA on Plating Thickness

Source	df	SS	MS
Between vendors	2	665.2	332.6
Within vendors	9	543.5	60.4
Totals	11	1208.7	

The F statistic equals 5.51 and is significant at the 5-percent significance level with 2 and 9 df. One might, therefore, conclude that three is a real difference in average plating thickness between these three vendors, and steps should be taken to select the most desirable vendor.

During a discussion of these results, it was pointed out that some of this difference between vendors might be due to unequal thickness in the brackets before plating. In fact, there might be a correlation between the thickness of the brackets before plating and the thickness of the plating. To see whether or not such an idea is at all reasonable, a scatterplot was made on the thickness of the bracket before plating X and the thickness of the zinc plating Y. Results are shown in Figure 16.1.

A glance at this scattergram would lead one to suspect that there is a positive correlation between the thickness of the bracket and the plating thickness. Since another variable, X, can be measured on each piece and may be related to the variable of interest Y, a covariance analysis may be run

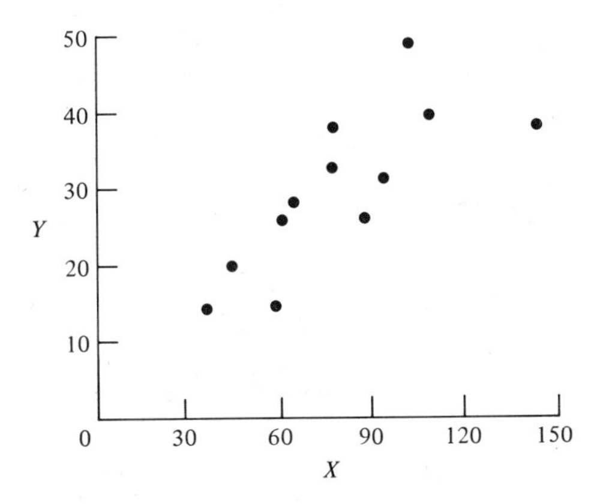

Figure 16.1 Scattergram of bracket thickness X versus plating thickness Y in 10^{-5} inches.

in order to remove the linear effect of X and Y when comparing vendors. A covariance analysis is used to remove the linear or higher order relationship of one or more independent variables from a dependent variable in order to assess the effect of a factor on the dependent variable.

Covariance

Since a positive correlation is suspected between X and Y above, a regression model might be written as

$$Y_{ij} = \mu + \beta(X_{ij} - \bar{X}) + \varepsilon_{ij} \tag{16.2}$$

where β is the true linear regression coefficient (or slope) between Y and X over all the data; $\bar{X}$ is the mean of the X values. In such a model, it is assumed that a measure of variable X can be made on each unit along with a corresponding measure of Y. For a covariance analysis, models (16.1) and (16.2) can be combined to give the covariance model

$$Y_{ij} = \mu + \beta(X_{ij} - \bar{X}) + \tau_j + \varepsilon_{ij} \tag{16.3}$$

In this model, the error term ε_{ij} should be smaller than in model (16.1) because of the removal of the effect of the covariate X_{ij}.

To determine how to adjust the analysis of variance to provide for the removal of this covariate, consider the regression model (16.2) in deviation form:

$$y = bx + e \tag{16.4}$$

where x and y are now deviations from their respective means ($x = X_{ij} - \bar{X}$, $y = Y_{ij} - \bar{Y}$) and b is the sample slope. By the method of least squares, it will be recalled that

$$b = \frac{\sum xy}{\sum x^2}$$

where summations are over all points. The sum of the squares of the errors of estimate, which have been minimized, is now, from Equation (16.4),

$$e = y - bx$$

$$\sum e^2 = \sum (y - bx)^2$$

$$= \sum y^2 - 2b \sum xy + b^2 \sum x^2$$

Substituting for b

$$b = \frac{\sum xy}{\sum x^2}$$

and

$$\sum e^2 = \sum y^2 - 2\frac{\sum xy}{\sum x^2} \cdot \sum xy + \left(\frac{\sum xy}{\sum x^2}\right)^2 \sum x^2$$

$$= \sum y^2 - \frac{(\sum xy)^2}{\sum x^2} \tag{16.5}$$

In this expression, the term $(\sum xy)^2/\sum x^2$ is the amount of reduction in the sum of squares of the y variable due to its linear regression on x. This term is then the sum of squares due to linear regression. It involves the sum of the cross products of the two variables ($\sum xy$) and the sum of squares of the x variable alone ($\sum x^2$). If a term of this type is subtracted from the sum of squares for the dependent variable y, the result will be a corrected or adjusted sum of squares for y.

The sums of squares and cross products can be computed on all the data (total), within each vendor or between vendors. In each case it will be helpful to recall that the sums of squares and cross products in terms of the original data are

$$\sum x^2 = \sum X^2 - \frac{(\sum X)^2}{N}$$

$$\sum y^2 = \sum Y^2 - \frac{(\sum Y)^2}{N}$$

$$\sum xy = \sum XY - \frac{(\sum X)(\sum Y)}{N}$$

Data on both variables appear with the appropriate totals in Table 16.3.

Table 16.3 Bracket Thickness X and Plating Thickness Y in 10^{-5} Inches from Three Vendors

	Vendor							
	A		*B*		*C*		Total	
	X	*Y*	*X*	*Y*	*X*	*Y*	*X*	*Y*
	110	40	60	25	62	27		
	75	38	75	32	90	24		
	93	30	38	13	45	20		
	97	47	140	35	59	13		
Totals	375	155	313	105	256	84	944	344

From Table 16.3 the sums of squares and cross products for the total data are[1]

$$T_{xx} = (110)^2 + (75)^2 + \cdots + (59)^2 - \frac{(944)^2}{12} = 9240.7$$

$$T_{yy} = (40)^2 + (38)^2 + \cdots + (13)^2 - \frac{(344)^2}{12} = 1208.7$$

$$T_{xy} = (110)(40) + (75)(38) + \cdots + (59)(13) - \frac{(944)(344)}{12} = 2332.7$$

[1] Here T, V, and E are used to denote sum of squares and cross products for totals, vendors, and error, respectively.

The between vendors sums of squares and cross products are computed by vendor totals as in an ANOVA:

$$V_{xx} = \frac{(375)^2 + (313)^2 + (256)^2}{4} - \frac{(944)^2}{12} = 1771.2$$

$$V_{yy} = \frac{(155)^2 + (105)^2 + (84)^2}{4} - \frac{(344)^2}{14} = 665.2$$

$$V_{xy} = \frac{(375)(155) + (313)(105) + (256)(84)}{4} - \frac{(944)(344)}{12} = 1062.2$$

By subtraction the error sums of squares are

$$E_{xx} = T_{xx} - V_{xx} = 9240.7 - 1771.2 = 7469.5$$

$$E_{yy} = T_{yy} - V_{yy} = 1208.7 - 665.2 = 543.5$$

$$E_{xy} = T_{xy} - V_{xy} = 2332.7 - 1062.2 = 1270.5$$

Using this information, the sums of squares of the dependent variable Y may be adjusted for regression on X. On the totals, the adjusted sum of squares is then

$$\text{adjusted} \sum y^2 = T_{yy} - \frac{T_{xy}^2}{T_{xx}} \quad \text{as in Equation (16.5)}$$

$$= 1208.7 - \frac{(2332.7)^2}{9240.7} = 619.8$$

and the adjusted sum of squares within vendors is

$$\text{adjusted} \sum y^2 = E_{yy} - \frac{(E_{xy})^2}{E_{xx}}$$

$$= 543.5 - \frac{(1270.5)^2}{7469.5} = 327.4$$

Then, by subtraction, the adjusted sum of squares between vendors is

$$619.8 - 327.4 = 292.4$$

These results are usually displayed as in Table 16.4.

With the adjusted sums of squares, the new F statistic is

$$F = \frac{146{,}2}{40.9} = 3.57$$

with 2 and 8 df. This is now not significant at the 5-percent significance level. This means that after removing the effect of bracket thickness, the vendors no longer differ in the average plating thickness on the brackets.

Table 16.4 Analysis of Covariance for Bracket Data

		SS and Products			Adjusted		
Source	df	$\sum y^2$	$\sum xy$	$\sum x^2$	$\sum y^2$	df	MS
Between vendors	2	665.2	1062.2	1771.2	—	—	—
Within vendors	9	543.5	1270.5	7469.5	327.4	8	40.9
Totals	11	1208.7	2332.7	9240.7	619.8	10	
Between vendors					292.4	2	146.2

Table 16.4 should have a word or two of explanation. The degrees of freedom on the adjusted sums of squares are reduced by 2 instead of 1 as estimates of both the mean and the slope are necessary in their computation. Adjustments are made here on the totals and the within sums of squares rather than on the between sum of squares as we are interested in making the adjustment based on an overall slope of Y on X and a within group average slope of Y on X as will be explained in more detail later.

In this analysis, three different types of regression can be identified: the "overall" regression of all the Y's on all the X's, the within vendors regression, and the regression of the three Y means on the three X means. This last regression could be quite different from the other two and is of little interest since we are attempting to adjust the Y's and the X's within the vendors. An estimate of this "average within vendor" slope is

$$b = \frac{\sum xy}{\sum x^2} = \frac{E_{xy}}{E_{xx}} = \frac{1270.5}{7469.5} = 0.17$$

If this slope or regression coefficient is used to adjust the observed Y means for the effect of X on Y, one finds

$$\text{adjusted } \bar{Y}_{.j} = \bar{Y}_{.j} - b(\bar{X}_{.j} - \bar{X}_{..})$$

For vendor A,

$$\text{adjusted } \bar{Y}_{.1} = \frac{155}{4} - (0.17)\left(\frac{375}{4} - \frac{944}{12}\right)$$

$$= 38.75 - (0.17)(93.75 - 78.67) = 36.19$$

For vendor B,

$$\text{adjusted } \bar{Y}_{.2} = 26.25 - (0.17)(78.25 - 78.67) = 26.31$$

For vendor C,

$$\text{adjusted } \bar{Y}_{.3} = 21.00 - (0.17)(64.00 - 78.67) = 23.49$$

A comparison of the last column above with the column of unadjusted means just to the right of the equal signs shows that these adjusted means are closer together than the unadjusted means which is confirmed in the preceding analysis. This can be seen graphically by plotting each vendor's data in a different symbol and "sliding" the means along lines parallel to this regression slope (Figure 16.2).

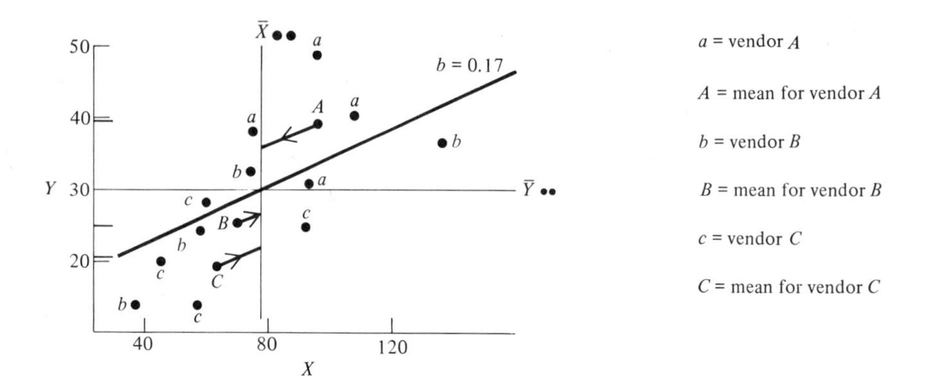

Figure 16.2 Plating thickness Y versus bracket thickness X plotted by vendors.

At the left-hand ordinate in Figure 16.2, the three unadjusted Y means are marked off. By moving these three means along lines parallel to the general slope ($b = 0.17$), a comparison is seen between the three means (on the $\bar{X}_{..}$ ordinate) after adjustment. This shows graphically the results of the covariance analysis. By examining each group of observations for a given vendor, it is seen that all three slopes within vendors are in the same direction and of similar magnitude. By considering all 12 points, the overall slope can be visualized. On the other hand, if the three means (indicated by the capital letters) are considered, the slope of these $\bar{Y}$'s on their corresponding $\bar{X}$'s is much steeper. Hence, adjustments are made on the basis of the overall slope and the "average" or "pooled" within groups slope.

One of the problems in covariance analysis is that there are many assumptions made in its application. These should be examined in some detail.

First, there are the usual assumptions of analysis of variance on the dependent variable Y: an additive model, normally and independently distributed error, and homogeneity of variance within the groups. This last assumption is often tested with Bartlett's or a similar test. In this problem, since the ranges of the three vendor readings are 17, 22, and 14, the assumption seems tenable.

In covariance analysis it is further assumed that the regression is linear, that the slope is not zero (covariance was necessary), that the regression coefficients within each group are homogeneous so that the "average" or "pooled" within groups regression can be used on all groups, and finally that the independent variable X is not affected by the treatments given to the groups.

The linearity assumption may be tested if the experiment is planned in such a way that there is more than one Y observation for each X. Since this was not done here, a look at the scattergram will have to suffice for this linearity assumption.

To test the hypothesis that the true slope β in Equation (16.3) is zero, consider whether or not the reduction in the error sum of squares is significant when compared to the error sum of squares. By an F test, the ratio would be

$$F_{1,N-k-1} = \frac{E_{xy}^2/E_{xx}}{\text{adjusted } E_{yy}/(N-k-1)}$$

$$F_{1,8} = \frac{216.1}{327.4/8} = \frac{216.1}{40.9} = 5.28$$

As the 5-percent F is 5.32, this result is nearly significant and certainly casts considerable doubt on the hypothesis that $\beta = 0$. Since the results after covariance adjustment were not significant as compared to the results before adjustment, it seems reasonable to conclude that covariance was necessary.

To test the hypothesis that all three regression coefficients are equal, let us compute each sample coefficient and adjust the sum of squares within each group by its own regression coefficient.

For vendor A,

$$b_A = \frac{\sum xy}{\sum x^2} = \frac{\sum XY - (\sum X)(\sum Y)/n}{\sum X^2 - (\sum X)^2/n}$$

Within group A,

$$b_A = \frac{14{,}599 - 58{,}125/4}{35{,}783 - 140{,}625/4} = 0.108$$

$$\text{adjusted} \sum y^2 = \sum y^2 - \frac{(\sum xy)^2}{\sum x^2} = 146.75 - 7.32 = 139.43$$

For vendor B,

$$b_B = \frac{9294 - 32{,}865/4}{30{,}269 - 24{,}492.25/4} = 0.187$$

$$\text{adjusted} \sum y^2 = 286.75 - 201.72 = 85.03$$

For vendor C,

$$b_C = \frac{5501 - 21{,}504/4}{17{,}450 - 65{,}536/4} = 0.117$$

$$\text{adjusted} \sum y^2 = 110.00 - 14.66 = 95.34$$

Summarizing, we obtain Table 16.5.

Table 16.5 Adjusted Sums of Squares within Each Vendor

Within Vendor	df	$\sum y^2$	b	Adjusted $\sum y^2$	df
A	3	146.75	0.108	139.43	2
B	3	286.75	0.187	85.03	2
C	3	110.00	0.117	95.34	2
Totals	9	543.50		319.80	6

Adjusting each within vendor sum of squares by its own regression coefficient reduces the degrees of freedom to 2 per vendor. Adding the new adjusted sums of squares, Table 16.5 shows a within vendor sum of squares of 319.80 based on 6 df. When the within vendor sum of squares was adjusted by a "pooled" within vendor regression, Table 16.4 showed an adjusted sum of squares within vendors of 327.4 based on 8 df. If there were significant differences in regression within the three vendors, this would show as a difference in these two figures: 327.4 − 319.8 = 7.6. An F test for this hypothesis would then be

$$F_{k-1,N-2k} = \frac{\left[\begin{array}{l}\text{adjusted} \sum y^2 \text{ (based on pooled within groups regression)} \\ \quad - \text{adjusted} \sum y^2 \text{ (based on regressions within each group)}\end{array}\right] \Big/ (k-1)}{\left[\text{adjusted} \sum y^2 \text{ (within each group)}\right]/(N-2k)}$$

and

$$F_{2,6} = \frac{(327.4 - 319.8)/2}{319.8/6} = \frac{3.8}{53.3} = <1$$

hence nonsignificant, and we conclude that the three regression coefficients are homogeneous.

The final assumption that the vendors do not affect the covariate X is tenable here on practical grounds as the vendor has nothing to do with the bracket thickness before plating. However, in some applications, this assumption needs to be checked by an F test on the X variable. For our data from

Table 16.4, such an F test would give

$$F_{2,9} = \frac{1771.2/2}{7469.5/9} = \frac{885.6}{829.9} = 1.07$$

which is obviously nonsignificant.

From this discussion of the assumptions, one notes that it is often a problem in the use of covariance analysis to satisfy all of the assumptions. This may be the reason many people avoid covariance although it is applicable whenever one suspects that another measured variable or variables affect the dependent variable.

The above techniques may be extended to handle several covariates, nonlinear regression, and several factors and interactions of interest. In handling randomized blocks, Latin squares, or experiments with several factors, Ostle [15] points out that the proper technique is to add the sum of squares and sum of cross products of the term of interest to the sum of squares and sum of cross products of the error, adjust this total, then adjust the error and determine the adjusted treatment effect by subtraction.

When several covariates are involved, many questions arise concerning how many covariates can be handled, how many really affect the dependent variable, how can a significant subset be found, and so on. All of these questions are of concern in any multiple regression problem and some discussion of these and other problems can be found in [4].

16.3 RESPONSE-SURFACE EXPERIMENTATION

Philosophy

The concept of a response surface involves a dependent variable Y, called the response variable, and several independent or controlled variables, $X_1, X_2, \cdots, X_k$. If all of these variables are assumed to be measurable, the response surface can be expressed as

$$Y = f(X_1, X_2, \cdots, X_k)$$

For the case of two independent variables such as temperature X_1 and time X_2, the yield Y of a chemical process can be expressed as

$$Y = f(X_1, X_2)$$

This surface can be plotted in three dimensions, with X_1 on the abscissa, X_2 on the ordinate, and Y plotted perpendicular to the X_1X_2 plane. If the values of X_1 and X_2 that yield the same Y are connected, we can picture the surface with a series of equal-yield lines, or contours. These are similar to the contours of equal height on topographic maps and the isobars on weather maps. Some response surfaces might be of the types in Figure 16.3. Two excellent references for understanding response-surface experimentation are [7] and [10].

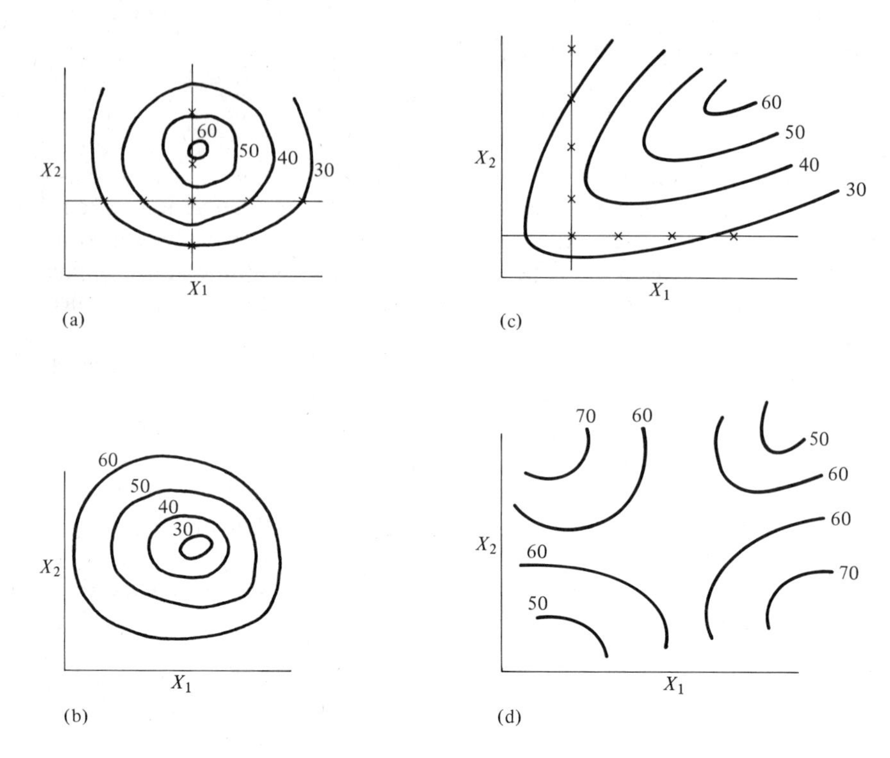

Figure 16.3 Some typical response surfaces in two dimensions. (*a*) Mound. (*b*) Depression. (*c*) Rising ridge. (*d*) Saddle.

The Twofold Problem

The problem involved in the use of response-surface experimentation is twofold, (1) to determine, on the basis of one experiment, where to move in the next experiment toward the optimal point on the underlying response surface; (2) having located the optimum or near optimum of the surface, to determine the equation of the response surface in an area near this optimum point.

One method of experimentation that seeks the optimal point of the response surface might be the traditional one-factor-at-a-time method. As shown in Figure 16.3(*a*), if X_2 is fixed and X_1 is varied, we find the X_1 optimal (or near optimal) value of response Y at the fixed value of X_2. Having found this X_1 value, experiments can now be run at this fixed X_1, and the X_2 for optimal response can be found. In the case of the mound in Figure 16.3(*a*), this method would lead eventually to the peak of the mound or near it. However, this same method, when applied to a surface such as the rising ridge in Figure 16.3(*c*), fails to lead to the maximum point on the response surface. In experimental work, the type of response surface is usually unknown, so that a better method is necessary if the optimum set of conditions is to be found for any surface.

The method developed by those who have worked in this area is called the *path of steepest ascent* method. The idea here is to run a simple experiment over a small area of the response surface where, for all practical purposes, the surface may be regarded as a plane. We then determine the equation of this plane and from it the direction we should take from this experiment in order to move toward the optimum of the surface. Since the next experiment should be in a direction in which we hope to scale the height the fastest, this is referred to as the *path of steepest ascent*. This technique does not determine how far away from the original experiment succeeding sequential experiments should be run, but it does indicate to the experimenter the direction along which the next experiment should be performed. A simple example will illustrate this method.

In order to determine the equation of the response surface, several special experimental designs have been developed which attempt to approximate this equation using the smallest number of experiments possible. In two dimensions, the simplest surface is a plane given by

$$Y = B_0X_0 + B_1X_1 + B_2X_2 + \varepsilon \tag{16.6}$$

where Y is the observed response, X_0 is taken as unity, and estimates of the B's are to be determined by the method of least squares which minimizes the sum of the squares of the errors ε. Such an equation is referred to as a first-order equation, since the power on each independent variable is unity.

If there is some evidence that the surface is not planar, a second-order equation in two dimensions may be a more suitable model:

$$Y = B_0X_0 + B_1X_1 + B_2X_2 + B_{11}X_1^2 + B_{12}X_1X_2 + B_{22}X_2^2 + \varepsilon \tag{16.7}$$

Here, the X_1X_2 term represents an interaction between the two variables X_1 and X_2.

If there are three independent or controlled variables, the first-order equation is again a plane or hyperplane

$$Y = B_0X_0 + B_1X_1 + B_2X_2 + B_3X_3 + \varepsilon \tag{16.8}$$

and the second-order equation is

$$\begin{aligned} Y = B_0X_0 &+ B_1X_1 + B_2X_2 + B_3X_3 + B_{11}X_1^2 + B_{22}X_2^2 \\ &+ B_{33}X_3^2 + B_{12}X_1X_2 + B_{13}X_1X_3 + B_{23}X_2X_3 + \varepsilon \end{aligned} \tag{16.9}$$

As the complexity of the surface increases, more coefficients must be estimated, and the number of experimental points must necessarily increase.

Several very clever designs have been developed that minimize the amount of work necessary to estimate these response-surface equations.

In order to determine the coefficients for these more complex surfaces and to interpret their geometric nature, both multiple regression techniques and the methods of solid analytical geometry are used. In the example which follows, only the simplest type of surface will be explored. The references quoted will give many more complex examples.

Example 16.2 Consider an example in which an experimenter is seeking the proper values for both concentration of filler to epoxy resin X_1 and position in the mold X_2 to minimize the abrasion on a plastic die. This abrasion or wear is measured as a decrease in thickness of the material after 10,000 cycles of abrasion. Since the maximum thickness is being sought, the first experiment should attempt to discover the direction in which succeeding experiments should be run in order to approach this maximum by the steepest path. Assuming the surface to be a plane in a small area, the first experiment will be used to determine the equation of this plane. The response surface is then

$$Y = B_0X_0 + B_1X_1 + B_2X_2 + \varepsilon$$

As there are three parameters to be estimated, B_0, B_1, and B_2, at least three experimental points must be taken to estimate these coefficients. Such a design might be an equilateral triangle, but since there are two factors, X_1 and X_2, each can be set at two levels and a 2^2 factorial may be used. Two concentrations were chosen, $\frac{1}{2}:1$ and $1:1$ (ratio of filler to resin), and two positions, 1 in. and 2 in. from a reference point, with responses Y, the thickness of the material in 10^{-4} in., as shown in Figure 16.4.

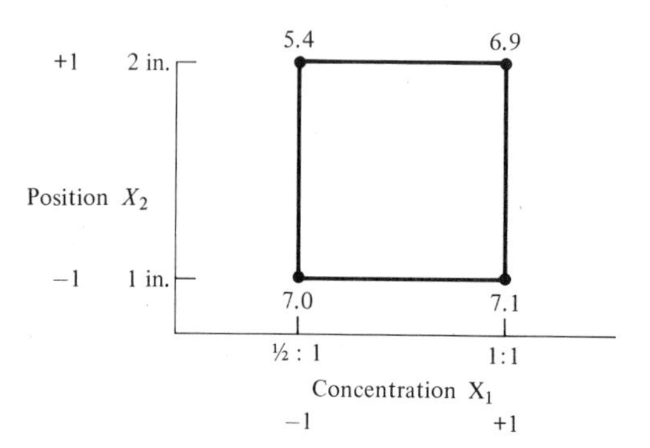

Figure 16.4 2^2 factorial example on response surface.

To determine the equation of the best fitting plane for these four points, consider the error in the prediction equation

$$\varepsilon = Y - B_0X_0 - B_1X_1 - B_2X_2$$

The sum of the squares for this error is

$$\sum \varepsilon^2 = \sum (Y - B_0X_0 - B_1X_1 - B_2X_2)^2$$

where the summation is over all points given in the design. To find the B's, we differentiate this expression with respect to each parameter and set these three expressions equal to zero. This provides three least squares normal equations, which can be solved for the best estimates of the B's. These estimates are designated as b's.

Differentiating the expression above gives

$$\frac{\partial(\sum \varepsilon^2)}{\partial B_0} = -2 \sum (Y - B_0X_0 - B_1X_1 - B_2X_2)X_0 = 0$$

$$\frac{\partial(\sum \varepsilon^2)}{\partial B_1} = -2 \sum (Y - B_0X_0 - B_1X_1 - B_2X_2)X_1 = 0$$

$$\frac{\partial(\sum \varepsilon^2)}{\partial B_2} = -2 \sum (Y - B_0X_0 - B_1X_1 - B_2X_2)X_2 = 0$$

and from these results we get

$$\begin{aligned} \sum X_0Y &= b_0 \sum X_0^2 + b_1 \sum X_0X_1 + b_2 \sum X_0X_2 \\ \sum X_1Y &= b_0 \sum X_0X_1 + b_1 \sum X_1^2 + b_2 \sum X_1X_2 \qquad (16.10) \\ \sum X_2Y &= b_0 \sum X_0X_2 + b_1 \sum X_1X_2 + b_2 \sum X_2^2 \end{aligned}$$

as the least squares normal equations.

By a proper choice of experimental variables, it is possible to reduce these equations considerably for simple solution. More complex models can be solved best by the use of matrix algebra. For the 2^2 factorial, the following coding scheme simplifies the solution of Equation (16.10). Set

$$X_1 = 4C - 3 \quad \text{where } C \text{ is the concentration}$$

$$X_2 = 2P - 1 \quad \text{where } P \text{ is the position}$$

X_0 is always taken as unity. For the experimental variables X_1 and X_2, the responses can be recorded as in Table 16.6. These experimental variables are also indicated on Figure 16.4.

Table 16.6 Orthogonal Layout for Figure 16.4

Y	X_0	X_1	X_2
7.0	1	−1	−1
5.4	1	−1	1
7.1	1	1	−1
6.9	1	1	1

An examination of the data as presented in Table 16.6 shows that X_0, X_1, and X_2 are all orthogonal to each other as

$$\sum X_1 = \sum X_2 = 0 \quad \text{and} \quad \sum X_1 X_2 = 0$$

Hence the least squares normal equations become

$$\begin{aligned} \sum X_0 Y &= b_0 n + b_1 \cdot 0 + b_2 \cdot 0 \\ \sum X_1 Y &= b_0 \cdot 0 + b_1 \cdot \sum X_1^2 + b_2 \cdot 0 \\ \sum X_2 Y &= b_0 \cdot 0 + b_1 \cdot 0 + b_2 \sum X_2^2 \end{aligned} \tag{16.11}$$

Solving gives

$$\begin{aligned} b_0 &= \sum X_0 Y / n \\ b_1 &= \sum X_1 Y / \sum X_1^2 \\ b_2 &= \sum X_2 Y / \sum X_2^2 \end{aligned} \tag{16.12}$$

For this problem

$$\begin{aligned} b_0 &= \frac{26.4}{4} = 6.60 \\ b_1 &= \frac{1.6}{4} = 0.40 \\ b_2 &= \frac{-1.8}{4} = -0.45 \end{aligned}$$

and the response surface can be approximated as

$$\hat{Y} = 6.60 + 0.40X_1 - 0.45X_2$$

To determine the sum of squares due to each of these terms in the model, we can use the results of Section 3.7 where the sum of squares due to b_i is

$$\text{SS}_{b_i} = b_i \cdot \sum X_i Y$$

Here

$$\begin{aligned} \text{SS}_{b_0} &= (6.60)(26.4) = 174.24 \\ \text{SS}_{b_1} &= (0.40)(1.6) = 0.64 \\ \text{SS}_{b_2} &= (-0.45)(-1.8) = 0.81 \end{aligned}$$

each carrying 1 df. The ANOVA table is shown as Table 16.7.

Table 16.7 ANOVA for 2^2 Factorial Response-Surface Example

Source	df	SS
b_0	1	174.24
b_1	1	0.64
b_2	1	0.81
Residual	1	0.49
Totals	4	176.18

Here, all 4 df are shown, since the b_0 term represents the degree of freedom usually associated with the mean, that is, the correction term in many examples. The total sum of squares is $\sum Y^2$ of the responses and the residual is what is left over. With this 1-df residual, no good test is available on the significance of each term in the model, nor is there any way to assess the adequacy of the planar model to describe the surface.

To decide on the direction for the next experiment, plot contours of equal response using the equation of the plane determined above

$$\hat{Y} = 6.60 + 0.40X_1 - 0.45X_2$$

Solve for X_2,

$$X_2 = \frac{6.60 - \hat{Y} + 0.40X_1}{0.45}$$

If $\hat{Y} = 5.5$,

$$X_2 = \frac{1.10 + 0.40X_1}{0.45}$$

when

$$X_1 = -1 \qquad X_2 = 1.56$$

$$X_1 = 1 \qquad X_2 = 3.33$$

If $\hat{Y} = 6.0$,

$$X_2 = \frac{0.60 + 0.40X_1}{0.45}$$

when

$$X_1 = -1 \qquad X_2 = 0.44$$

$$X_1 = 1 \qquad X_2 = 2.22$$

If $\hat{Y} = 6.5$,

$$X_2 = \frac{0.10 + 0.40X_1}{0.45}$$

when

$$X_1 = -1 \qquad X_2 = -0.67$$

$$X_1 = 1 \qquad X_2 = 1.11$$

If $\hat{Y} = 7.0$,

$$X_2 = \frac{-0.40 + 0.40X_1}{0.45}$$

when

$$X_1 = -1 \qquad X_2 = -1.67$$

$$X_1 = 1 \qquad X_2 = 0$$

If $\hat{Y} = 7.5$,

$$X_2 = \frac{-0.90 + 0.40X_1}{0.45}$$

when

$$X_1 = -1 \qquad X_2 = -2.89$$

$$X_1 = 1 \qquad X_2 = -1.11$$

Plotting these five contours on the original diagram gives the pattern shown in Figure 16.5.

By moving in a direction normal to these contours and "up" the surface, we can anticipate larger values of response until the peak is reached. To decide

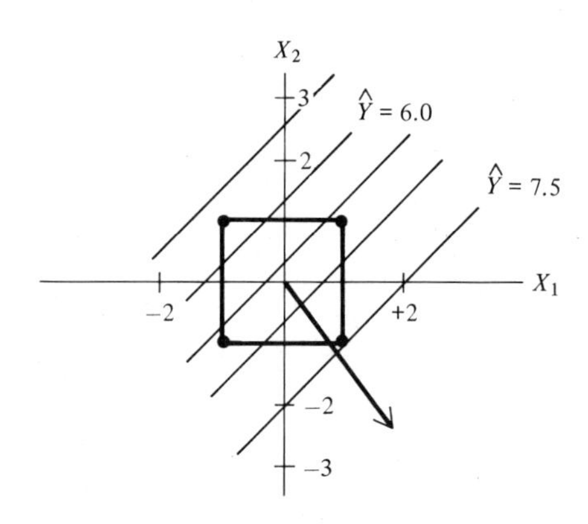

Figure 16.5 Contours on 2^2 factorial response-surface example.

on a possible set of conditions for the next experiment, consider the equation of a normal to these contours through the point (0, 0). The contours are

$$X_2 = \frac{6.60 - \hat{Y} + 0.40X_1}{0.45}$$

and their slope is $0.40/0.45 = 8/9$. The normal will have a slope $= -9/8$, and its equation is

$$X_2 - 0 = \frac{-9}{8}(X_1 - 0)$$

$$X_2 = \frac{-9}{8}X_1 \qquad \text{(see arrow in Figure 16.5)}$$

This technique will not tell how far to go in this direction, but we might run the next experiment with the center of the factorial at $(+1, -9/8)$. The four points would be

X_1	X_2
0	$-\frac{1}{8}$
0	$-2\frac{1}{8}$
2	$-\frac{1}{8}$
2	$-2\frac{1}{8}$

These points are expressed in terms of the experimental variables, but must be decoded to see where to set the concentration and position. Using the coding,

$$X_1 = 4C - 3 \quad \text{or} \quad C = \frac{X_1 + 3}{4}$$

$$X_2 = 2P - 1 \quad \text{or} \quad P = \frac{X_2 + 1}{2}$$

the four new points would be

X_1	X_2	C	P
0	$-\frac{1}{8}$	$\frac{3}{4}:1$	$\frac{7}{16}$ in
0	$-2\frac{1}{8}$	$\frac{3}{4}:1$	$-\frac{9}{16}$ in
2	$-\frac{1}{8}$	$1\frac{1}{4}:1$	$\frac{7}{16}$ in
2	$-2\frac{1}{8}$	$1\frac{1}{4}:1$	$-\frac{9}{16}$ in

if these settings are possible. Responses may now be taken at these four points on a new 2^2 factorial, and after analysis we can again decide the direction of steepest ascent. This prodecure continues until the optimum is obtained.

The above design is sufficient to indicate the direction for subsequent experiments, but it does not provide a good measure of experimental error to test the significance of b_0, b_1, and b_2, nor is there any test of how well the plane approximates the surface. One way to improve on this design is to take two or more points at the center of the square. By replication at the same point, an estimate of experimental error is obtained and the average of the center-point responses will provide an estimate of "goodness of fit" of the plane. If the experiment is near the maximum response, this center point might be somewhat above the four surrounding points, which would indicate the need of a more complex model.

To see how this design would help, consider two observations of response at the center of the example above. Using the same responses at the vertices of the square, the results might be those shown in Figure 16.6.

Y	X_0	X_1	X_2
7.0	1	−1	−1
5.4	1	−1	1
7.1	1	1	−1
6.9	1	1	1
6.6	1	0	0
6.8	1	0	0

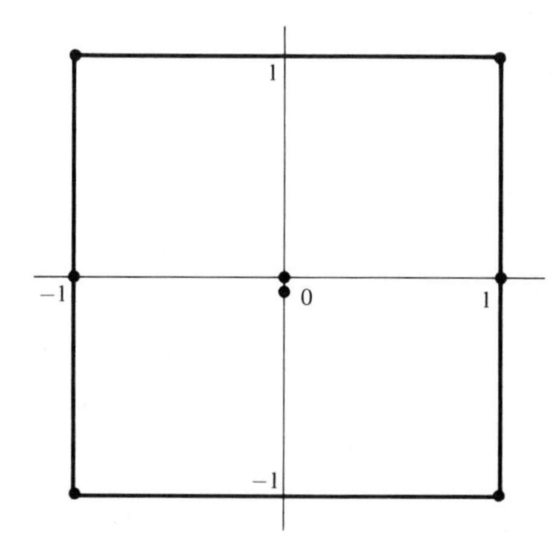

Figure 16.6 2^2 factorial with two points in center.

The coefficients in the model are still estimated by Equation (16.12):

$$b_0 = \frac{\sum X_0 Y}{n} = \frac{39.8}{6} = 6.63$$

$$b_1 = \frac{\sum X_1 Y}{\sum X_1^2} = \frac{1.6}{4} = 0.40$$

$$b_2 = \frac{\sum X_2 Y}{\sum X_2^2} = \frac{-1.8}{4} = -0.45$$

and

$$\text{SS}_{b_0} = 6.63(39.8) = 263.87$$
$$\text{SS}_{b_1} = 0.40(1.6) = 0.64$$
$$\text{SS}_{b_2} = -0.45(-1.8) = 0.81$$

For the sum of squares of the error at (0, 0)

$$\text{SS}_e = (6.6)^2 + (6.8)^2 - \frac{(13.4)^2}{2} = 89.80 - 89.78 = 0.02$$

The analysis is shown in Table 16.8.

Table 16.8 ANOVA for 2^2 + Two Center Points

Source	df	SS	MS
Total	6	265.98	
b_0	1	263.87	236.87
b_1	1	0.64	0.64
b_2	1	0.81	0.81
Residual	3	0.66	—
Error	1	0.02	0.02
Lack of fit	2	0.64	0.32

Testing these effects against error gives

$$b_0\colon F_{1,1} = \frac{236.87}{0.02} = 11{,}843.5$$

$$b_1\colon F_{1,1} = \frac{0.64}{0.02} = 32$$

$$b_2\colon F_{1,1} = \frac{0.81}{0.02} = 40.5$$

$$\text{lack of fit: } F_{2,1} = \frac{0.32}{0.02} = 16$$

With such a small number of degrees of freedom, only b_0 shows significance at the 5-percent level. However, the tests on the b terms are larger than the test on lack of fit, which may indicate that the plane

$$\hat{Y} = 6.63 + 0.40X_1 - 0.45X_2$$

is a fair approximation to the surface where this first experiment was run.

More Complex Surfaces

For more factors—controlled variables—the model for the response variable is more complex, but several designs have been found useful in estimating the coefficients of these surfaces.

When three variables are involved, a first approximation is again a plane or hyperplane of the form

$$Y = B_0X_0 + B_1X_1 + B_2X_2 + B_3X_3 + \varepsilon$$

For this first-order surface, at least four points must be taken to estimate the four B's. Since three dimensions are involved, we might consider a 2^3 factorial. As this design has eight experimental conditions at its vertices, the design often used is a one-half replication of a 2^3 factorial with two or more points at the center of the cube. These six points (two at the center) are sufficient to estimate all four B's and to test for lack of fit of this plane to the surface in three dimensions. After this initial one-half replication of the 2^3, the other half might be run, giving more information for a better fit.

If a plane is not a good fit in two dimensions, a second-order model might be tried. Such a response surface has the form

$$Y = B_0X_0 + B_1X_1 + B_2X_2 + B_{11}X_1^2 + B_{12}X_1X_2 + B_{22}X_2^2 + \varepsilon$$

Here six B's are to be estimated. The simplest design for this model would be a pentagon (five points) plus center points. This would yield six or more responses, and all six B's could be estimated.

In developing these designs, Box and others have found that the calculations can be simplified if the design can be rotated. A *rotable design* is one which has equal predictability in all directions from the center, and the points are at a constant distance from the center. All first-order designs are rotatable, as the square and one-half replication of the cube in the cases above. The simplest second-order design that is rotatable is the pentagon with a point at the center, as given above.

In three dimensions, a second-order surface is given by

$$Y = B_0X_0 + B_1X_1 + B_2X_2 + B_3X_3 + B_{11}X_1^2 + B_{22}X_2^2 + B_{33}X_3^2 + B_{12}X_1X_2 + B_{13}X_1X_3 + B_{23}X_2X_3 + \varepsilon \quad (16.13)$$

which has 10 unknown coefficients. The cube for a three-dimensional model has only eight points, so a special design has been developed, called a *central composite* design. It is a 2^3 factorial with points along each axis at a distance from the center equal to the distance to each vertex. This gives $8 + 6 = 14$ points plus a point at the center makes 15, which is adequate for estimating the B's in Equation (16.13). This design is pictured in Figure 16.7.

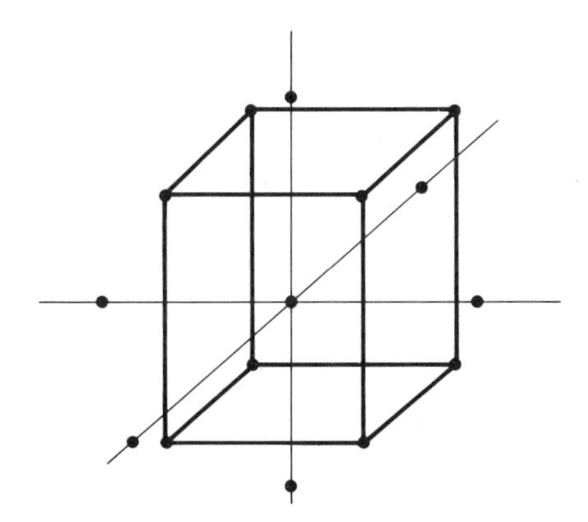

Figure 16.7 Central composite design.

In attempting to determine the equation of the response surface, some concepts in solid analytical geometry are often helpful. If a second-order model in two dimensions is

$$Y = B_0X_0 + B_1X_1 + B_2X_2 + B_{11}X_1^2 + B_{12}X_1X_2 + B_{22}X_2^2 + \varepsilon$$

the shape of this surface can be determined by reducing this equation to what is called *canonical form*. This would be

$$Y = B'_{11}X'^2_1 + B'_{22}X'^2_2$$

This is accomplished by a translation of axes to remove terms in X_1 and X_2, then a rotation of axes to remove the X_1X_2 term. From B'_{11} and B'_{22}, we can determine the shape of the surface, whether it is a sphere, ellipsoid, paraboloid, hyperboloid, or other. In higher dimensions this becomes more complicated, but it can still be useful in describing the response surface.

16.4 EVOLUTIONARY OPERATION (EVOP)

Philosophy

Evolutionary operation is a method of process operation that has a built-in procedure to increase productivity. The technique was developed by Box [5].

This book and Barnett's article [1] should be read in order to understand how the method works. Many chemical companies have reported considerable success using EVOP.

The procedure consists of running a simple experiment, usually a factorial, within the range of operability of a process as it is currently running. It is assumed that the variables to be controlled are measureable and can be set within a short distance of the current settings without disturbing production quality. The idea is to gather data on a response variable, usually yield, at the various points of an experimental design. When one set of data has been taken at all the points, one *cycle* is said to have been completed. One cycle is usually not sufficient to detect any shift in the response, so a second cycle is taken. This continues until the effect of one or more control variables, their interactions, or a change in the mean shows up as significant when compared with a measure of experimental error. This estimate of error is obtained from the cycle data, thus making the experiment self-contained. After a significant increase in yield has been detected, one *phase* is said to have been completed, and at this point a decision is usually made to change the basic operating conditions in a direction that should improve the yield. Several cycles may be necessary before a shift can be detected. The objective here, as with response surfaces, is to move in the direction of an optimum response. Response surface experimentation is primarily a laboratory or research technique; evolutionary operation is a production-line method.

In order to facilitate the EVOP procedure, a simple form has been developed to be used on the production line for each cycle of a 2^2 factorial with a point at the center. In the sections that follow, an example will be run using these forms, and later the details of the form will be developed.

Example 16.3 To illustrate the EVOP procedure, consider a chemical process in which temperature and pressure are varied over short ranges, and the resulting chemical yield is recorded. Since two controlled variables affect the yield, a 2^2 factorial should indicate the effect of each factor as well as a possible interaction between them. By taking a point at the center of a 2^2 factorial, we can also check on a change in the mean (CIM) by comparing this point at the center with the four points around the vertices of the square. If the process should be straddling a maximum, the center point should eventually (after several cycles) be significantly above the peripheral points at the vertices. The standard form for EVOP locates the five points in this design as indicated in Figure 16.8.

By comparing the responses (or average responses) at points 3 and 4 with those at 2 and 5, an effect of variable X_1 may be detected. Likewise, by comparing responses at 3 and 5 with those at 2 and 4, we may assess the X_2 effect. Comparing the responses at 2 and 3 with those at 4 and 5 will indicate an interaction effect, and comparing the responses at 2, 3, 4, and 5 with those at

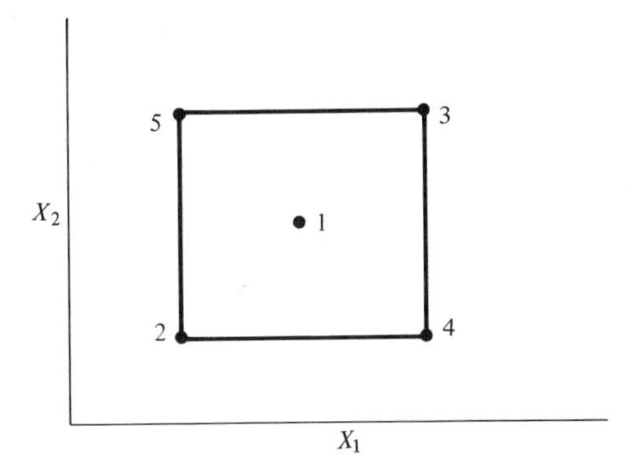

Figure 16.8

1 will indicate a change in the mean, if present. The forms shown in Table 16.9 are filled in with data to show their use, which is self-explanatory.

Not much can be learned from the first cycle unless some separate estimate of standard deviation is available. The second cycle (Table 16.10) will begin to show the method for testing the effects.

Since, in the second cycle, none of the effects are numerically larger than their error limits, the true effect could easily be zero. In this case another cycle must be run. The only items that might need explaining are under the calculation of standard deviation. The range referred to here is the range of the differences (iv), and $f_{k,n}$ is found in a table where $k = 5$ for these five-point designs, and n is the cycle number. Part of such a table shows

$n =$	2	3	4	5	6	7	8
$f_{5,n} =$	0.30	0.35	0.37	0.38	0.39	0.40	0.40

The third cycle is shown in Table 16.11. Here the temperature effect is seen to be significant, with an increase in temperature giving a higher yield. Once a significant effect has been found, the first *phase* of the EVOP procedure has been completed. The average results at this point are usually displayed on an EVOP bulletin board as in Figure 16.9.

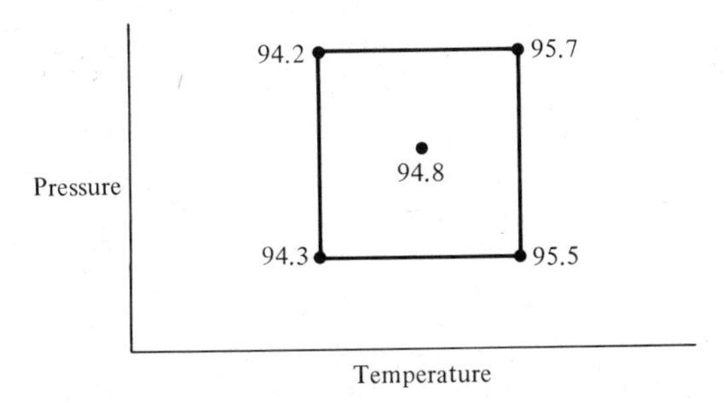

Figure 16.9

Table 16.9 EVOP Work Sheet—First Cycle

5 3 / 1 / 2 4	Cycle: $n = 1$ Response: Yield	Project: 424 Phase: 1 Date: 10/12

Calculation of Averages

Operating Conditions	(1)	(2)	(3)	(4)	(5)
(i) Previous cycle sum					
(ii) Previous cycle average					
(iii) New observations	94.0	94.5	96.5	94.5	94.5
(iv) Differences [(ii) less (iii)]					
(v) New sums	94.0	94.5	96.5	94.5	94.5
(vi) New averages $\bar{Y}_i$	94.0	94.5	96.5	94.5	94.5

Calculation of Standard Deviation

Previous sum $s =$

Previous average $s =$

Range $=$

New $s =$ range $\times f_{k,n} =$

New sum $s =$

New average $s = \dfrac{\text{new sum } s}{n - 1}$

Calculation of Effects

Temperature effect $= \frac{1}{2}(\bar{Y}_3 + \bar{Y}_4 - \bar{Y}_2 - \bar{Y}_5) = 1.00$

Pressure effect $= \frac{1}{2}(\bar{Y}_3 + \bar{Y}_5 - \bar{Y}_2 - \bar{Y}_4) = 1.00$

$T \times P$ interaction effect $= \frac{1}{2}(\bar{Y}_2 + \bar{Y}_3 - \bar{Y}_4 - \bar{Y}_5) = 1.00$

Change in mean effect $= \frac{1}{5}(\bar{Y}_2 + \bar{Y}_3 + \bar{Y}_4 + \bar{Y}_5 - 4\bar{Y}_1) = 0.80$

Calculations of Error Limits

For new average $= \dfrac{2}{\sqrt{n}} s =$

For new effects $= \dfrac{2}{\sqrt{n}} s =$

For change in mean $= \dfrac{1.78}{\sqrt{n}} s =$

Table 16.10 EVOP Work Sheet—Second Cycle

5		3
	1	
2		4

Cycle: $n = 2$
Response: Yield

Project: 424
Phase: 1
Date: 10/12

Calculations of Averages

Operating Conditions	(1)	(2)	(3)	(4)	(5)
(i) Previous cycle sum	94.0	94.5	96.5	94.5	94.5
(ii) Previous cycle average	94.0	94.5	96.5	94.5	94.5
(iii) New observations	96.0	95.0	95.0	96.5	94.0
(iv) Differences [(ii) less (iii)]	−2.0	−0.5	1.5	−2.0	−0.5
(v) New sums	190.0	189.5	191.5	191.0	188.5
(vi) New averages $\bar{Y}_i$	95.0	94.7	95.7	95.5	94.2

Calculation of Standard Deviation

Previous sum s =
Previous average s =
Range = 3.5
New s = range × $f_{k,n}$ = 1.05
New sum s = 1.05

$$\text{New average } s = \frac{\text{new sum } s}{n - 1} = 1.05$$

Calculation of Effects

$$\text{Temperature effect} = \tfrac{1}{2}(\bar{Y}_3 + \bar{Y}_4 - \bar{Y}_2 - \bar{Y}_5) = 1.15$$

$$\text{Pressure effect} = \tfrac{1}{2}(\bar{Y}_3 + \bar{Y}_5 - \bar{Y}_2 - \bar{Y}_4) = -0.15$$

$$T \times P \text{ interaction effect} = \tfrac{1}{2}(\bar{Y}_2 + \bar{Y}_3 - \bar{Y}_4 - \bar{Y}_5) = 0.35$$

$$\text{Change in mean effect} = \tfrac{1}{5}(\bar{Y}_2 + \bar{Y}_3 + \bar{Y}_4 + \bar{Y}_5 - 4\bar{Y}_1) = 0.02$$

Calculation of Error Limits

$$\text{For new average} = \frac{2}{\sqrt{n}}\, s = 1.48$$

$$\text{For new effects} = \frac{2}{\sqrt{n}}\, s = 1.48$$

$$\text{For change in mean} = \frac{1.78}{\sqrt{n}}\, s = 1.32$$

Table 16.11 EVOP Work Sheet—Third Cycle

5 3 1 2 4	Cycle: $n = 3$ Response: Yield	Project: 424 Phase: 1 Date: 10/12

Calculations of Averages

Operating Conditions	(1)	(2)	(3)	(4)	(5)
(i) Previous cycle sum	190.0	189.5	191.5	191.0	188.5
(ii) Previous cycle average	95.0	94.7	95.7	95.5	94.2
(iii) New observations	94.5	93.5	96.0	97.0	94.0
(iv) Differences [(ii) less (iii)]	0.5	1.2	−0.3	−1.5	0.2
(v) New sums	284.5	283.0	287.2	286.5	282.7
(vi) New averages $\bar{Y}_i$	94.8	94.3	95.7	95.5	94.2

Calculation of Standard Deviation

Previous sum $s = 1.05$

Previous average $s = 1.05$

Range $= 2.7$

New s = range $\times f_{k,n} = 0.95$

New sum $s = 2.00$

New average $s = \dfrac{\text{new sum } s}{n - 1} = 1.00$

Calculation of Effects

Temperature effect $= \frac{1}{2}(\bar{Y}_3 + \bar{Y}_4 - \bar{Y}_2 - \bar{Y}_5) = 1.35$*

Pressure effect $= \frac{1}{2}(\bar{Y}_3 + \bar{Y}_5 - \bar{Y}_2 - \bar{Y}_4) = 0.05$

$T \times P$ interaction effect $= \frac{1}{2}(\bar{Y}_2 + \bar{Y}_3 - \bar{Y}_4 - \bar{Y}_5) = -0.15$

Change in mean effect $= \frac{1}{5}(\bar{Y}_2 + \bar{Y}_3 + \bar{Y}_4 + \bar{Y}_5 - 4\bar{Y}_1) = 0.10$

Calculation of Error Limits

For new average $\dfrac{2}{\sqrt{n}} s = 1.16$

For new effects $= \dfrac{2}{\sqrt{n}} s = 1.16$

For change in mean $= \dfrac{1.78}{\sqrt{n}} s = 1.02$

* Significant at 5-percent level.

Now an EVOP committee usually reviews these data and decides whether or not to reset the operating conditions. If they change the operating conditions (point 1), then EVOP is reinstated around this new point and the second phase is begun. EVOP is continued again until significant changes are detected. In fact, it goes on continually, seeking to optimize a process.

This is a very simple example with just two independent variables. The references should be consulted for more complex situations.

EVOP Form Rationale

Most of the steps in the EVOP form above are quite clear. It may be helpful to see where some of the constants come from.

In estimating the standard deviation from the range of differences in step (iv), consider a general expression for these differences

$$D_p = \frac{X_{p1} + X_{p2} + \cdots + X_{p,n-1}}{n-1} - X_{p,n}$$

where D_p represents the difference at any point p of the design. X_{pi} represents the observation at point p in the ith cycle. n is, of course, the number of cycles. Since the variance of a sum equals the sum of the variances for independent variables, and since the variance of a constant times a random variable equals the square of the constant times the variance of the random variable,

$$\sigma_{D_p}^2 = \frac{1}{(n-1)^2}\left[\sigma_{xp1}^2 + \sigma_{xp2}^2 + \cdots + \sigma_{xp,n-1}^2\right] + \sigma_{xp,n}^2$$

Since all these x's represent the same population, their variances are all alike, and then

$$\sigma_D^2 = \frac{1}{(n-1)^2}\left[(n-1)\sigma_x^2\right] + \sigma_x^2 = \frac{n}{n-1}\sigma_x^2$$

and

$$\sigma_D = \sqrt{\frac{n}{n-1}}\,\sigma_x$$

The standard deviation of the population can then be written in terms of the standard deviation of these differences

$$\sigma_x = \sqrt{\frac{n-1}{n}}\,\sigma_D$$

σ_D can now be estimated from the range of these differences R_d. From the quality control field

$$\sigma_D = \frac{R_d}{d_2}$$

where d_2 depends on the number of differences in the range, which is 5 on this form. Here $d_2 = 2.326$ for samples of 5,

$$\sigma_D = \frac{R_d}{2.326}$$

and the standard deviation of the population is estimated by

$$\sigma_x = \sqrt{\frac{n-1}{n}}\frac{R_d}{2.326}$$

The quantity

$$\sqrt{\frac{n-1}{n}}\frac{1}{2.326}$$

is called $f_{k,n}$ in the EVOP form where $k = 5$.

Note that

$$f_{5,2} = \sqrt{\frac{1}{2}}\frac{R_d}{2.326} = 0.30R_d$$

$$f_{5,6} = \sqrt{\frac{5}{6}}\frac{R_d}{2.326} = 0.39R_d$$

which tallies with the table values given.

To determine the error limits for the effects, two standard deviation limits are used, as they represent the approximate 95 percent confidence limits on the parameter being estimated.

For any effect such as

$$E = \tfrac{1}{2}(\bar{Y}_3 + \bar{Y}_4 - \bar{Y}_2 - \bar{Y}_5)$$

its variance would be

$$V_E = \tfrac{1}{4}(\sigma^2_{\bar{Y}_3} + \sigma^2_{\bar{Y}_4} + \sigma^2_{\bar{Y}_2} + \sigma^2_{\bar{Y}_5}) = \tfrac{1}{4}(4\sigma^2_{\bar{Y}}) = \sigma^2_{\bar{Y}} = \frac{\sigma^2_{Y}}{n}$$

and two standard deviation limits on an effect would be

$$\pm 2\frac{\sigma_Y}{\sqrt{n}} \quad \text{estimated by} \quad \pm 2\frac{s}{\sqrt{n}}$$

For the change in mean effect

$$\text{CIM} = \tfrac{1}{5}(\bar{Y}_2 + \bar{Y}_3 + \bar{Y}_4 + \bar{Y}_5 - 4\bar{Y}_1)$$

$$V_{\text{CIM}} = \tfrac{1}{25}(\sigma^2_{\bar{Y}_2} + \sigma^2_{\bar{Y}_3} + \sigma^2_{\bar{Y}_4} + \sigma^2_{\bar{Y}_5} + 16\sigma^2_{\bar{Y}_1})$$

$$= \frac{20}{25}(\sigma^2_{\bar{Y}}) = \frac{20}{25}\frac{\sigma^2_Y}{n}$$

$$\sigma_{\text{CIM}} = \sqrt{\frac{4}{5}}\cdot\frac{\sigma_Y}{\sqrt{n}}$$

and two standard deviation limits would be

$$\pm 2 \cdot \sqrt{\frac{4}{5}} \frac{s}{\sqrt{n}} = \pm 1.78 \frac{s}{\sqrt{n}}$$

as given on the EVOP form.

Example 16.4 This EVOP procedure often has three factors each at two levels. The $2^3 = 8$ experimental points are run as well as two points in the center of the cube, giving ten experimental points per cycle. Usually this design is run in two blocks where the *ABC* interaction is confounded with blocks. The standard form for the location of the ten points is shown in Figure 16.10.

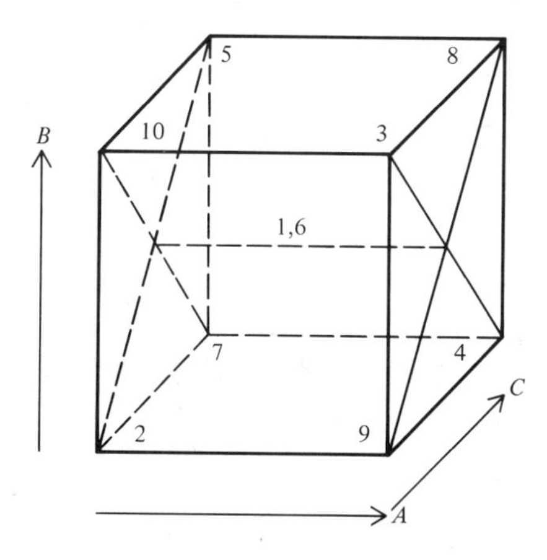

Figure 16.10

The first five points (1 through 5) are run as the first block and the last five points (6 through 10) are run as the second block. Tables 16.12 and 16.13 illustrate this EVOP technique for the second cycle of an example on the mean surface factor for grinder slivers where the factors are drop-out temperature *A*, back-zone temperature *B*, and atmosphere *C*. The worksheets should be self-explanatory as they are based on the design principles of confounding a 2^3 factorial in two blocks of four observations. One could terminate this process with the second cycle as it shows drop-out temperature *A*, back-zone temperature *B*, and their *AB* interaction to be significant in producing changes in the mean surface factor for grinder slivers.

Table 16.12 EVOP Worksheet—Block 1 in 2^3

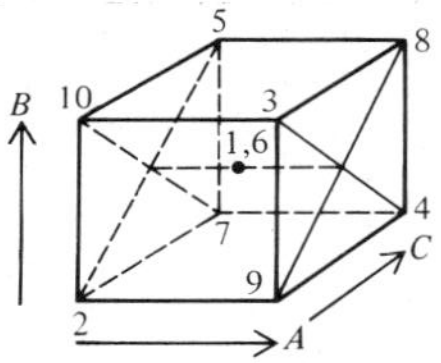

Cycle: $n = 2$
Response: Mean surface factor for grinder slivers
Factors: A = Drop-out temperature 2145° and 2165°
B = Back-zone temperature 2140° and 2160°
C = Atmosphere: Reducing and oxidizing

Project: 478
Phase: 1
Date: 11/6

Calculation of Averages—Block 1

Operating Conditions (Block 1)	(1)	(2)	(3)	(4)	(5)
(i) Previous cycle sum	3.9	0.0	1.6	9.8	4.3
(ii) Previous cycle sum	3.9	0.0	1.6	9.8	4.3
(iii) New observations	5.6	4.0	6.7	15.0	3.4
(iv) Differences [(ii) less (iii)]	−1.7	−4.0	−5.1	−5.2	−0.9
(v) New sums	9.5	4.0	8.3	24.8	7.7
(vi) New averages $\bar{Y}_i$	4.8	2.0	4.2	12.4	3.8

Calculation of Standard Deviation

Previous sum s = (all blocks)
Previous average s =
Range = 4.3
New sum s = range × $f_{k,n}$ = 1.29
New sum s = 1.29 (all blocks)
New average $s = \dfrac{\text{new sum } s}{2n - 3} = 1.29$

Calculation of Effects—Block 1

$A - BC = \frac{1}{2}(\bar{Y}_3 + \bar{Y}_4 - \bar{Y}_2 - \bar{Y}_5) = 5.40$
$B - AC = \frac{1}{2}(\bar{Y}_3 + \bar{Y}_5 - \bar{Y}_2 - \bar{Y}_4) = -3.20$
$C - AB = \frac{1}{2}(\bar{Y}_4 + \bar{Y}_5 - \bar{Y}_2 - \bar{Y}_3) = 5.00$
Change in mean effect $= \frac{1}{5}(\bar{Y}_2 + \bar{Y}_3 + \bar{Y}_4 + \bar{Y}_5 - 4\bar{Y}_1) = 0.64$

Table 16.13 EVOP Worksheet—Block 2 in 2^3

Cycle: $n = 2$

Calculation of Averages—Block 2						Calculation of Standard Deviation
Operating Conditions (Block 2)	(6)	(7)	(8)	(9)	(10)	
(i) Previous cycle sum	4.2	4.4	2.8	10.1	0.0	Previous sum s = 1.29 (all blocks)
(ii) Previous cycle average	4.2	4.4	2.8	10.1	0.0	Previous average s = 1.29
(iii) New observations	3.2	3.4	6.6	15.0	1.9	Range = 5.9
(iv) Differences [(ii) less (iii)]	1.0	0.9	−3.8	−4.9	−1.9	New sum s = range × $f_{k,n}$ = 1.77
(v) New sums	7.4	7.7	9.4	25.1	1.9	New sum s = 3.06 (all blocks)
(vi) New averages $\bar{Y}_i$	3.7	3.8	4.7	12.6	1.0	New average $s = \dfrac{\text{new sum } s}{2n - 2} = 1.53$

Calculation of Effects—Block 2

$$A + BC = \tfrac{1}{2}(\bar{Y}_8 + \bar{Y}_9 - \bar{Y}_7 - \bar{Y}_{10}) = 6.25$$

$$B + AC = \tfrac{1}{2}(\bar{Y}_8 + \bar{Y}_{10} - \bar{Y}_7 - \bar{Y}_9) = -5.35$$

$$C + AB = \tfrac{1}{2}(\bar{Y}_7 + \bar{Y}_8 - \bar{Y}_9 - \bar{Y}_{10}) = -2.55$$

$$\text{Change in mean effect} = \tfrac{1}{5}(\bar{Y}_7 + \bar{Y}_8 + \bar{Y}_9 + \bar{Y}_{10} - 4\bar{Y}_6) = 1.46$$

Calculation of Error Limits

$$\text{For new averages} = \frac{2s}{\sqrt{n}} = \pm 2.16$$

$$\text{For new effects} = \frac{1.42s}{\sqrt{n}} = \pm 1.53$$

$$\text{For change in mean} = \frac{1.26s}{\sqrt{n}} = \pm 1.36$$

Calculation of Effects—Both Blocks

$$A = \tfrac{1}{2}[(A + BC) + (A - BC)] = 5.82^*$$
$$B = \tfrac{1}{2}[(B + AC) + (B - AC)] = -4.28^*$$
$$C = \tfrac{1}{2}[(C + AB) + (C - AB)] = 1.23$$
$$\text{Change in mean effect} = \tfrac{1}{2}(\text{CIM}_1 + \text{CIM}_2) = 1.05$$

$$AB = \tfrac{1}{2}[(C + AB) - (C - AB)] = -3.78^*$$
$$AC = \tfrac{1}{2}[(B + AC) - (B - AC)] = -1.08$$
$$BC = \tfrac{1}{2}[(A + BC) - (A - BC)] = -0.42$$

* One asterisk indicates significance at the 5 percent level.

16.5 ANALYSIS OF ATTRIBUTE DATA

Philosophy

Two assumptions used in the application of the analysis of variance technique are that (1) the response variable is normally distributed and (2) the variances of the experimental errors are equal throughout the experiment. In practice it is often necessary to deal with attribute data where the response variable is either 0 or 1. In such cases one often records the number of occurrences of a particular phenomenon or the percentage of such occurrences.

It is well known that the number of occurrences per unit such as defects per piece, errors per page, customers per unit time, often follow a Poisson distribution where such response variables are not only nonnormal but variances and means are equal. When proportions or percentages are used as the response variable, the data are binomial and again variances are related to means and the basic assumptions of ANOVA do not hold. Some studies have shown that lack of normality is not too serious in applying the ANOVA technique, but most statisticians recommend that a transformation be made on the original data if it is known to be nonnormal. Recommendations for proper transformations are given in Davies [7].

Another technique, called *factorial chi-square*, has been found very useful in treating attribute data in industrial problems. This technique is described by Batson [2] who gives several examples of its application to actual problems. While this factorial chi-square technique may not be as precise as a regular ANOVA on transformed data, its simplicity makes it well worth consideration for many applied problems.

To illustrate the techniques mentioned above, consider the following example from a consulting-firm study.

Example 16.4 Interest centered on undesireable marks on steel samples from a grinding operation. The response variable was simply whether or not these marks occurred. Four factors were believed to affect these marks: blade size A, centering B, leveling C, and speed D. Each of these four factors was set at two levels giving a 2^4 factorial experiment. Each of the 16 treatment combinations was run in a completely randomized order, and 20 steel samples were produced under each of the 16 experimental conditions. The number of damaged samples in each group of 20 is the recorded variable in Table 16.14.

If the above responses are each divided by 20, the proportions of damaged samples can be determined. Since proportions follow a binomial distribution, an arc sine transformation is appropriate (see [7]). Taking the arc sin $\sqrt{p}$ as the response variable, Yates' method can be used to determine the sums of squares as shown in Table 16.15.

If the three- and four-way interactions are assumed to be nonexistent, the resulting ANOVA table is shown as Table 16.16.

Table 16.14 Damaged Steel Samples in 2^4 Experiment

		Factor A			
		0		1	
		Factor B		Factor B	
Factor C	Factor D	0	1	0	1
0	0	0	0	16	20
	1	0	0	10	20
1	0	0	0	10	14
	1	1	0	12	20

Table 16.15 Yates Method on Transformed Data of Steel Sample Experiment

Treatment	Proportion p	Transformed Variable $X \equiv \arcsin \sqrt{p}$	(1)	(2)	(3)	(4)	SS
(1)	0.00	0.00	1.11	2.68	4.46	9.51	—
a	0.80	1.11	1.57	1.78	5.05	9.05	5.12
b	0.00	0.00	0.79	2.36	4.46	1.89	0.22
ab	1.00	1.57	0.99	2.69	4.59	2.35	0.35
c	0.00	0.00	0.79	2.68	0.66	−1.23	0.09
ac	0.50	0.79	1.57	1.78	1.23	−1.03	0.06
bc	0.00	0.00	1.12	2.36	0.66	−0.59	0.03
abc	0.70	0.99	1.57	2.23	1.69	−0.13	0.01
d	0.00	0.00	1.11	0.46	−0.90	0.59	0.03
ad	0.50	0.79	1.57	0.20	−0.33	0.13	0.01
bd	0.00	0.00	0.79	0.78	−0.90	0.57	0.02
abd	1.00	1.57	0.99	0.45	−0.13	1.03	0.06
cd	0.05	0.23	0.79	0.46	−0.26	0.57	0.02
acd	0.60	0.89	1.57	0.20	−0.33	0.77	0.04
bcd	0.00	0.00	0.66	0.78	−0.26	−0.07	0.00
$abcd$	1.00	1.57	1.57	0.91	0.13	0.39	0.01

This analysis shows the highly significant blade size effect which is obvious from a glance at Table 16.14. It also shows a significant centering effect and a blade size–centering interaction which is not as obvious from Table 16.14.

To apply the factorial chi-square technique to this same example, Table 16.17 is filled in. The plus and minus signs simply give the proper weights for the main effects and two-way interactions and T is the value of the corresponding contrast. D is the total number of samples in the total experiment;

Table 16.16 ANOVA for Steel Sample Experiment

Source	df	SS	MS
A	1	5.12	5.12***
B	1	0.22	0.22*
AB	1	0.35	0.35**
C	1	0.09	0.09
AC	1	0.06	0.06
BC	1	0.03	0.03
D	1	0.03	0.03
AD	1	0.01	0.01
BD	1	0.02	0.02
CD	1	0.02	0.02
Error, pooled	5	0.12	0.02

* One asterisk indicates significance at the 5-percent level; two, the 1-percent level; and three, the 0.1-percent level.

here 20 per cell or $20 \times 2^4 = 320$ samples. The chi-square (χ^2) value is determined by the formula

$$\chi^2_{[1]} = \frac{N^2}{(S)(F)} \times \frac{T^2}{D} \tag{16.14}$$

Here N is the total number of items in the whole experiment and S is the number of occurrences (successes) in the N. F is the number of non-occurrences (or failures). In the example $N = 320$ samples, $S = 123$ with undesirable marks. For factor A, then

$$\chi^2_{[1]} = \frac{(320)^2}{(123)(197)}(45.75)$$

$$= (4.23)(45.75) = 193.52$$

The 4.23 is constant for these data. The resulting chi-square values are then compared to χ^2 with 1 df at the significance level desired. For the 5-percent level, $\chi^2_{[1]} = 3.84$ and again factors A, B, and AB are significant.

As seen in Table 16.17, the calculations for this technique are quite simple and the results are consistent with those using a more precise method. Extensions to problems other than 2^n factorials can be found in Batson [2].

Table 16.17 Factorial Chi-Square Work Sheet for Table 16.14

Blade *A*	0								1														
Center *B*	0				1				0				1										
Level *C*	0		1		0		1		0		1		0		1								
Speed *D*	0	1	0	1	0	1	0	1	0	1	0	1	0	1	0	1	Σ +	Σ −	T	T^2	D	T^2/D	χ^2
$S/20$	0	0	0	1	0	0	0	0	16	10	10	12	20	20	14	20							
Main effects																							
Blade *A*	−	−	−	−	−	−	−	−	+	+	+	+	+	+	+	+	122	−1	121	14,641	320	45.75	193.52*
Center *B*	−	−	−	−	+	+	+	+	−	−	−	−	+	+	+	+	74	−49	25	625	320	1.95	8.25*
Level *C*	−	−	+	+	−	−	+	+	−	−	+	+	−	−	+	+	57	−66	−9	81	320	0.25	1.06
Speed *D*	−	+	−	+	−	+	−	+	−	+	−	+	−	+	−	+	63	−60	3	9	320	0.03	0.12
Interactions																							
AB	+	+	+	+	−	−	−	−	−	−	−	−	+	+	+	+	75	−48	27	729	320	2.28	9.65*
AC	+	+	−	−	+	+	−	−	−	−	+	+	−	−	+	+	56	−67	−9	81	320	0.25	1.06
AD	+	−	+	−	+	−	+	−	−	+	−	+	−	+	−	+	62	−61	1	1	320	0.01	0.04
BC	+	+	−	−	−	−	+	+	+	+	−	−	−	−	+	+	60	−63	−3	9	320	0.03	0.12
BD	+	−	+	−	−	+	−	+	+	−	+	−	−	+	−	+	66	−57	9	81	320	0.25	1.06
CD	+	−	−	+	+	−	−	+	+	−	−	+	+	−	−	+	69	−54	+15	225	320	0.70	2.96

* One asterisk indicates significance at the 5-percent level.

PROBLEMS

16.1 One study reports on the breaking strength Y of samples of seven different starch films where the thickness of each film X was also recorded as a covariate. Results rounded off considerably are as follows:

Source	df	$\sum y^2$	$\sum xy$	$\sum x^2$	Adjusted $\sum y^2$	df	MS
Between starches		6,000,000	50,000	500			
Within starches	100						
Totals		8,000,000	60,000	600			

1. Complete this table and set up a test for the effect of starch film on breaking strength both before and after adjustment for thickness.
2. Write the mathematical model for this experiment, briefly describing each term.
3. State the assumptions implied in the tests of hypotheses made above and describe briefly how you would test their validity.

16.2 In studying the effect of two different teaching methods, the gain score of each pupil in a public school test was used as the response variable. The pretest scores are also available. Run a covariance analysis on the data below to determine whether or not the new teaching method did show an improvement in average achievement as measured by this public school test.

Method I			Method II		
Pupil	Pretest Score	Gain	Pupil	Pretest Score	Gain
1	28	5	1	24	7
2	43	−8	2	40	9
3	36	4	3	41	4
4	35	1	4	33	4
5	31	4	5	45	5
6	34	1	6	41	−2
7	33	4	7	41	6
8	38	8	8	33	11
9	39	1	9	41	−1
10	44	−1	10	30	15
11	36	−7	11	45	4
12	21	2	12	50	2
13	34	−1	13	42	9
14	33	1	14	47	3
15	25	7	15	43	−2

Method I			Method II		
Pupil	Pretest Score	Gain	Pupil	Pretest Score	Gain
16	43	1	16	34	2
17	34	3	17	44	7
18	40	5	18	43	4
19	36	4	19	28	3
20	30	9	20	46	−1
21	35	−6	21	43	2
22	38	0	22	46	−4
23	31	−14	23	21	4
24	41	4			
25	35	10			
26	27	10			
27	45	1			

16.3 Data below are for five individuals who have been subjected to four conditions or treatments represented by the groups or lots 1–4. Y represents some measure of an individual supposedly affected by the variations in the treatments of the four lots. X represents another measure of an individual which may affect the value of Y even in the absence of the four treatments. The problem is to determine whether or not the Y means of the four groups differ significantly from each other after the effects of the X variable have been removed.

Groups or Lots							
1		2		3		4	
X	Y	X	Y	X	Y	X	Y
29	22	15	30	16	12	5	23
20	22	9	32	31	8	25	25
14	20	1	26	26	13	16	28
21	24	6	25	35	25	10	26
6	12	19	37	12	7	24	23

16.4 Consider a two-way classified design with more than one observation per cell.
1. Show the F test for one of the main effects in a covariance analysis if both factors are fixed.
2. Show the F test for the interaction in a covariance analysis if one factor is fixed and the other one is random.

16.5 For Example 16.4 in the text, another response variable was measured—the mean surface factor for heat slivers. Results of the first three cycles at the ten experimental points gave

Cycle	(1)	(2)	(3)	(4)	(5)	(6)	(7)	(8)	(9)	(10)
1	4.0	6.8	5.9	4.5	4.3	4.2	7.8	8.3	11.0	5.9
2	5.2	7.2	6.1	3.7	4.0	4.5	7.0	8.1	12.5	6.3
3	4.8	8.0	7.0	3.6	3.8	4.3	6.5	7.8	12.7	7.0

Set up EVOP worksheets for these data and comment on your results after three cycles.

16.6 Data on yield in pounds for varying percent concentration at levels 29, 30, and 31 percent, and varying power at levels 400, 450, and 500 watts gave the following results in four cycles at the five points in a 2^2 design:

Cycle	(1)	(2)	(3)	(4)	(5)
1	477	472	411	476	372
2	469	452	430	468	453
3	465	396	375	468	292
4	451	469	363	432	460

Analyze by EVOP methods.

16.7 The response variable in the data below is the number of defective knife handles in samples of 72 knives where the factors are machines, lumber grades, and replications. Analyze after making a suitable transformation of the data.

Lumber Grade	Replication	Machine 1	Machine 2
A	I	8	4
	II	4	6
B	I	9	3
	II	10	4

16.8 Analyze the data of Problem 16.7 using factorial chi square. Compare with the results of Problem 16.7.

17
Summary

Throughout this book the three phases of an experiment have been emphasized: the experiment, the design, and the analysis. At the end of most chapters an outline of the designs considered up to that point has been presented. It may prove useful to the reader to have a complete outline as a summary of the designs presented in Chapters 1–15.

This is not the only way such an outline could be constructed, but it represents an attempt to see each design as part of the overall picture of designed experiments. A look at the outline will reveal that the same design is often used for an entirely different experiment, which simply illustrates the fact that much care must be taken to spell out the experiment and its design or method of randomization before an experiment is performed.

For each experiment, the chapter reference in this book is given.

Experiment	*Design*	*Analysis*	*Chapter Reference*
I. Single factor			
	1. Completely randomized	1. One-way ANOVA	Chapter 3
	$Y_{ij} = \mu + \tau_j + \varepsilon_{ij}$		
	2. Randomized block	2.	
	$Y_{ij} = \mu + \beta_i + \tau_j + \varepsilon_{ij}$		
	a. Complete	*a.* Two-way ANOVA	Chapter 4
	b. Incomplete, balanced	*b.* Special ANOVA	
	c. Incomplete, general	*c.* General regression method	
	3. Latin square	3.	
	$Y_{ijk} = \mu + \beta_i + \tau_j + \gamma_k + \varepsilon_{ijk}$		
	a. Complete	*a.* Three-way ANOVA	Chapter 5
	b. Incomplete, Youden square	*b.* Special ANOVA (like 2*b*)	
	4. Graeco-Latin square	4. Four-way ANOVA	Chapter 5
	$Y_{ijkm} = \mu + \beta_i + \tau_j + \gamma_k + \omega_m + \varepsilon_{ijkm}$		
II. Two or more factors			
A. Factorial (crossed)			
	1. Completely randomized	1.	
	$Y_{ijk} = \mu + A_i + B_j + AB_{ij} + \varepsilon_{k(ij)}, \cdots$ for more factors		
	a. General case	*a.* ANOVA with interactions	Chapter 6
	b. 2^n case	*b.* Yates method or general ANOVA; use: (1), *a*, *b*, *ab*, $\cdots$	Chapter 7

Experiment	Design	Analysis	Chapter Reference
	c. 3^n case	*c*. General ANOVA; use 00, 10, 20, 01, 011, $\cdots$ and $A \times B = AB + AB^2, \cdots$ for interaction	Chapter 9
	2. Randomized block	2.	
	a. Complete $Y_{ijk} = \mu + R_k + A_i + B_j + AB_{ij} + \varepsilon_{ijk}$	*a*. Factorial ANOVA with replications R_k	Chapter 12
	b. Incomplete, confounding:	*b*.	
	i. Main effect—split-plot $Y_{ijk} = \mu + \underbrace{R_i + A_j + RA_{ij}}_{\text{whole plot}} + \underbrace{B_k + RB_{ik} + AB_{jk} + RAB_{ijk}}_{\text{split-plot}}$	i. Split-plot ANOVA	Chapter 13
	ii. Interactions in 2^n and 3^n	ii.	Chapter 14
	(1) Several replications $Y_{ijkq} = \mu + R_i + \beta_j + R\beta_{ij} + A_k + B_q + AB_{kq} + \varepsilon_{ijkq}$	(1) Factorial ANOVA with replications R_i and blocks β_j or confounded interaction	

Experiment	Design	Analysis	Chapter Reference
	(2) One replication only $Y_{ijk} = \mu + \beta_i + A_j + B_k + AB_{jk}, \cdots$	(2) Factorial ANOVA with blocks β_i or confounded interaction	Chapter 14
	(3) Fractional replication—aliases $Y_{ijk} = \mu + A_i + B_j + AB_{ij}, \cdots$	(3) Factorial ANOVA with aliases or aliased interactions	Chapter 15
	(a) in 2^n		
	(b) in 3^n; $n = 3$: Latin square		
	3. Latin square	3.	
	a. Complete $Y_{ijkm} = \mu + R_k + \gamma_m + A_i + B_j + AB_{ij} + \varepsilon_{ijkm}$	a. Factorial ANOVA with replications and positions	Chapter 12
B. Nested (hierarchical)			
	1. Completely randomized $Y_{ijk} = \mu + A_i + B_{j(i)} + \varepsilon_{k(ij)}$	1. Nested ANOVA	Chapter 11
	2. Randomized block	2.	
	a. Complete $Y_{ijk} = \mu + R_k + A_i + B_{j(i)} + \varepsilon_{ijk}$	a. Nested ANOVA with blocks R_k	Chapters 4 and 11
	3. Latin square	3.	
	a. Complete $X_{ijkm} = \mu + R_k + \gamma_m + A_i + B_{j(i)} + \varepsilon_{ijkm}$	a. Nested ANOVA with blocks and positions	Chapters 5 and 11

Experiment	*Design*	*Analysis*	*Chapter Reference*
C. Nested factorial			
	1. Completely randomized $Y_{ijkm} = \mu + A_i + B_{j(i)} + C_k + AC_{ik} + BC_{kj(i)} + \varepsilon_{m(ijk)}$	1. Nested factorial ANOVA	Chapter 11
	2. Randomized block	2.	
	a. Complete $Y_{ijkm} = \mu + R_k + A_i + B_{j(i)} + C_m + AC_{im} + BC_{mj(i)} + \varepsilon_{ijkm}$	*a*. Nested factorial ANOVA with blocks R_k	Chapters 4 and 11
	3. Latin square	3.	
	a. Complete $Y_{ijkmq} = \mu + R_k + \gamma_m + A_i + B_{j(i)} + C_q + AC_{iq} + BC_{qj(i)} + \varepsilon_{ijkmq}$	*a*. Nested factorial ANOVA with blocks and positions	Chapters 5 and 11

Glossary of Terms

Alias. An effect in a fractionally replicated design which "looks like" or cannot be distinguished from another effect.

Alpha (α). Size of the type I error or probability of rejecting a hypothesis when true.

Beta (β). Size of the type II error or probability of accepting a hypothesis when some alternative hypothesis is true.

Canonical form. Form of a second-degree response-surface equation which allows determination of the type of surface.

Completely randomized design. A design in which all treatments are assigned to the experimental units in a completely random manner.

Confidence limits. Two values between which a parameter is said to lie with a specified degree of confidence.

Confounding. An experimental arrangement in which certain effects cannot be distinguished from others. One such effect is usually blocks.

Consistent estimator. An estimator of a parameter whose value comes closer to the parameter as the sample size is increased.

Contrast. A linear combination of treatment totals or averages where the sum of the coefficients is zero.

Correlation coefficient (Pearson r). The square root of the proportion of total variation accounted for by linear regression.

Correlation index R. The square root of the proportion of total variation accounted for by the regression equation of the degree being fitted to the data.

Covariance analysis. Technique for removing the effect of one or more variables on the response variable before assessing the effect of treatments on the response variable.

Critical region. A set of values of a test statistic where the hypothesis under test is rejected.

Defining contrast. An expression that indicates which effects are confounded with blocks in a factorial design that is confounded.

Effect of a factor. The change in response produced by a change in level of the factor.

Errors. Type I: Rejecting a hypothesis when true. Type II: Accepting a hypothesis when false.

Eta squared η^2. Proportion of total variation accounted for by a regression line through all average responses for each value of the independent variable.

Evolutionary operation. An experimental procedure for collecting information to improve a process without disturbing production.

Expected value of a statistic. The average value of a statistic if it were calculated from an infinite number of equal-sized samples from a given population.

Factorial chi square. Technique for handling attribute data in an analysis of variance.

Factorial experiment. An experiment in which all levels of each factor in the experiment are combined with all levels of every other factor.

Fractional replication. An experimental design in which only a fraction of a complete factorial is run.

Graeco-Latin square. An experimental design in which four factors are so arranged that each level of each factor is combined only once with each level of the other three factors.

Incomplete block design. A randomized block design in which not all treatment combinations can be included in one block. Such a design is called a *balanced* design if each pair of treatments occurs together the same number of times.

Interaction. An interaction between two factors means that a change in response between levels of one factor is not the same for all levels of the other factor.

Latin square. An experimental design in which each level of each factor is combined only once with each level of two other factors.

Least squares method. Method of assigning estimates of parameters in a model such that the sum of the squares of the errors is minimized.

Mean. Of a sample

$$\bar{Y} = \sum_{i=1}^{n} Y_i/n$$

Of a population

$$\mu = E(Y).$$

Mean square. An unbiased estimate of a population variance. Determined by dividing a sum of squares by its degree of freedom.

Minimum variance estimate. If two or more estimates $u_1, u_2, \cdots, u_k$ are made of the same parameter θ, the estimate with the smallest variance is called the minimum variance estimate.

Mixed model. The model of a factorial experiment in which one or more factors are at fixed levels and at least one factor is at random levels.

Nested experiment. An experiment in which the levels of one factor are chosen within the levels of another factor.

Nested factorial experiment. An experiment in which some factors are factorial or crossed with others and some factors are nested within others.

Operating characteristic curve. A plot of the probability of acceptance of a hypothesis under test versus the true parameter of the population.

Orthogonal contrasts. Two contrasts are said to be orthogonal if the products of their corresponding coefficients add to zero.

Orthogonal polynomials. Sets of polynomials used in regression analysis such that each polynomial in the set is orthogonal to all other polynomials in the set.

Parameter. A characteristic of a population, such as the population mean or variance.

Point estimate. A single statistic used to estimate a parameter.

Power of a test. A plot of the probability of rejection of a hypothesis under test versus the true parameter. It is the complement of the operating characteristic curve.

Principal block. The block in a confounded design that contains the treatment combination in which all factors are at their lowest level.

Random model. A model of a factorial experiment in which the levels of the factor are chosen at random.

Random sample. A sample in which each member of the population sampled has an equal chance of being selected in this sample.

Randomized block design. An experimental design in which the treatment combinations are randomized within a block and several blocks are run.

Rotatable design. An experimental design that has equal predictive power in all directions from a center point, and in which all experimental points are equidistant from this center point.

Split-plot design. An experimental design in which a main effect is confounded with blocks due to the practical necessities of the order of experimentation.

Standard error of estimate. The standard deviation of errors of estimate around a least squares fitted regression model.

Statistic u. A measure computed from a sample.

Statistical hypothesis H_0. An assumption about a population being sampled.

Statistical inference. Inferring something about a population of measures from a sample of that population.

Statistics. A tool for decision making in the light of uncertainty.

Steepest ascent method. A method of sequential experiments that will direct the experimenter toward the optimum of a response surface.

Sum of squares SS. The sum of the squares of deviations of a random variable from its mean,

$$\text{SS} = \sum_{i=1}^{n} (Y_i - \bar{Y})^2$$

Test of a hypothesis. A rule by which a hypothesis is accepted or rejected.

Test statistic. A statistic used to test a hypothesis.

Treatment combination. A given combination showing the levels of all factors to be run for that set of experimental conditions.

Unbiased statistic. A statistic whose expected value equals the parameter it is estimating.

Variance. Of a sample

$$s^2 = \frac{\sum_{i=1}^{n} (Y_i - \bar{Y})^2}{n - 1} = \frac{\text{SS}}{df}$$

Of a population

$$\sigma^2 = E(Y - \mu)^2.$$

Youden square. An incomplete Latin square.

References

[1] Barnett, E. H., "Introduction to Evolutionary Operation," *Industrial and Engineering Chemistry*, vol. 52 (June 1960), p. 500.

[2] Batson, H. C., "Applications of Factorial Analysis to Experiments in Chemistry," *National Convention Transactions of the American Society for Quality Control* (1956), pp. 9–23.

[3] Bennett, C. A., and N. L. Franklin, *Statistical Analysis in Chemistry and the Chemical Industry*. New York: John Wiley & Sons, Inc. (1954).

[4] *Biometrics*, vol. 13 (Sept. 1957), pp. 261–405.

[5] Box, G. E. P., and N. R. Draper, *Evolutionary Operation: A Statistical Method for Process Improvement*. New York: Wiley, 1969.

[6] Burr, I. W., *Engineering Statistics and Quality Control*. New York: McGraw-Hill Book Company (1953).

[7] Davies, O. L., *Design and Analysis of Industrial Experiments*. New York: Hafner Publishing Company (1954).

[8] Dixon, W. J., and F. J. Massey, *An Introduction to Statistical Analysis* (2nd Ed.). New York: McGraw-Hill Book Company (1957).

[9] Fisher, R. A., and F. Yates, *Statistical Tables for Biological, Agricultural and Medical Research* (4th Ed.). Edinburgh and London: Oliver & Boyd, Ltd. (1953).

[10] Hunter, J. S., "Determination of Optimum Operating Conditions by Experimental Methods," *Industrial Quality Control* (Dec.–Feb. 1958–1959).

[11] Kempthorne, O., *The Design and Analysis of Experiments*. New York: John Wiley & Sons, Inc. (1952).

[12] Keuls, M., "The Use of the Studentized Range in Connection with an Analysis of Variance," *Euphytica*, vol. 1 (1952), pp. 112–122.

[13] McCall, C. H., Jr., "Linear Contrasts, Parts I, II, and III," *Industrial Quality Control* (July–Sept., 1960).

[14] Miller, L. D., "An Investigation of the Machinability of Malleable Iron Using Ceramic Tools," unpublished MSIE thesis, Purdue University (1959).

[15] Ostle, B., *Statistics in Research* (2nd Ed.). Iowa State University Press (1963).

[16] Owen, D. B., *Handbook of Statistical Tables.* Boston: Addison-Wesley Publishing Company, Inc. (1962).

[17] Scheffe, H., "A Method for Judging All Contrasts in the Analysis of Variance," *Biometrics*, vol. XL (June, 1953).

[18] Snedecor, G. W., and W. C. Cochran, *Statistical Methods.* 6th Edition. Iowa State University Press (1967).

[19] Wortham, A. W., and T. E. Smith, *Practical Statistics in Experimental Design.* Dallas: Dallas Publishing House (1960).

[20] Yates, F., *Design and Analysis of Factorial Experiments.* London: Imperial Bureau of Soil Sciences (1937).

Table A Areas under the Normal Curve* (Proportion of Total Area under the Curve from $-\infty$ to Designated Z Value)

Z	0.09	0.08	0.07	0.06	0.05	0.04	0.03	0.02	0.01	0.00
−3.5	0.00017	0.00017	0.00018	0.00019	0.00019	0.00020	0.00021	0.00022	0.00022	0.00023
−3.4	0.00024	0.00025	0.00026	0.00027	0.00028	0.00029	0.00030	0.00031	0.00033	0.00034
−3.3	0.00035	0.00036	0.00038	0.00039	0.00040	0.00042	0.00043	0.00045	0.00047	0.00048
−3.2	0.00050	0.00052	0.00054	0.00056	0.00058	0.00060	0.00062	0.00064	0.00066	0.00069
−3.1	0.00071	0.00074	0.00076	0.00079	0.00082	0.00085	0.00087	0.00090	0.00094	0.00097
−3.0	0.00100	0.00104	0.00107	0.00111	0.00114	0.00118	0.00122	0.00126	0.00131	0.00135
−2.9	0.0014	0.0014	0.0015	0.0015	0.0016	0.0016	0.0017	0.0017	0.0018	0.0019
−2.8	0.0019	0.0020	0.0021	0.0021	0.0022	0.0023	0.0023	0.0024	0.0025	0.0026
−2.7	0.0026	0.0027	0.0028	0.0029	0.0030	0.0031	0.0032	0.0033	0.0034	0.0035
−2.6	0.0036	0.0037	0.0038	0.0039	0.0040	0.0041	0.0043	0.0044	0.0045	0.0047
−2.5	0.0048	0.0049	0.0051	0.0052	0.0054	0.0055	0.0057	0.0059	0.0060	0.0062
−2.4	0.0064	0.0066	0.0068	0.0069	0.0071	0.0073	0.0075	0.0078	0.0080	0.0082
−2.3	0.0084	0.0087	0.0089	0.0091	0.0094	0.0096	0.0099	0.0102	0.0104	0.0107
−2.2	0.0110	0.0113	0.0116	0.0119	0.0122	0.0125	0.0129	0.0132	0.0136	0.0139
−2.1	0.0143	0.0146	0.0150	0.0154	0.0158	0.0162	0.0166	0.0170	0.0174	0.0179
−2.0	0.0183	0.0188	0.0192	0.0197	0.0202	0.0207	0.0212	0.0217	0.0222	0.0228
−1.9	0.0233	0.0239	0.0244	0.0250	0.0256	0.0262	0.0268	0.0274	0.0281	0.0287
−1.8	0.0294	0.0301	0.0307	0.0314	0.0322	0.0329	0.0336	0.0344	0.0351	0.0359
−1.7	0.0367	0.0375	0.0384	0.0392	0.0401	0.0409	0.0418	0.0427	0.0436	0.0446
−1.6	0.0455	0.0465	0.0475	0.0485	0.0495	0.0505	0.0516	0.0526	0.0537	0.0548
−1.5	0.0559	0.0571	0.0582	0.0594	0.0606	0.0618	0.0630	0.0643	0.0655	0.0668
−1.4	0.0681	0.0694	0.0708	0.0721	0.0735	0.0749	0.0764	0.0778	0.0793	0.0808
−1.3	0.0823	0.0838	0.0853	0.0869	0.0885	0.0901	0.0918	0.0934	0.0951	0.0968
−1.2	0.0985	0.1003	0.1020	0.1038	0.1057	0.1075	0.1093	0.1112	0.1131	0.1151
−1.1	0.1170	0.1190	0.1210	0.1230	0.1251	0.1271	0.1292	0.1314	0.1335	0.1357
−1.0	0.1379	0.1401	0.1423	0.1446	0.1469	0.1492	0.1515	0.1539	0.1562	0.1587
−0.9	0.1611	0.1635	0.1660	0.1685	0.1711	0.1736	0.1762	0.1788	0.1814	0.1841
−0.8	0.1867	0.1894	0.1922	0.1949	0.1977	0.2005	0.2033	0.2061	0.2090	0.2119
−0.7	0.2148	0.2177	0.2207	0.2236	0.2266	0.2297	0.2327	0.2358	0.2389	0.2420
−0.6	0.2451	0.2483	0.2514	0.2546	0.2578	0.2611	0.2643	0.2676	0.2709	0.2743
−0.5	0.2776	0.2810	0.2843	0.2877	0.2912	0.2946	0.2981	0.3015	0.3050	0.3085
−0.4	0.3121	0.3156	0.3192	0.3228	0.3264	0.3300	0.3336	0.3372	0.3409	0.3446
−0.3	0.3483	0.3520	0.3557	0.3594	0.3632	0.3669	0.3707	0.3745	0.3783	0.3821
−0.2	0.3859	0.3897	0.3936	0.3974	0.4013	0.4052	0.4090	0.4129	0.4168	0.4207
−0.1	0.4247	0.4286	0.4325	0.4364	0.4404	0.4443	0.4483	0.4522	0.4562	0.4602
−0.0	0.4641	0.4681	0.4721	0.4761	0.4801	0.4840	0.4880	0.4920	0.4960	0.5000

* Adapted from E. L. Grant, *Statistical Quality Control* (2nd Ed.). New York: McGraw-Hill Book Company, Inc. (1952), Table A, pp. 510–511. Reproduced by permission of the publisher.

Table A (*Continued*)

z	0.00	0.01	0.02	0.03	0.04	0.05	0.06	0.07	0.08	0.09
+0.0	0.5000	0.5040	0.5080	0.5120	0.5160	0.5199	0.5239	0.5279	0.5319	0.5359
+0.1	0.5398	0.5438	0.5478	0.5517	0.5557	0.5596	0.5636	0.5675	0.5714	0.5753
+0.2	0.5793	0.5832	0.5871	0.5910	0.5948	0.5987	0.6026	0.6064	0.6103	0.6141
+0.3	0.6179	0.6217	0.6255	0.6293	0.6331	0.6368	0.6406	0.6443	0.6480	0.6517
+0.4	0.6554	0.6591	0.6628	0.6664	0.6700	0.6736	0.6772	0.6808	0.6844	0.6879
+0.5	0.6915	0.6950	0.6985	0.7019	0.7054	0.7088	0.7123	0.7157	0.7190	0.7224
+0.6	0.7257	0.7291	0.7324	0.7357	0.7389	0.7422	0.7454	0.7486	0.7517	0.7549
+0.7	0.7580	0.7611	0.7642	0.7673	0.7704	0.7734	0.7764	0.7794	0.7823	0.7852
+0.8	0.7881	0.7910	0.7939	0.7967	0.7995	0.8023	0.8051	0.8079	0.8106	0.8133
+0.9	0.8159	0.8186	0.8212	0.8238	0.8264	0.8289	0.8315	0.8340	0.8365	0.8389
+1.0	0.8413	0.8438	0.8461	0.8485	0.8508	0.8531	0.8554	0.8577	0.8599	0.8621
+1.1	0.8643	0.8665	0.8686	0.8708	0.8729	0.8749	0.8770	0.8790	0.8810	0.8830
+1.2	0.8849	0.8869	0.8888	0.8907	0.8925	0.8944	0.8962	0.8980	0.8997	0.9015
+1.3	0.9032	0.9049	0.9066	0.9082	0.9099	0.9115	0.9131	0.9147	0.9162	0.9177
+1.4	0.9192	0.9207	0.9222	0.9236	0.9251	0.9265	0.9279	0.9292	0.9306	0.9319
+1.5	0.9332	0.9345	0.9357	0.9370	0.9382	0.9394	0.9406	0.9418	0.9429	0.9441
+1.6	0.9452	0.9463	0.9474	0.9484	0.9495	0.9505	0.9515	0.9525	0.9535	0.9545
+1.7	0.9554	0.9564	0.9573	0.9582	0.9591	0.9599	0.9608	0.9616	0.9625	0.9633
+1.8	0.9641	0.9649	0.9656	0.9664	0.9671	0.9678	0.9686	0.9693	0.9699	0.9706
+1.9	0.9713	0.9719	0.9726	0.9732	0.9738	0.9744	0.9750	0.9756	0.9761	0.9767
+2.0	0.9773	0.9778	0.9783	0.9788	0.9793	0.9798	0.9803	0.9808	0.9812	0.9817
+2.1	0.9821	0.9826	0.9830	0.9834	0.9838	0.9842	0.9846	0.9850	0.9854	0.9857
+2.2	0.9861	0.9864	0.9868	0.9871	0.9875	0.9878	0.9881	0.9884	0.9887	0.9890
+2.3	0.9893	0.9896	0.9898	0.9901	0.9904	0.9906	0.9909	0.9911	0.9913	0.9916
+2.4	0.9918	0.9920	0.9922	0.9925	0.9927	0.9929	0.9931	0.9932	0.9934	0.9936
+2.5	0.9938	0.9940	0.9941	0.9943	0.9945	0.9946	0.9948	0.9949	0.9951	0.9952
+2.6	0.9953	0.9955	0.9956	0.9957	0.9959	0.9960	0.9961	0.9962	0.9963	0.9964
+2.7	0.9965	0.9966	0.9967	0.9968	0.9969	0.9970	0.9971	0.9972	0.9973	0.9974
+2.8	0.9974	0.9975	0.9976	0.9977	0.9977	0.9978	0.9979	0.9979	0.9980	0.9981
+2.9	0.9981	0.9982	0.9983	0.9983	0.9984	0.9984	0.9985	0.9985	0.9986	0.9986
+3.0	0.99865	0.99869	0.99874	0.99878	0.99882	0.99886	0.99889	0.99893	0.99896	0.99900
+3.1	0.99903	0.99906	0.99910	0.99913	0.99915	0.99918	0.99921	0.99924	0.99926	0.99929
+3.2	0.99931	0.99934	0.99936	0.99938	0.99940	0.99942	0.99944	0.99946	0.99948	0.99950
+3.3	0.99952	0.99953	0.99955	0.99957	0.99958	0.99960	0.99961	0.99962	0.99964	0.99965
+3.4	0.99966	0.99967	0.99969	0.99970	0.99971	0.99972	0.99973	0.99974	0.99975	0.99976
+3.5	0.99977	0.99978	0.99978	0.99979	0.99980	0.99981	0.99981	0.99982	0.99983	0.99983

Table B Student's t Distribution

df	Percentile point						
	70	80	90	95	97.5	99	99.5
1	.73	1.38	3.08	6.31	12.71	31.82	63.66
2	.62	1.06	1.89	2.92	4.30	6.96	9.92
3	.58	.98	1.64	2.35	3.18	4.54	5.84
4	.57	.94	1.53	2.13	2.78	3.75	4.60
5	.56	.92	1.48	2.01	2.57	3.36	4.03
6	.55	.91	1.44	1.94	2.45	3.14	3.71
7	.55	.90	1.42	1.90	2.36	3.00	3.50
8	.55	.89	1.40	1.86	2.31	2.90	3.36
9	.54	.88	1.38	1.83	2.26	2.82	3.25
10	.54	.88	1.37	1.81	2.23	2.76	3.17
11	.54	.88	1.36	1.80	2.20	2.72	3.11
12	.54	.87	1.36	1.78	2.18	2.68	3.06
13	.54	.87	1.35	1.77	2.16	2.65	3.01
14	.54	.87	1.34	1.76	2.14	2.62	2.98
15	.54	.87	1.34	1.75	2.13	2.60	2.95
16	.54	.86	1.34	1.75	2.12	2.58	2.92
17	.53	.86	1.33	1.74	2.11	2.57	2.90
18	.53	.86	1.33	1.73	2.10	2.55	2.88
19	.53	.86	1.33	1.73	2.09	2.54	2.86
20	.53	.86	1.32	1.72	2.09	2.53	2.84
21	.53	.86	1.32	1.72	2.08	2.52	2.83
22	.53	.86	1.32	1.72	2.07	2.51	2.82
23	.53	.86	1.32	1.71	2.07	2.50	2.81
24	.53	.86	1.32	1.71	2.06	2.49	2.80
25	.53	.86	1.32	1.71	2.06	2.48	2.79
26	.53	.86	1.32	1.71	2.06	2.48	2.78
27	.53	.86	1.31	1.70	2.05	2.47	2.77
28	.53	.86	1.31	1.70	2.05	2.47	2.76
29	.53	.85	1.31	1.70	2.04	2.46	2.76
30	.53	.85	1.31	1.70	2.04	2.46	2.75
40	.53	.85	1.30	1.68	2.02	2.42	2.70
50	.53	.85	1.30	1.67	2.01	2.40	2.68
60	.53	.85	1.30	1.67	2.00	2.39	2.66
80	.53	.85	1.29	1.66	1.99	2.37	2.64
100	.53	.84	1.29	1.66	1.98	2.36	2.63
200	.52	.84	1.29	1.65	1.97	2.34	2.60
500	.52	.84	1.28	1.65	1.96	2.33	2.59
∞	.52	.84	1.28	1.64	1.96	2.33	2.58

Table C Chi Square*

ν	Probability										
	0.99	0.98	0.95	0.90	0.80	0.20	0.10	0.05	0.02	0.01	0.001
1	0.0^3157	0.0^3628	0.00393	0.0158	0.0642	1.642	2.706	3.841	5.412	6.635	10.827
2	0.0201	0.0404	0.103	0.211	0.446	3.219	4.605	5.991	7.824	9.210	13.815
3	0.115	0.185	0.352	0.584	1.005	4.642	6.251	7.815	9.837	11.341	16.268
4	0.297	0.429	0.711	1.064	1.649	5.989	7.779	9.488	11.668	13.277	18.465
5	0.554	0.752	1.145	1.610	2.343	7.289	9.236	11.070	13.388	15.086	20.517
6	0.872	1.134	1.635	2.204	3.070	8.558	10.645	12.592	15.033	16.812	22.457
7	1.239	1.564	2.167	2.833	3.822	9.803	12.017	14.067	16.622	18.475	24.322
8	1.646	2.032	2.733	3.490	4.594	11.030	13.362	15.507	18.168	20.090	26.125
9	2.088	2.532	3.325	4.168	5.380	12.242	14.684	16.919	19.679	21.666	27.877
10	2.558	3.059	3.940	4.865	6.179	13.442	15.987	18.307	21.161	23.209	29.588
11	3.053	3.609	4.575	5.578	6.989	14.631	17.275	19.675	22.618	24.725	31.264
12	3.571	4.178	5.226	6.304	7.807	15.812	18.549	21.026	24.054	26.217	32.909
13	4.107	4.765	5.892	7.042	8.634	16.985	19.812	22.362	25.472	27.688	34.528
14	4.660	5.368	6.571	7.790	9.467	18.151	21.064	23.685	26.873	29.141	36.123
15	5.229	5.985	7.261	8.547	10.307	19.311	22.307	24.996	28.259	30.578	37.697
16	5.812	6.614	7.962	9.312	11.152	20.465	23.542	26.296	29.633	32.000	39.252
17	6.408	7.255	8.672	10.085	12.002	21.615	24.769	27.587	30.995	33.409	40.790
18	7.015	7.906	9.390	10.865	12.857	22.760	25.989	28.869	32.346	34.805	42.312
19	7.633	8.567	10.117	11.651	13.716	23.900	27.204	30.144	33.687	36.191	43.820
20	8.260	9.237	10.851	12.443	14.578	25.038	28.412	31.410	35.020	37.566	45.315
21	8.897	9.915	11.591	13.240	15.445	26.171	29.615	32.671	36.343	38.932	46.797
22	9.542	10.600	12.338	14.041	16.314	27.301	30.813	33.924	37.659	40.289	48.268
23	10.196	11.293	13.091	14.848	17.187	28.429	32.007	35.172	38.968	41.638	49.728
24	10.856	11.992	13.848	15.659	18.062	29.553	33.196	36.415	40.270	42.980	51.179
25	11.524	12.697	14.611	16.473	18.940	30.675	34.382	37.652	41.566	44.314	52.620
26	12.198	13.409	15.379	17.292	19.820	31.795	35.563	38.885	42.856	45.642	54.052
27	12.879	14.125	16.151	18.114	20.703	32.912	36.741	40.113	44.140	46.963	55.476
28	13.565	14.847	16.928	18.939	21.588	34.027	37.916	41.337	45.419	48.278	56.893
29	14.256	15.574	17.708	19.768	22.475	35.139	39.087	42.557	46.693	49.588	58.302
30	14.953	16.306	18.493	20.599	23.364	36.250	40.256	43.773	47.962	50.892	59.703

* Abridged from R. A. Fisher and F. Yates, *Statistical Tables for Biological, Agricultural and Medical Research.* Edinburgh and London: Oliver & Boyd, Ltd (1953), Table IV. Reproduced by permission of the authors and publishers. For larger values of ν, the expression $\sqrt{2\chi^2} - \sqrt{2\nu - 1}$ may be used as a normal deviate with unit variance, remembering that the probability for χ^2 corresponds with that of a single tail of a normal curve.

Table D *F* Distribution*

df for denom.	$1-\alpha$	df for numerator											
		1	2	3	4	5	6	7	8	9	10	11	12
1	.75	5.83	7.50	8.20	8.58	8.82	8.98	9.10	9.19	9.26	9.32	9.36	9.41
	.90	39.9	49.5	53.6	55.8	57.2	58.2	58.9	59.4	59.9	60.2	60.5	60.7
	.95	161	200	216	225	230	234	237	239	241	242	243	244
2	.75	2.57	3.00	3.15	3.23	3.28	3.31	3.34	3.35	3.37	3.38	3.39	3.39
	.90	8.53	9.00	9.16	9.24	9.29	9.33	9.35	9.37	9.38	9.39	9.40	9.41
	.95	18.5	19.0	19.2	19.2	19.3	19.3	19.4	19.4	19.4	19.4	19.4	19.4
	.99	98.5	99.0	99.2	99.2	99.3	99.3	99.4	99.4	99.4	99.4	99.4	99.4
3	.75	2.02	2.28	2.36	2.39	2.41	2.42	2.43	2.44	2.44	2.44	2.45	2.45
	.90	5.54	5.46	5.39	5.34	5.31	5.28	5.27	5.25	5.24	5.23	5.22	5.22
	.95	10.1	9.55	9.28	9.12	9.10	8.94	8.89	8.85	8.81	8.79	8.76	8.74
	.99	34.1	30.8	29.5	28.7	28.2	27.9	27.7	27.5	27.3	27.2	27.1	27.1
4	.75	1.81	2.00	2.05	2.06	2.07	2.08	2.08	2.08	2.08	2.08	2.08	2.08
	.90	4.54	4.32	4.19	4.11	4.05	4.01	3.98	3.95	3.94	3.92	3.91	3.90
	.95	7.71	6.94	6.59	6.39	6.26	6.16	6.09	6.04	6.00	5.96	5.94	5.91
	.99	21.2	18.0	16.7	16.0	15.5	15.2	15.0	14.8	14.7	14.5	14.4	14.4
5	.75	1.69	1.85	1.88	1.89	1.89	1.89	1.89	1.89	1.89	1.89	1.89	1.89
	.90	4.06	3.78	3.62	3.52	3.45	3.40	3.37	3.34	3.32	3.30	3.28	3.27
	.95	6.61	5.79	5.41	5.19	5.05	4.95	4.88	4.82	4.77	4.74	4.71	4.68
	.99	16.3	13.3	12.1	11.4	11.0	10.7	10.5	10.3	10.2	10.1	9.96	9.89
6	.75	1.62	1.76	1.78	1.79	1.79	1.78	1.78	1.77	1.77	1.77	1.77	1.77
	.90	3.78	3.46	3.29	3.18	3.11	3.05	3.01	2.98	2.96	2.94	2.92	2.90
	.95	5.99	5.14	4.76	4.53	4.39	4.28	4.21	4.15	4.10	4.06	4.03	4.00
	.99	13.7	10.9	9.78	9.15	8.75	8.47	8.26	8.10	7.98	7.87	7.79	7.72
7	.75	1.57	1.70	1.72	1.72	1.71	1.71	1.70	1.70	1.69	1.69	1.69	1.68
	.90	3.59	3.26	3.07	2.96	2.88	2.83	2.78	2.75	2.72	2.70	2.68	2.67
	.95	5.59	4.74	4.35	4.12	3.97	3.87	3.79	3.73	3.68	3.64	3.60	3.57
	.99	12.2	9.55	8.45	7.85	7.46	7.19	6.99	6.84	6.72	6.62	6.54	6.47
8	.75	1.54	1.66	1.67	1.66	1.66	1.65	1.64	1.64	1.64	1.63	1.63	1.62
	.90	3.46	3.11	2.92	2.81	2.73	2.67	2.62	2.59	2.56	2.54	2.52	2.50
	95	5.32	4.46	4.07	3.84	3.69	3.58	3.50	3.44	3.39	3.35	3.31	3.28
	.99	11.3	8.65	7.59	7.01	6.63	6.37	6.18	6.03	5.91	5.81	5.73	5.67
9	.75	1.51	1.62	1.63	1.63	1.62	1.61	1.60	1.60	1.59	1.59	1.58	1.58
	.90	3.36	3.01	2.81	2.69	2.61	2.55	2.51	2.47	2.44	2.42	2.40	2.38
	.95	5.12	4.26	3.86	3.63	3.48	3.37	3.29	3.23	3.18	3.14	3.10	3.07
	.99	10.6	8.02	6.99	6.42	6.06	5.80	5.61	5.47	5.35	5.26	5.18	5.11
10	.75	1.49	1.60	1.60	1.59	1.59	1.58	1.57	1.56	1.56	1.55	1.55	1.54
	.90	3.28	2.92	2.73	2.61	2.52	2.46	2.41	2.38	2.35	2.32	2.30	2.28
	.95	4.96	4.10	3.71	3.48	3.33	3.22	3.14	3.07	3.02	2.98	2.94	2.91
	.99	10.0	7.56	6.55	5.99	5.64	5.39	5.20	5.06	4.94	4.85	4.77	4.71
11	.75	1.47	1.58	1.58	1.57	1.56	1.55	1.54	1.53	1.53	1.52	1.52	1.51
	.90	3.23	2.86	2.66	2.54	2.45	2.39	2.34	2.30	2.27	2.25	2.23	2.21
	.95	4.84	3.98	3.59	3.36	3.20	3.09	3.01	2.95	2.90	2.85	2.82	2.79
	.99	9.65	7.21	6.22	5.67	5.32	5.07	4.89	4.74	4.63	4.54	4.46	4.40
12	.75	1.46	1.56	1.56	1.55	1.54	1.53	1.52	1.51	1.51	1.50	1.50	1.49
	.90	3.18	2.81	2.61	2.48	2.39	2.33	2.28	2.24	2.21	2.19	2.17	2.15
	.95	4.75	3.89	3.49	3.26	3.11	3.00	2.91	2.85	2.80	2.75	2.72	2.69
	.99	9.33	6.93	5.95	5.41	5.06	4.82	4.64	4.50	4.39	4.30	4.22	4.16

Table D (*Continued*)

df for numerator													
15	20	24	30	40	50	60	100	120	200	500	∞	1 − α	df for denom.
9.49	9.58	9.63	9.67	9.71	9.74	9.76	9.78	9.80	9.82	9.84	9.85	.75	
61.2	61.7	62.0	62.3	62.5	62.7	62.8	63.0	63.1	63.2	63.3	63.3	.90	1
246	248	249	250	251	252	252	253	253	254	254	254	.95	
3.41	3.43	3.43	3.44	3.45	3.45	3.46	3.47	3.47	3.48	3.48	3.48	.75	
9.42	9.44	9.45	9.46	9.47	9.47	9.47	9.48	9.48	9.49	9.49	9.49	.90	2
19.4	19.4	19.5	19.5	19.5	19.5	19.5	19.5	19.5	19.5	19.5	19.5	.95	
99.4	99.4	99.5	99.5	99.5	99.5	99.5	99.5	99.5	99.5	99.5	99.5	.99	
2.46	2.46	2.46	2.47	2.47	2.47	2.47	2.47	2.47	2.47	2.47	2.47	.75	
5.20	5.18	5.18	5.17	5.16	5.15	5.15	5.14	5.14	5.14	5.14	5.13	.90	3
8.70	8.66	8.64	8.62	8.59	8.58	8.57	8.55	8.55	8.54	8.53	8.53	.95	
26.9	26.7	26.6	26.5	26.4	26.4	26.3	26.2	26.2	26.2	26.1	26.1	.99	
2.08	2.08	2.08	2.08	2.08	2.08	2.08	2.08	2.08	2.08	2.08	2.08	.75	
3.87	3.84	3.83	3.82	3.80	3.80	3.79	3.78	3.78	3.77	3.76	3.76	.90	
5.86	5.80	5.77	5.75	5.72	5.70	5.69	5.66	5.66	5.65	5.64	5.63	.95	4
14.2	14.0	13.9	13.8	13.7	13.7	13.7	13.6	13.6	13.5	13.5	13.5	.99	
1.89	1.88	1.88	1.88	1.88	1.88	1.87	1.87	1.87	1.87	1.87	1.87	.75	
3.24	3.21	3.19	3.17	3.16	3.15	3.14	3.13	3.12	3.12	3.11	3.10	.90	5
4.62	4.56	4.53	4.50	4.46	4.44	4.43	4.41	4.40	4.39	4.37	4.36	.95	
9.72	9.55	9.47	9.38	9.29	9.24	9.20	9.13	9.11	9.08	9.04	9.02	.99	
1.76	1.76	1.75	1.75	1.75	1.75	1.74	1.74	1.74	1.74	1.74	1.74	.75	
2.87	2.84	2.82	2.80	2.78	2.77	2.76	2.75	2.74	2.73	2.73	2.72	.90	6
3.94	3.87	3.84	3.81	3.77	3.75	3.74	3.71	3.70	3.69	3.68	3.67	.95	
7.56	7.40	7.31	7.23	7.14	7.09	7.06	6.99	6.97	6.93	6.90	6.88	.99	
1.68	1.67	1.67	1.66	1.66	1.66	1.65	1.65	1.65	1.65	1.65	1.65	.75	
2.63	2.59	2.58	2.56	2.54	2.52	2.51	2.50	2.49	2.48	2.48	2.47	.90	7
3.51	3.44	3.41	3.38	3.34	3.32	3.30	3.27	3.27	3.25	3.24	3.23	.95	
6.31	6.16	6.07	5.99	5.91	5.86	5.82	5.75	5.74	5.70	5.67	5.65	.99	
1.62	1.61	1.60	1.60	1.59	1.59	1.59	1.58	1.58	1.58	1.58	1.58	.75	
2.46	2.42	2.40	2.38	2.36	2.35	2.34	2.32	2.32	2.31	2.30	2.29	.90	8
3.22	3.15	3.12	3.08	3.04	3.02	3.01	2.97	2.97	2.95	2.94	2.93	.95	
5.52	5.36	5.28	5.20	5.12	5.07	5.03	4.96	4.95	4.91	4.88	4.86	.99	
1.57	1.56	1.56	1.55	1.55	1.54	1.54	1.53	1.53	1.53	1.53	1.53	.75	
2.34	2.30	2.28	2.25	2.23	2.22	2.21	2.19	2.18	2.17	2.17	2.16	.90	9
3.01	2.94	2.90	2.86	2.83	2.80	2.79	2.76	2.75	2.73	2.72	2.71	.95	
4.96	4.81	4.73	4.65	4.57	4.52	4.48	4.42	4.40	4.36	4.33	4.31	.99	
1.53	1.52	1.52	1.51	1.51	1.50	1.50	1.49	1.49	1.49	1.48	1.48	.75	
2.24	2.20	2.18	2.16	2.13	2.12	2.11	2.09	2.08	2.07	2.06	2.06	.90	10
2.85	2.77	2.74	2.70	2.66	2.64	2.62	2.59	2.58	2.56	2.55	2.54	.95	
4.56	4.41	4.33	4.25	4.17	4.12	4.08	4.01	4.00	3.96	3.93	3.91	.99	
1.50	1.49	1.49	1.48	1.47	1.47	1.47	1.46	1.46	1.46	1.45	1.45	.75	
2.17	2.12	2.10	2.08	2.05	2.04	2.03	2.00	2.00	1.99	1.98	1.97	.90	11
2.72	2.65	2.61	2.57	2.53	2.51	2.49	2.46	2.45	2.43	2.42	2.40	.95	
4.25	4.10	4.02	3.94	3.86	3.81	3.78	3.71	3.69	3.66	3.62	3.60	.99	
1.48	1.47	1.46	1.45	1.45	1.44	1.44	1.43	1.43	1.43	1.42	1.42	.75	
2.10	2.06	2.04	2.01	1.99	1.97	1.96	1.94	1.93	1.92	1.91	1.90	.90	12
2.62	2.54	2.51	2.47	2.43	2.40	2.38	2.35	2.34	2.32	2.31	2.30	.95	
4.01	3.86	3.78	3.70	3.62	3.57	3.54	3.47	3.45	3.41	3.38	3.36	.99	

Table D (*Continued*)

df for denom.	$1-\alpha$	df for numerator 1	2	3	4	5	6	7	8	9	10	11	12
	.75	1.45	1.54	1.54	1.53	1.52	1.51	1.50	1.49	1.49	1.48	1.47	1.47
13	.90	3.14	2.76	2.56	2.43	2.35	2.28	2.23	2.20	2.16	2.14	2.12	2.10
	.95	4.67	3.81	3.41	3.18	3.03	2.92	2.83	2.77	2.71	2.67	2.63	2.60
	.99	9.07	6.70	5.74	5.21	4.86	4.62	4.44	4.30	4.19	4.10	4.02	3.96
	.75	1.44	1.53	1.53	1.52	1.51	1.50	1.48	1.48	1.47	1.46	1.46	1.45
14	.90	3.10	2.73	2.52	2.39	2.31	2.24	2.19	2.15	2.12	2.10	2.08	2.05
	.95	4.60	3.74	3.34	3.11	2.96	2.85	2.76	2.70	2.65	2.60	2.57	2.53
	.99	8.86	6.51	5.56	5.04	4.69	4.46	4.28	4.14	4.03	3.94	3.86	3.80
	.75	1.43	1.52	1.52	1.51	1.49	1.48	1.47	1.46	1.46	1.45	1.44	1.44
15	.90	3.07	2.70	2.49	2.36	2.27	2.21	2.16	2.12	2.09	2.06	2.04	2.02
	.95	4.54	3.68	3.29	3.06	2.90	2.79	2.71	2.64	2.59	2.54	2.51	2.48
	.99	8.68	6.36	5.42	4.89	4.56	4.32	4.14	4.00	3.89	3.80	3.73	3.67
	.75	1.42	1.51	1.51	1.50	1.48	1.48	1.47	1.46	1.45	1.45	1.44	1.44
16	.90	3.05	2.67	2.46	2.33	2.24	2.18	2.13	2.09	2.06	2.03	2.01	1.99
	.95	4.49	3.63	3.24	3.01	2.85	2.74	2.66	2.59	2.54	2.49	2.46	2.42
	.99	8.53	6.23	5.29	4.77	4.44	4.20	4.03	3.89	3.78	3.69	3.62	3.55
	.75	1.42	1.51	1.50	1.49	1.47	1.46	1.45	1.44	1.43	1.43	1.42	1.41
17	.90	3.03	2.64	2.44	2.31	2.22	2.15	2.10	2.06	2.03	2.00	1.98	1.96
	.95	4.45	3.59	3.20	2.96	2.81	2.70	2.61	2.55	2.49	2.45	2.41	2.38
	.99	8.40	6.11	5.18	4.67	4.34	4.10	3.93	3.79	3.68	3.59	3.52	3.46
	.75	1.41	1.50	1.49	1.48	1.46	1.45	1.44	1.43	1.42	1.42	1.41	1.40
18	.90	3.01	2.62	2.42	2.29	2.20	2.13	2.08	2.04	2.00	1.98	1.96	1.93
	.95	4.41	3.55	3.16	2.93	2.77	2.66	2.58	2.51	2.46	2.41	2.37	2.34
	.99	8.29	6.01	5.09	4.58	4.25	4.01	3.84	3.71	3.60	3.51	3.43	3.37
	.75	1.41	1.49	1.49	1.47	1.46	1.44	1.43	1.42	1.41	1.41	1.40	1.40
19	.90	2.99	2.61	2.40	2.27	2.18	2.11	2.06	2.02	1.98	1.96	1.94	1.91
	.95	4.38	3.52	3.13	2.90	2.74	2.63	2.54	2.48	2.42	2.38	2.34	2.31
	.99	8.18	5.93	5.01	4.50	4.17	3.94	3.77	3.63	3.52	3.43	3.36	3.30
	.75	1.40	1.49	1.48	1.46	1.45	1.44	1.42	1.42	1.41	1.40	1.39	1.39
20	.90	2.97	2.59	2.38	2.25	2.16	2.09	2.04	2.00	1.96	1.94	1.92	1.89
	.95	4.35	3.49	3.10	2.87	2.71	2.60	2.51	2.45	2.39	2.35	2.31	2.28
	.99	8.10	5.85	4.94	4.43	4.10	3.87	3.70	3.56	3.46	3.37	3.29	3.23
	.75	1.40	1.48	1.47	1.45	1.44	1.42	1.41	1.40	1.39	1.39	1.38	1.37
22	.90	2.95	2.56	2.35	2.22	2.13	2.06	2.01	1.97	1.93	1.90	1.88	1.86
	.95	4.30	3.44	3.05	2.82	2.66	2.55	2.46	2.40	2.34	2.30	2.26	2.23
	.99	7.95	5.72	4.82	4.31	3.99	3.76	3.59	3.45	3.35	3.26	3.18	3.12
	.75	1.39	1.47	1.46	1.44	1.43	1.41	1.40	1.39	1.38	1.38	1.37	1.36
24	.90	2.93	2.54	2.33	2.19	2.10	2.04	1.98	1.94	1.91	1.88	1.85	1.83
	.95	4.26	3.40	3.01	2.78	2.62	2.51	2.42	2.36	2.30	2.25	2.21	2.18
	.99	7.82	5.61	4.72	4.22	3.90	3.67	3.50	3.36	3.26	3.17	3.09	3.03
	.75	1.38	1.46	1.45	1.44	1.42	1.41	1.40	1.39	1.37	1.37	1.36	1.35
26	.90	2.91	2.52	2.31	2.17	2.08	2.01	1.96	1.92	1.88	1.86	1.84	1.81
	.95	4.23	3.37	2.98	2.74	2.59	2.47	2.39	2.32	2.27	2.22	2.18	2.15
	.99	7.72	5.53	4.64	4.14	3.82	3.59	3.42	3.29	3.18	3.09	3.02	2.96
	.75	1.38	1.46	1.45	1.43	1.41	1.40	1.39	1.38	1.37	1.36	1.35	1.34
28	.90	2.89	2.50	2.29	2.16	2.06	2.00	1.94	1.90	1.87	1.84	1.81	1.79
	.95	4.20	3.34	2.95	2.71	2.56	2.45	2.36	2.29	2.24	2.19	2.15	2.12
	.99	7.64	5.45	4.57	4.07	3.75	3.53	3.36	3.23	3.12	3.03	2.96	2.90

Table D (*Continued*)

df for numerator												$1-\alpha$	df for denom.
15	20	24	30	40	50	60	100	120	200	500	∞		
1.46	1.45	1.44	1.43	1.42	1.42	1.42	1.41	1.41	1.40	1.40	1.40	.75	
2.05	2.01	1.98	1.96	1.93	1.92	1.90	1.88	1.88	1.86	1.85	1.85	.90	13
2.53	2.46	2.42	2.38	2.34	2.31	2.30	2.26	2.25	2.23	2.22	2.21	.95	
3.82	3.66	3.59	3.51	3.43	3.38	3.34	3.27	3.25	3.22	3.19	3.17	.99	
1.44	1.43	1.42	1.41	1.41	1.40	1.40	1.39	1.39	1.39	1.38	1.38	.75	
2.01	1.96	1.94	1.91	1.89	1.87	1.86	1.83	1.83	1.82	1.80	1.80	.90	
2.46	2.39	2.35	2.31	2.27	2.24	2.22	2.19	2.18	2.16	2.14	2.13	.95	14
3.66	3.51	3.43	3.35	3.27	3.22	3.18	3.11	3.09	3.06	3.03	3.00	.99	
1.43	1.41	1.41	1.40	1.39	1.39	1.38	1.38	1.37	1.37	1.36	1.36	.75	
1.97	1.92	1.90	1.87	1.85	1.83	1.82	1.79	1.79	1.77	1.76	1.76	.90	15
2.40	2.33	2.29	2.25	2.20	2.18	2.16	2.12	2.11	2.10	2.08	2.07	.95	
3.52	3.37	3.29	3.21	3.13	3.08	3.05	2.98	2.96	2.92	2.89	2.87	.99	
1.41	1.40	1.39	1.38	1.37	1.37	1.36	1.36	1.35	1.35	1.34	1.34	.75	
1.94	1.89	1.87	1.84	1.81	1.79	1.78	1.76	1.75	1.74	1.73	1.72	.90	16
2.35	2.28	2.24	2.19	2.15	2.12	2.11	2.07	2.06	2.04	2.02	2.01	.95	
3.41	3.26	3.18	3.10	3.02	2.97	2.93	2.86	2.84	2.81	2.78	2.75	.99	
1.40	1.39	1.38	1.37	1.36	1.35	1.35	1.34	1.34	1.34	1.33	1.33	.75	
1.91	1.86	1.84	1.81	1.78	1.76	1.75	1.73	1.72	1.71	1.69	1.69	.90	17
2.31	2.23	2.19	2.15	2.10	2.08	2.06	2.02	2.01	1.99	1.97	1.96	.95	
3.31	3.16	3.08	3.00	2.92	2.87	2.83	2.76	2.75	2.71	2.68	2.65	.99	
1.39	1.38	1.37	1.36	1.35	1.34	1.34	1.33	1.33	1.32	1.32	1.32	.75	
1.89	1.84	1.81	1.78	1.75	1.74	1.72	1.70	1.69	1.68	1.67	1.66	.90	18
2.27	2.19	2.15	2.11	2.06	2.04	2.02	1.98	1.97	1.95	1.93	1.92	.95	
3.23	3.08	3.00	2.92	2.84	2.78	2.75	2.68	2.66	2.62	2.59	2.57	.99	
1.38	1.37	1.36	1.35	1.34	1.33	1.33	1.32	1.32	1.31	1.31	1.30	.75	
1.86	1.81	1.79	1.76	1.73	1.71	1.70	1.67	1.67	1.65	1.64	1.63	.90	19
2.23	2.16	2.11	2.07	2.03	2.00	1.98	1.94	1.93	1.91	1.89	1.88	.95	
3.15	3.00	2.92	2.84	2.76	2.71	2.67	2.60	2.58	2.55	2.51	2.49	.99	
1.37	1.36	1.35	1.34	1.33	1.33	1.32	1.31	1.31	1.30	1.30	1.29	.75	
1.84	1.79	1.77	1.74	1.71	1.69	1.68	1.65	1.64	1.63	1.62	1.61	.90	20
2.20	2.12	2.08	2.04	1.99	1.97	1.95	1.91	1.90	1.88	1.86	1.84	.95	
3.09	2.94	2.86	2.78	2.69	2.64	2.61	2.54	2.52	2.48	2.44	2.42	.99	
1.36	1.34	1.33	1.32	1.31	1.31	1.30	1.30	1.30	1.29	1.29	1.28	.75	
1.81	1.76	1.73	1.70	1.67	1.65	1.64	1.61	1.60	1.59	1.58	1.57	.90	22
2.15	2.07	2.03	1.98	1.94	1.91	1.89	1.85	1.84	1.82	1.80	1.78	.95	
2.98	2.83	2.75	2.67	2.58	2.53	2.50	2.42	2.40	2.36	2.33	2.31	.99	
1.35	1.33	1.32	1.31	1.30	1.29	1.29	1.28	1.28	1.27	1.27	1.26	.75	
1.78	1.73	1.70	1.67	1.64	1.62	1.61	1.58	1.57	1.56	1.54	1.53	.90	24
2.11	2.03	1.98	1.94	1.89	1.86	1.84	1.80	1.79	1.77	1.75	1.73	.95	
2.89	2.74	2.66	2.58	2.49	2.44	2.40	2.33	2.31	2.27	2.24	2.21	.99	
1.34	1.32	1.31	1.30	1.29	1.28	1.28	1.26	1.26	1.26	1.25	1.25	.75	
1.76	1.71	1.68	1.65	1.61	1.59	1.58	1.55	1.54	1.53	1.51	1.50	.90	26
2.07	1.99	1.95	1.90	1.85	1.82	1.80	1.76	1.75	1.73	1.71	1.69	.95	
2.81	2.66	2.58	2.50	2.42	2.36	2.33	2.25	2.23	2.19	2.16	2.13	.99	
1.33	1.31	1.30	1.29	1.28	1.27	1.27	1.26	1.25	1.25	1.24	1.24	.75	
1.74	1.69	1.66	1.63	1.59	1.57	1.56	1.53	1.52	1.50	1.49	1.48	.90	28
2.04	1.96	1.91	1.87	1.82	1.79	1.77	1.73	1.71	1.69	1.67	1.65	.95	
2.75	2.60	2.52	2.44	2.35	2.30	2.26	2.19	2.17	2.13	2.09	2.06	.99	

Table D (*Continued*)

df for denom.	1 − α	df for numerator 1	2	3	4	5	6	7	8	9	10	11	12
30	.75	1.38	1.45	1.44	1.42	1.41	1.39	1.38	1.37	1.36	1.35	1.35	1.34
	.90	2.88	2.49	2.28	2.14	2.05	1.98	1.93	1.88	1.85	1.82	1.79	1.77
	.95	4.17	3.32	2.92	2.69	2.53	2.42	2.33	2.27	2.21	2.16	2.13	2.09
	.99	7.56	5.39	4.51	4.02	3.70	3.47	3.30	3.17	3.07	2.98	2.91	2.84
40	.75	1.36	1.44	1.42	1.40	1.39	1.37	1.36	1.35	1.34	1.33	1.32	1.31
	.90	2.84	2.44	2.23	2.09	2.00	1.93	1.87	1.83	1.79	1.76	1.73	1.71
	.95	4.08	3.23	2.84	2.61	2.45	2.34	2.25	2.18	2.12	2.08	2.04	2.00
	.99	7.31	5.18	4.31	3.83	3.51	3.29	3.12	2.99	2.89	2.80	2.73	2.66
60	.75	1.35	1.42	1.41	1.38	1.37	1.35	1.33	1.32	1.31	1.30	1.29	1.29
	.90	2.79	2.39	2.18	2.04	1.95	1.87	1.82	1.77	1.74	1.71	1.68	1.66
	.95	4.00	3.15	2.76	2.53	2.37	2.25	2.17	2.10	2.04	1.99	1.95	1.92
	.99	7.08	4.98	4.13	3.65	3.34	3.12	2.95	2.82	2.72	2.63	2.56	2.50
120	.75	1.34	1.40	1.39	1.37	1.35	1.33	1.31	1.30	1.29	1.28	1.27	1.26
	.90	2.75	2.35	2.13	1.99	1.90	1.82	1.77	1.72	1.68	1.65	1.62	1.60
	.95	3.92	3.07	2.68	2.45	2.29	2.17	2.09	2.02	1.96	1.91	1.87	1.83
	.99	6.85	4.79	3.95	3.48	3.17	2.96	2.79	2.66	2.56	2.47	2.40	2.34
200	.75	1.33	1.39	1.38	1.36	1.34	1.32	1.31	1.29	1.28	1.27	1.26	1.25
	.90	2.73	2.33	2.11	1.97	1.88	1.80	1.75	1.70	1.66	1.63	1.60	1.57
	.95	3.89	3.04	2.65	2.42	2.26	2.14	2.06	1.98	1.93	1.88	1.84	1.80
	.99	6.76	4.71	3.88	3.41	3.11	2.89	2.73	2.60	2.50	2.41	2.34	2.27
∞	.75	1.32	1.39	1.37	1.35	1.33	1.31	1.29	1.28	1.27	1.25	1.24	1.24
	.90	2.71	2.30	2.08	1.94	1.85	1.77	1.72	1.67	1.63	1.60	1.57	1.55
	.95	3.84	3.00	2.60	2.37	2.21	2.10	2.01	1.94	1.88	1.83	1.79	1.75
	.99	6.63	4.61	3.78	3.32	3.02	2.80	2.64	2.51	2.41	2.32	2.25	2.18

Table D* (*Continued*)

df for numerator													
15	20	24	30	40	50	60	100	120	200	500	∞	1 − α	df for denom.
1.32	1.30	1.29	1.28	1.27	1.26	1.26	1.25	1.24	1.24	1.23	1.23	.75	
1.72	1.67	1.64	1.61	1.57	1.55	1.54	1.51	1.50	1.48	1.47	1.46	.90	30
2.01	1.93	1.89	1.84	1.79	1.76	1.74	1.70	1.68	1.66	1.64	1.62	.95	
2.70	2.55	2.47	2.39	2.30	2.25	2.21	2.13	2.11	2.07	2.03	2.01	.99	
1.30	1.28	1.26	1.25	1.24	1.23	1.22	1.21	1.21	1.20	1.19	1.19	.75	
1.66	1.61	1.57	1.54	1.51	1.48	1.47	1.43	1.42	1.41	1.39	1.38	.90	40
1.92	1.84	1.79	1.74	1.69	1.66	1.64	1.59	1.58	1.55	1.53	1.51	.95	
2.52	2.37	2.29	2.20	2.11	2.06	2.02	1.94	1.92	1.87	1.83	1.80	.99	
1.27	1.25	1.24	1.22	1.21	1.20	1.19	1.17	1.17	1.16	1.15	1.15	.75	
1.60	1.54	1.51	1.48	1.44	1.41	1.40	1.36	1.35	1.33	1.31	1.29	.90	60
1.84	1.75	1.70	1.65	1.59	1.56	1.53	1.48	1.47	1.44	1.41	1.39	.95	
2.35	2.20	2.12	2.03	1.94	1.88	1.84	1.75	1.73	1.68	1.63	1.60	.99	
1.24	1.22	1.21	1.19	1.18	1.17	1.16	1.14	1.13	1.12	1.11	1.10	.75	
1.55	1.48	1.45	1.41	1.37	1.34	1.32	1.27	1.26	1.24	1.21	1.19	.90	120
1.75	1.66	1.61	1.55	1.50	1.46	1.43	1.37	1.35	1.32	1.28	1.25	.95	
2.19	2.03	1.95	1.86	1.76	1.70	1.66	1.56	1.53	1.48	1.42	1.38	.99	
1.23	1.21	1.20	1.18	1.16	1.14	1.12	1.11	1.10	1.09	1.08	1.06	.75	
1.52	1.46	1.42	1.38	1.34	1.31	1.28	1.24	1.22	1.20	1.17	1.14	.90	200
1.72	1.62	1.57	1.52	1.46	1.41	1.39	1.32	1.29	1.26	1.22	1.19	.95	
2.13	1.97	1.89	1.79	1.69	1.63	1.58	1.48	1.44	1.39	1.33	1.28	.99	
1.22	1.19	1.18	1.16	1.14	1.13	1.12	1.09	1.08	1.07	1.04	1.00	.75	
1.49	1.42	1.38	1.34	1.30	1.26	1.24	1.18	1.17	1.13	1.08	1.00	.90	∞
1.67	1.57	1.52	1.46	1.39	1.35	1.32	1.24	1.22	1.17	1.11	1.00	.95	
2.04	1.88	1.79	1.70	1.59	1.52	1.47	1.36	1.32	1.25	1.15	1.00	.99	

* Abridged from E. S. Pearson and H. O. Hartley (Eds.), *Biometrika Tables for Statisticians*, vol. 1 (2nd Ed.). New York: Cambridge University Press (1958), Table 18. Reproduced by permission of E. S. Pearson and the trustees of *Biometrika*.

Table E.1 Upper 5-Percent Points of Studentized Range q*

	p**																		
n_2	2	3	4	5	6	7	8	9	10	11	12	13	14	15	16	17	18	19	20
1	18.0	26.7	32.8	37.2	40.5	43.1	45.4	47.3	49.1	50.6	51.9	53.2	54.3	55.4	56.3	57.2	58.0	58.8	59.6
2	6.09	8.28	9.80	10.89	11.73	12.43	13.03	13.54	13.99	14.39	14.75	15.08	15.38	15.65	15.91	16.14	16.36	16.57	16.77
3	4.50	5.88	6.83	7.51	8.04	8.47	8.85	9.18	9.46	9.72	9.95	10.16	10.35	10.52	10.69	10.84	10.98	11.12	11.24
4	3.93	5.00	5.76	6.31	6.73	7.06	7.35	7.60	7.83	8.03	8.21	8.37	8.52	8.67	8.80	8.92	9.03	9.14	9.24
5	3.61	4.54	5.18	5.64	5.99	6.28	6.52	6.74	6.93	7.10	7.25	7.39	7.52	7.64	7.75	7.86	7.95	8.04	8.13
6	3.46	4.34	4.90	5.31	5.63	5.89	6.12	6.32	6.49	6.65	6.79	6.92	7.04	7.14	7.24	7.34	7.43	7.51	7.59
7	3.34	4.16	4.68	5.06	5.35	5.59	5.80	5.99	6.15	6.29	6.42	6.54	6.65	6.75	6.84	6.93	7.01	7.08	7.16
8	3.26	4.04	4.53	4.89	5.17	5.40	5.60	5.77	5.92	6.05	6.18	6.29	6.39	6.48	6.57	6.65	6.73	6.80	6.87
9	3.20	3.95	4.42	4.76	5.02	5.24	5.43	5.60	5.74	5.87	5.98	6.09	6.19	6.28	6.36	6.44	6.51	6.58	6.65
10	3.15	3.88	4.33	4.66	4.91	5.12	5.30	5.46	5.60	5.72	5.83	5.93	6.03	6.12	6.20	6.27	6.34	6.41	6.47
11	3.11	3.82	4.26	4.58	4.82	5.03	5.20	5.35	5.49	5.61	5.71	5.81	5.90	5.98	6.06	6.14	6.20	6.27	6.33
12	3.08	3.77	4.20	4.51	4.75	4.95	5.12	5.27	5.40	5.51	5.61	5.71	5.80	5.88	5.95	6.02	6.09	6.15	6.21
13	3.06	3.73	4.15	4.46	4.69	4.88	5.05	5.19	5.32	5.43	5.53	5.63	5.71	5.79	5.86	5.93	6.00	6.06	6.11
14	3.03	3.70	4.11	4.41	4.64	4.83	4.99	5.13	5.25	5.36	5.46	5.56	5.64	5.72	5.79	5.86	5.92	5.98	6.03
15	3.01	3.67	4.08	4.37	4.59	4.78	4.94	5.08	5.20	5.31	5.40	5.49	5.57	5.65	5.72	5.79	5.85	5.91	5.96
16	3.00	3.65	4.05	4.34	4.56	4.74	4.90	5.03	5.15	5.26	5.35	5.44	5.52	5.59	5.66	5.73	5.79	5.84	5.90
17	2.98	3.62	4.02	4.31	4.52	4.70	4.86	4.99	5.11	5.21	5.31	5.39	5.47	5.55	5.61	5.68	5.74	5.79	5.84
18	2.97	3.61	4.00	4.28	4.49	4.67	4.83	4.96	5.07	5.17	5.27	5.35	5.43	5.50	5.57	5.63	5.69	5.74	5.79
19	2.96	3.59	3.98	4.26	4.47	4.64	4.79	4.92	5.04	5.14	5.23	5.32	5.39	5.46	5.53	5.59	5.65	5.70	5.75
20	2.95	3.58	3.96	4.24	4.45	4.62	4.77	4.90	5.01	5.11	5.20	5.28	5.36	5.43	5.50	5.56	5.61	5.66	5.71
24	2.92	3.53	3.90	4.17	4.37	4.54	4.68	4.81	4.92	5.01	5.10	5.18	5.25	5.32	5.38	5.44	5.50	5.55	5.59
30	2.89	3.48	3.84	4.11	4.30	4.46	4.60	4.72	4.83	4.92	5.00	5.08	5.15	5.21	5.27	5.33	5.38	5.43	5.48
40	2.86	3.44	3.79	4.04	4.23	4.39	4.52	4.63	4.74	4.82	4.90	4.98	5.05	5.11	5.17	5.22	5.27	5.32	5.36
60	2.83	3.40	3.74	3.98	4.16	4.31	4.44	4.55	4.65	4.73	4.81	4.88	4.94	5.00	5.06	5.11	5.15	5.20	5.24
120	2.80	3.36	3.69	3.92	4.10	4.24	4.36	4.47	4.56	4.64	4.71	4.78	4.84	4.90	4.95	5.00	5.04	5.09	5.13
∞	2.77	3.32	3.63	3.86	4.03	4.17	4.29	4.39	4.47	4.55	4.62	4.68	4.74	4.80	4.84	4.89	4.93	4.97	5.01

* From J. M. May, "Extended and Corrected Tables of the Upper Percentage Points of the Studentized Range," *Biometrika*, vol. 39 (1952), pp. 192–193. Reproduced by permission of the trustees of *Biometrika*.

** p is the number of quantities (for example, means) whose range is involved. n_2 is the degrees of freedom in the error estimate.

Table E.2 Upper 1-Percent Points of Studentized Range q

n_2*	p* 2	3	4	5	6	7	8	9	10	11	12	13	14	15	16	17	18	19	20
1	90.0	135	164	186	202	216	227	237	246	253	260	266	272	227	282	286	290	294	298
2	14.0	19.0	22.3	24.7	26.6	28.2	29.5	30.7	31.7	32.6	33.4	34.1	34.8	35.4	36.0	36.5	37.0	37.5	37.9
3	8.26	10.6	12.2	13.3	14.2	15.0	15.6	16.2	16.7	17.1	17.5	17.9	18.2	18.5	18.8	19.1	19.3	19.5	19.8
4	6.51	8.12	9.17	9.96	10.6	11.1	11.5	11.9	12.3	12.6	12.8	13.1	13.3	13.5	13.7	13.9	14.1	14.2	14.4
5	5.70	6.97	7.80	8.42	8.91	9.32	9.67	9.97	10.24	10.48	10.70	10.89	11.08	11.24	11.40	11.55	11.68	11.81	11.93
6	5.24	6.33	7.03	7.56	7.97	8.32	8.61	8.87	9.10	9.30	9.49	9.65	9.81	9.95	10.08	10.21	10.32	10.43	10.54
7	4.95	5.92	6.54	7.01	7.37	7.68	7.94	8.17	8.37	8.55	8.71	8.86	9.00	9.12	9.24	9.35	9.46	9.55	9.65
8	4.74	5.63	6.20	6.63	6.96	7.24	7.47	7.68	7.87	8.03	8.18	8.31	8.44	8.55	8.66	8.76	8.85	8.94	9.03
9	4.60	5.43	5.96	6.35	6.66	6.91	7.13	7.32	7.49	7.65	7.78	7.91	8.03	8.13	8.23	9.32	8.41	8.49	8.57
10	4.48	5.27	5.77	6.14	6.43	6.67	6.87	7.05	7.21	7.36	7.48	7.60	7.71	7.81	7.91	7.99	8.07	8.15	8.22
11	4.39	5.14	5.62	5.97	6.25	6.48	6.67	6.84	6.99	7.13	7.25	7.36	7.46	7.56	7.65	7.73	7.81	7.88	7.95
12	4.32	5.04	5.50	5.84	6.10	6.32	6.51	6.67	6.81	6.94	7.06	7.17	7.26	7.36	7.44	7.52	7.59	7.66	7.73
13	4.26	4.96	5.40	5.73	5.98	6.19	6.37	6.53	6.67	6.79	6.90	7.01	7.10	7.19	7.27	7.34	7.42	7.48	7.55
14	4.21	4.89	5.32	5.63	5.88	6.08	6.26	6.41	6.54	6.66	6.77	6.87	6.96	7.05	7.12	7.20	7.27	7.33	7.39
15	4.17	4.83	5.25	5.56	5.80	5.99	6.16	6.31	6.44	6.55	6.66	6.76	6.84	6.93	7.00	7.07	7.14	7.20	7.26
16	4.13	4.78	5.19	5.49	5.72	5.92	6.08	6.22	6.35	6.46	6.56	6.66	6.74	6.82	6.90	6.97	7.03	7.09	7.15
17	4.10	4.74	5.14	5.43	5.66	5.85	6.01	6.15	6.27	6.38	6.48	6.57	6.66	6.73	6.80	6.87	6.94	7.00	7.05
18	4.07	4.70	5.09	5.38	5.60	5.79	5.94	6.08	6.20	6.31	6.41	6.50	6.58	6.65	6.72	6.79	6.85	6.91	6.96
19	4.05	4.67	5.05	5.33	5.55	5.73	5.89	6.02	6.14	6.25	6.34	6.43	6.51	6.58	6.65	6.72	6.78	6.84	6.89
20	4.02	4.64	5.02	5.29	5.51	5.69	5.84	5.97	6.09	6.19	6.29	6.37	6.45	6.52	6.59	6.65	6.71	6.76	6.82
24	3.96	4.54	4.91	5.17	5.37	5.54	5.69	5.81	5.92	6.02	6.11	6.19	6.26	6.33	6.39	6.45	6.51	6.56	6.61
30	3.89	4.45	4.80	5.05	5.24	5.40	5.54	5.65	5.76	5.85	5.93	6.01	6.08	6.14	6.20	6.26	6.31	6.36	6.41
40	3.82	4.37	4.70	4.93	5.11	5.27	5.39	5.50	5.60	5.69	5.77	5.84	5.90	5.96	6.02	6.07	6.12	6.17	6.21
60	3.76	4.28	4.60	4.82	4.99	5.13	5.25	5.36	5.45	5.53	5.60	5.67	5.73	5.79	5.84	5.89	5.93	5.98	6.02
120	3.70	4.20	4.50	4.71	4.87	5.01	5.12	5.21	5.30	5.38	5.44	5.51	5.56	5.61	5.66	5.71	5.75	5.79	5.83
∞	3.64	4.12	4.40	4.60	4.76	4.88	4.99	5.08	5.16	5.23	5.29	5.35	5.40	5.45	5.49	5.54	5.57	5.61	5.65

* p is the number of quantities (for example, means) whose range is involved. n_2 is the degrees of freedom in the error estimate.

Table F Coefficients of Orthogonal Polynomials

		X											
k	Polynomial	1	2	3	4	5	6	7	8	9	10	$\Sigma\,\xi_i^2$	λ
3	Linear	−1	0	1								2	1
	Quadratic	1	−2	1								6	3
4	Linear	−3	−1	1	3							20	2
	Quadratic	1	−1	−1	1							4	1
	Cubic	−1	3	−3	1							20	$\frac{10}{3}$
5	Linear	−2	−1	0	1	2						10	1
	Quadratic	2	−1	−2	−1	2						14	1
	Cubic	−1	2	0	−2	1						10	$\frac{5}{6}$
	Quartic	1	−4	6	−4	1						70	$\frac{35}{12}$
6	Linear	−5	−3	−1	1	3	5					70	2
	Quadratic	5	−1	−4	−4	−1	5					84	$\frac{3}{2}$
	Cubic	−5	7	4	−4	−7	5					180	$\frac{5}{3}$
	Quartic	1	−3	2	2	−3	1					28	$\frac{7}{12}$
7	Linear	−3	−2	−1	0	1	2	3				28	1
	Quadratic	5	0	−3	−4	−3	0	5				84	1
	Cubic	−1	1	1	0	−1	−1	1				6	$\frac{1}{6}$
	Quartic	3	−7	1	6	1	−7	3				154	$\frac{7}{12}$
8	Linear	−7	−5	−3	−1	1	3	5	7			168	2
	Quadratic	7	1	−3	−5	−5	−3	1	7			168	1
	Cubic	−7	5	7	3	−3	−7	−5	7			264	$\frac{2}{3}$
	Quartic	7	−13	−3	9	9	−3	−13	7			616	$\frac{7}{12}$
	Quintic	−7	23	−17	−15	15	17	−23	7			2184	$\frac{7}{10}$
9	Linear	−4	−3	−2	−1	0	1	2	3	4		60	1
	Quadratic	28	7	−8	−17	−20	−17	−8	7	28		2772	3
	Cubic	−14	7	13	9	0	−9	−13	−7	14		990	$\frac{5}{6}$
	Quartic	14	−21	−11	9	18	9	−11	−21	14		2002	$\frac{7}{12}$
	Quintic	−4	11	−4	−9	0	9	4	−11	4		468	$\frac{3}{20}$
10	Linear	−9	−7	−5	−3	−1	1	3	5	7	9	330	2
	Quadratic	6	2	−1	−3	−4	−4	−3	−1	2	6	132	$\frac{1}{2}$
	Cubic	−42	14	35	31	12	−12	−31	−35	−14	42	8580	$\frac{5}{3}$
	Quartic	18	−22	−17	3	18	18	3	−17	−22	18	2860	$\frac{5}{12}$
	Quintic	−6	14	−1	−11	−6	6	11	1	−14	6	780	$\frac{1}{10}$

Answers to Odd-Numbered Problems

CHAPTER 2

2.1 $\bar{Y} = 18{,}472.9$, $s^2 = 41.7$.

2.3 For a two-sided alternative, some points are
$\mu = 18{,}473; 18{,}472; 18{,}471; 18{,}470; 18{,}469; 18{,}468; 18{,}467$
$P_a = \beta = 0.08, 0.39, 0.80, 0.95, 0.80, 0.39, 0.08.$

2.5 $n = 7$.

2.7 Do not reject hypothesis, since $\chi^2 = 17.05$ with 11 df.

2.9 Do not reject hypothesis, since $F = 1.8$.

2.11 Reject hypothesis at 1-percent level, since $|t| = 13.4$.

2.13 1. Reject hypothesis, since $z = 4.0$.
2. Do not reject hypothesis, since $z = 1.71$.
3. Do not reject hypothesis z, since $z < 1$.

2.15 For $\mu = 13 \quad 15 \quad 17 \quad 19$
Power $(1 - \beta) = \quad 0.04 \quad 0.27 \quad 0.70 \quad 0.95$.

2.17 Reject hypothesis at $\alpha = 0.05$, since $t = 8.32$.

2.19 Reject hypothesis of equal variances at $\alpha = 0.05$, since $F = 17.4$. Do not reject hypothesis of equal means at $\alpha = 0.05$, since t' is < 1.

CHAPTER 3

3.1

Source	df	SS	MS
Between A levels	4	253.04	63.26
Error	20	76.80	3.84
Totals	24	329.84	

Significant at the 1-percent level.

3.3 Two such contrasts might be

	SS
$C_1 = T_A - T_C = 31$	60.06
$C_2 = T_A - 2T_B + T_C = -79$	130.02
	190.08

Neither is significant at the 5-percent level.

3.5

	B	A	C
$\bar{Y}_{.j} =$	25.4	22.4	18.5

None significantly different.

3.7 Two sets and their sums of squares might be

			SS
Set 1:	$C_1 = 2T_1 - 2T_5$	$= -28$	49.0
	$C_2 = 4T_2 - 6T_4$	$= -108$	48.6
	$C_3 = 11T_1 + 11T_2 - 14T_3 + 11T_4 + 11T_5$	$= 71$	1.3
	$C_4 = 10T_1 - 4T_2 - 4T_4 + 10T_5$	$= 8$	0.1
			99.0

			SS
Set 2:	$C_1 = 6T_1 - 2T_2$	$= -18$	3.4
	$C_2 = 11T_1 + 11T_2 - 8T_3$	$= -237$	33.6
	$C_3 = 4T_1 + 4T_2 + 4T_3 - 19T_4$	$= -252$	36.3
	$C_4 = 2T_1 + 2T_2 + 2T_3 + 2T_4 - 23T_5$	$= -172$	25.7
			99.0

3.9 Proofs.

3.13 If $C_1 = 3T_{.1} - T_{.2} = -9$ which is numerically less than 17.90. Nonsignificant.
If $C_2 = 11T_{.1} + 11T_{.2} - 8T_{.3} = -237$ which is numerically greater than 149.30. Significant.

3.15 1.

Source	df	SS	MS
Between subgrades	3	120	40
Error	16	160	10
Totals	19	280	

2. A fixed model.
3. $F_{3,16} = 4.0$, reject at 5-percent level.
4. One set:

$$C_1 = T_{.1} - T_{.2}$$
$$C_2 = T_{.1} + T_{.2} - 2T_{.3}$$
$$C_3 = T_{.1} + T_{.2} + T_{.3} - 3T_{.4}$$

5. Rejection levels are 4.23, 5.15, 5.71.
6. $\bar{Y}_{.2} \pm 2.47$.

CHAPTER 4

4.1

Source	df	SS	MS
Between coater types	3	1.53	0.51*
Between days	2	0.21	0.10
Error	6	0.54	0.09
Totals	11	2.28	

* Significant at the 5-percent level.

4.3

	K	A	L	M
$\bar{Y}_{.j}$:	5.27	4.87	4.73	4.27

K and M are significantly different at the 5-percent level.

4.5 (5.57)

Source	df	SS	MS
Coater types	3	1.93	0.64*
Days	2	0.32	0.16
Error	5	0.46	0.09
Totals	10	2.71	

* Significant at the 5-percent level.

4.7 Three possible contrasts are

		SS
$C_1 = Q_1 - Q_2$	$= 4$	0.33
$C_2 = Q_1 + Q_2 - 2Q_3$	$= 6$	0.25
$C_3 = Q_1 + Q_2 + Q_3 - 3Q_4$	$= 36$	4.50
		5.08

None significant.

4.9 $SS_{error} = 50.25$ (as before).

4.11 Four contrasts might be

$C_1 = T_M - T_A$ Nonsignificant
$C_2 = T_M - T_K$ Significant at 5 percent by Scheffé
$C_3 = T_M - T_L$ Nonsignificant
$C_4 = T_A - T_K$ Nonsignificant

4.13 If seven were used,

Source	df
Treatments	1
Blocks (men)	6
Error	6
Total	13

4.17 If treatment A is X alone, B is Y alone, C is the control, D is X then Y, E is Y then X, F is X then Y after one day, and G is Y then X after one day, one orthogonal set might be

$$C_1 = T_A - T_B$$
$$C_2 = T_D - T_E$$
$$C_3 = T_F - T_G$$
$$C_4 = T_A + T_B + T_D + T_E + T_F + T_G - 6T_C$$
$$C_5 = T_A + T_B - T_D - T_E$$
$$C_6 = T_A + T_B + T_D + T_E - 2T_F - 2T_G$$

CHAPTER 5

5.1

Source	df	SS	MS
Electrodes	4	1.04	0.26
Strips	4	3.44	0.86
Positions	4	1.84	0.46
Error	12	9.92	0.83
Totals	24	16.24	

None significant.

5.3

	B	C	E	D	A
$\bar{Y}_{.j}$:	2.26	2.54	2.76	3.14	3.42

B is thus better (less time) than E, D, or A. It is not significantly better than C.

5.5

Source	df	SS	MS
Electrodes	4	3.82	0.96**
Positions	3	0.55	0.18
Strips (unadjusted)	4	0.26	—
Error	8	0.42	0.05
Totals	19	5.05	

** Significant at the 1-percent level.

5.7 5 × 5 Graeco-Latin square.

	1	2	3	4	5
I	$A\alpha$	$B\beta$	$C\gamma$	$D\delta$	$E\varepsilon$
II	$E\delta$	$A\varepsilon$	$B\alpha$	$C\beta$	$D\gamma$
III	$D\beta$	$E\gamma$	$A\delta$	$B\varepsilon$	$C\alpha$
IV	$C\varepsilon$	$D\alpha$	$E\beta$	$A\gamma$	$B\delta$
V	$B\gamma$	$C\delta$	$D\varepsilon$	$E\alpha$	$A\beta$

5.9

Charge stock	Day 1	2	3
1	*A*	*B*	*D*
2	*B*	*C*	*E*
3	*C*	*D*	*F*
4	*D*	*E*	*G*
5	*E*	*F*	*A*
6	*F*	*G*	*B*
7	*G*	*A*	*C*

CHAPTER 6

6.1

Source	df	SS	MS
Exhaust index	2	4608.17	2304.08***
Pump heater voltage	1	96.34	96.34
$P \times E$ interaction	2	283.16	141.58*
Error	6	139.00	23.17
Totals	11	5126.67	

* Significant.
*** Highly significant.
(Readings were first multiplied by 1000.)

6.3 On exhaust index means only

$\bar{Y}_{.j}$: 9.25 21.25 55.50 (all × 10^{-3})

All are significantly different from one another.
By cell means (because of interaction)

$\bar{Y}_{ij.}$: 7.5 11 12 30.5 53 58

6.5

Source	df	SS	MS
Humidity	2	9.07	4.53
Temperature	2	8.66	4.33
$T \times H$ interaction	4	6.07	1.52
Error	27	28.50	1.06
Totals	35	52.30	

6.7 At the 5-percent significance level, reject H_2 and H_3 but not H_1.

6.9 Plot of cell totals versus feed for the two material lines are not parallel (interaction), and both material and feed effect are obvious.

6.11

Source	df	SS	MS
Temperature	2	600.09	300.04**
Mix	2	1.59	0.80*
$T \times M$	4	0.98	0.25*
Laboratory	3	3.85	1.28**
$T \times L$	6	1.54	0.26*
$M \times L$	6	0.76	0.13*
$T \times M \times L$	12	0.90	0.07*
Error	36	0.67	0.02
Totals	71	610.38	

* One asterisk indicates significance of the 5-percent level; two, the 1-percent level; three, the 0.1-percent level.

6.13 How randomize over laboratories? Or, over temperatures in laboratories?

6.15 Two-dimensional stimulus more variable than three dimensional. Means by analysts:

Analyst:	1	4	3	5	2
Mean:	1.16	1.23	1.42	1.44	1.64

J by A means:

1.04 1.14 1.17 1.20 1.24 1.42 1.58 1.59 1.70 1.70

6.17

Source	df	SS	MS
Thickness A	1	846.81	846.81**
Temperature B	1	5041.00	5041.00**
AB	1	509.63	509.63**
Drying condition C	1	5.88	5.88
AC	1	1.44	1.44
BC	1	15.21	15.21*
ABC	1	0.14	0.14
Length of wash L	3	69.76	23.25*
AL	3	15.79	5.26
BL	3	3.04	1.01
ABL	3	11.45	3.82
CL	3	9.65	3.22
ACL	3	8.07	2.69
BCL	3	6.43	2.14
$ABCL$	3	5.66	1.89
Error	32	87.21	2.73
Totals	63	6637.17	

* One asterisk indicates significance at the 5-percent level; two, the 1-percent level.

CHAPTER 7

7.1

Source	df	SS	MS
Factor A	1	0.08	0.08
Factor B	1	70.08	70.08**
$A \times B$ interaction	1	24.08	24.08
Error	8	36.68	4.58
Totals	11	130.92	

** Significant at the 1-percent level.

7.3

Source	df	SS	MS
A	1	2704.00	2704.00
B	1	26,732.25	26,732.25
AB	1	7744.00	7744.00
C	1	24,025.00	24,025.00
AC	1	16,256.25	16,256.25
BC	1	64,516.00	64,516.00
ABC	1	420.25	420.25
Error	8	246,284.00	30,785.50
Totals	15	388,681.75	

Nothing significant at the 5-percent level.

7.5 No plots, since there are no significant effects at the 5-percent level.

7.7

Source	df	SS	MS
A	1	13,736.5	
B	1	6188.3	
AB	1	22,102.5	Same
C	1	22.7	as
AC	1	22,525.0	SS
BC	1	12,051.3	
ABC	1	20,757.0	
D	1	81,103.8	
AD	1	145,665.0*	
BD	1	9214.3	
ABD	1	126,630.3*	
CD	1	148.8	
ACD	1	6757.0	
BCD	1	294.0	
ABCD	1	19,453.8	
Error	16	431,599.4	26,975.0
Totals	31	918,249.7	

* Significant at 5-percent level.

7.9 See Problem 6.17 answer.

7.11 2^6 factorial. Probably use four-, five-, and six-way interactions as error term with 22 df.

CHAPTER 8

8.1

Source	df	SS	MS
Linear *A*	1	16.82	16.82*
Quadratic *A*	1	202.30	202.30**
Cubic *A*	1	28.88	28.88*
Quartic *A*	1	5.04	5.04
Error	20	76.80	3.84
Totals	24	329.84	

* One asterisk indicates significance at the 5-percent level; two, the 1-percent level.

8.3 $r^2 = 0.0510$, $\eta^2 = 0.7676$, $R^2_{\text{cubic}} = 0.7519$.

8.5

Source	df	SS	MS
Lacquer	3	63.79	21.26**
Times	2	126.59	63.29**
$L \times T$ interaction	6	38.07	6.35
Error	12	34.51	2.88
Totals	23	262.96	

** Significant at 1-percent level.

8.7 Lacquer cubic: $Y_x' = 13.429 + 4.922u - 0.375u^2 - 2.36u^3$.
Time quadratic: $Y_x' = 15.002 - 2.188u - 3.063u^2$.

CHAPTER 9

9.1

Source	df	SS	MS
Temperature	2	348.44	174.22
Humidity	2	333.77	166.88
$T \times H$ interaction	4	358.23	89.56
Error	9	680.00	75.66
Totals	17	1720.44	

None significant at the 5-percent level.

9.3 $\left.\begin{aligned} AB &= 320.11 \\ AB^2 &= 38.11 \end{aligned}\right\} 358.22$

9.5 After coding data by subtracting 2.6 and multiplying by 10, we have

Source	df	SS	MS
Surface thickness A	2	2544.71	1272.35
Base thickness B	2	4787.37	2393.68
$A \times B$ interaction	4	185.18	46.29
Subbase thickness C	2	4165.15	2082.57
$A \times C$ interaction	4	189.72	47.48
$B \times C$ interaction	4	27.74	6.93
$A \times B \times C$ interaction	8	100.06	12.51
Error	27	71.50	2.65

All except $B \times C$ are significant at the 5-percent level, but the main effects predominate.

9.7

Source	df	SS	Total
AB	2	66.93	185.19
AB^2	2	118.26	
AC	2	58.93	189.75
AC^2	2	130.82	
BC	2	5.82	27.75
BC^2	2	21.93	
ABC	2	17.60	100.05
ABC^2	2	8.26	
AB^2C	2	53.48	
AB^2C^2	2	20.71	

9.9 Plots show strong linear effects of A, B, and C. They show AB and AC interaction—but it is slight compared to the main effects. A slight quadratic trend in A is also noted.

9.11

Source	df	SS	MS
Resin type R	1	0.050	0.050*
Weight fraction S	2	0.540	
S_L	1	0.540	0.540**
S_Q	1	0.000	0.000
Gate setting G	2	30.300	
G_L	1	30.150	30.150**
G_Q	1	0.150	0.150*
RS	2	0.040	0.020
RG	2	0.060	0.030*
SG	4	0.343	
S_LG_L	1	0.280	0.280**
S_LG_Q	1	0.002	0.002
S_QG_L	1	0.010	0.010
S_QG_Q	1	0.051	0.051*
RSG	4	0.050	0.013
Error	18	0.150	0.0083
Totals	35	31.530	

* One asterisk indicates significance at the 5-percent level; two, the 1-percent level.

9.13 $Y'_x = 3.20$, $Y'_x = 3.02$.

CHAPTER 10

10.1

Source	EMS
O_i	$\sigma_\varepsilon^2 + 10\sigma_O^2$
A_j	$\sigma_\varepsilon^2 + 2\sigma_{OA}^2 + 6\phi_A$
OA_{ij}	$\sigma_\varepsilon^2 + 2\sigma_{OA}^2$
$\varepsilon_{k(ij)}$	σ_ε^2

Tests are indicated by arrows. None significant at the 5-percent level.

10.3

Source	EMS
A_i	$\sigma_\varepsilon^2 + nc\sigma_{AB}^2 + nbc\sigma_A^2$
B_j	$\sigma_\varepsilon^2 + nc\sigma_{AB}^2 + nac\sigma_B^2$
AB_{ij}	$\sigma_\varepsilon^2 + nc\sigma_{AB}^2$
C_k	$\sigma_\varepsilon^2 + n\sigma_{ABC}^2 + na\sigma_{BC}^2 + nb\sigma_{AC}^2 + nab\phi_C$
AC_{ik}	$\sigma_\varepsilon^2 + n\sigma_{ABC}^2 + nb\sigma_{AC}^2$
BC_{jk}	$\sigma_\varepsilon^2 + n\sigma_{ABC}^2 + na\sigma_{BC}^2$
ABC_{ijk}	$\sigma_\varepsilon^2 + n\sigma_{ABC}^2$
$\varepsilon_{m(ijk)}$	σ_ε^2

Tests are obvious.
No direct test on C.

10.5

Source	EMS
A_i	$\sigma_\varepsilon^2 + nb\sigma_{ACD}^2 + nbc\sigma_{AD}^2 + nbd\sigma_{AC}^2 + nbcd\phi_A$
B_j	$\sigma_\varepsilon^2 + na\sigma_{BCD}^2 + nac\sigma_{BD}^2 + nad\sigma_{BC}^2 + nacd\phi_B$
AB_{ij}	$\sigma_\varepsilon^2 + n\sigma_{ABCD}^2 + nc\sigma_{ABD}^2 + nd\sigma_{ABC}^2 + ncd\phi_{AB}$
C_k	$\sigma_\varepsilon^2 + nab\sigma_{CD}^2 + nabd\sigma_C^2$
AC_{ik}	$\sigma_\varepsilon^2 + nb\sigma_{ACD}^2 + nbd\sigma_{AC}^2$
BC_{jk}	$\sigma_\varepsilon^2 + na\sigma_{BCD}^2 + nad\sigma_{BC}^2$
ABC_{ijk}	$\sigma_\varepsilon^2 + n\sigma_{ABCD}^2 + nd\sigma_{ABC}^2$
D_m	$\sigma_\varepsilon^2 + nab\sigma_{CD}^2 + nabc\sigma_D^2$
AD_{im}	$\sigma_\varepsilon^2 + nb\sigma_{ACD}^2 + nbc\sigma_{AD}^2$
BD_{jm}	$\sigma_\varepsilon^2 + na\sigma_{BCD}^2 + nac\sigma_{BD}^2$
ABD_{ijm}	$\sigma_\varepsilon^2 + n\sigma_{ABCD}^2 + nc\sigma_{ABD}^2$
CD_{km}	$\sigma_\varepsilon^2 + nab\sigma_{CD}^2$
ACD_{ikm}	$\sigma_\varepsilon^2 + nb\sigma_{ACD}^2$
BCD_{jkm}	$\sigma_\varepsilon^2 + na\sigma_{BCD}^2$
$ABCD_{ijkm}$	$\sigma_\varepsilon^2 + n\sigma_{ABCD}^2$
$\varepsilon_{q(ijkm)}$	σ_ε^2

No direct test on A, B, or AB.

10.7 $N_0 = 4.44$, $s_\varepsilon^2 = 1.16$, $s_A^2 = 5.32$.

10.11 To test A, $\text{MS} = \text{MS}_{AB} + \text{MS}_{AC} - \text{MS}_{ABC}$
B, $\text{MS} = \text{MS}_{AB} + \text{MS}_{BC} - \text{MS}_{ABC}$
C, $\text{MS} = \text{MS}_{AC} + \text{MS}_{BC} - \text{MS}_{ABC}$

CHAPTER 11

11.1

Source	df	SS	MS	EMS
L_i	2	27.42	13.71	$\sigma_\varepsilon^2 + 3\sigma_R^2 + 12\phi_L$
$R_{j(i)}$	9	36.38	4.04	$\sigma_\varepsilon^2 + 3\sigma_R^2$
$\varepsilon_{k(ij)}$	24	21.80	0.91	σ_ε^2
Totals	35	85.60		

Difference between rolls within lots is significant at the 1-percent level.

11.3

Source	EMS
A_i	$\sigma_\varepsilon^2 + 2\sigma_C^2 + 6\sigma_B^2 + 24\phi_A$
$B_{j(i)}$	$\sigma_\varepsilon^2 + 2\sigma_C^2 + 6\sigma_B^2$
$C_{k(ij)}$	$\sigma_\varepsilon^2 + 2\sigma_C^2$
$\varepsilon_{m(ijk)}$	σ_ε^2

Tests are obvious.

11.5

Source	df	SS	MS	EMS
T_i	1	5,489,354	5,489,354	$\sigma_\varepsilon^2 + 6\sigma_M^2 + 12\phi_T$
$M_{j(i)}$	2	408,250	204,125	$\sigma_\varepsilon^2 + 6\sigma_M^2$
S_k	1	8971	8971	$\sigma_\varepsilon^2 + 3\sigma_{MS}^2 + 12\phi_S$
TS_{ik}	1	37,445	37,445	$\sigma_\varepsilon^2 + 3\sigma_{MS}^2 + 6\phi_{TS}$
$MS_{kj(i)}$	2	31,520	15,760	$\sigma_\varepsilon^2 + 3\sigma_{MS}^2$
$\varepsilon_{m(ijk)}$	16	316,930	19,808	σ_ε^2

11.7

Source	EMS
T_i	$\sigma_\varepsilon^2 + 18\sigma_m^2 + 36\phi_t$
$M_{j(i)}$	$\sigma_\varepsilon^2 + 18\sigma_m^2$
S_k	$\sigma_\varepsilon^2 + 9\sigma_{ms}^2 + 36\phi_s$
TS_{ik}	$\sigma_\varepsilon^2 + 9\sigma_{ms}^2 + 18\phi_{ts}$
$MS_{kj(i)}$	$\sigma_\varepsilon^2 + 9\sigma_{ms}^2$
P_m	$\sigma_\varepsilon^2 + 6\sigma_{mp}^2 + 24\phi_p$
TP_{im}	$\sigma_\varepsilon^2 + 6\sigma_{mp}^2 + 12\phi_{tp}$
$MP_{mj(i)}$	$\sigma_\varepsilon^2 + 6\sigma_{mp}^2$
SP_{km}	$\sigma_\varepsilon^2 + 3\sigma_{msp}^2 + 12\phi_{sp}$
TSP_{ikm}	$\sigma_\varepsilon^2 + 3\sigma_{msp}^2 + 6\phi_{tsp}$
$MSP_{mkj(i)}$	$\sigma_\varepsilon^2 + 3\sigma_{msp}^2$
$\varepsilon_{q(ijkm)}$	σ_ε^2

11.9

Source	df	SS	MS	EMS
F_i	4	3886	971**	$\sigma_\varepsilon^2 + 9\sigma_s^2 + 54\phi_f$
C_j	1	109	109**	$\sigma_\varepsilon^2 + 9\sigma_s^2 + 135\phi_c$
FC_{ij}	4	721	180**	$\sigma_\varepsilon^2 + 9\sigma_s^2 + 27\phi_{fc}$
$S_{k(ij)}$	20	39	1.95**	$\sigma_\varepsilon^2 + 9\sigma_s^2$
P_m	2	161	80.5**	$\sigma_\varepsilon^2 + 3\sigma_{sp}^2 + 90\phi_p$
FP_{im}	8	114	14.3**	$\sigma_\varepsilon^2 + 3\sigma_{sp}^2 + 18\phi_{fp}$
CP_{jm}	2	3	1.5	$\sigma_\varepsilon^2 + 3\sigma_{sp}^2 + 45\phi_{cp}$
FCP_{ijm}	8	15	1.9	$\sigma_\varepsilon^2 + 3\sigma_{sp}^2 + 9\phi_{fcp}$
$SP_{mk(ij)}$	40	33	0.82	$\sigma_\varepsilon^2 + 3\sigma_{sp}^2$
$\varepsilon_{q(ijkm)}$	180	63	0.35	σ_ε^2
Totals	269	5144		

** Significant at the 1-percent level.

11.11 $Y_{ijkm} = \mu + G_i + C_j + GC_{ij} + S_{k(ij)} + T_m + GT_{im} + CT_{jm} + GCT_{ijm} + TS_{mk(ij)}$

11.13

Source	df	EMS
D_i	2	$\sigma_S^2 + 22\sigma_D^2$
G_j	1	$\sigma_{SG}^2 + 11\sigma_{DG}^2 + 33\phi_G$
DG_{ij}	2	$\sigma_{SG}^2 + 11\sigma_{DG}^2$
$S_{k(i)}$	30	σ_S^2
$SG_{jk(i)}$	30	σ_{SG}^2

11.15

Source	df	EMS
R_i	1	$\sigma_\varepsilon^2 + 6\sigma_H^2 + 36\sigma_R^2$
$H_{j(i)}$	10	$\sigma_\varepsilon^2 + 6\sigma_H^2$
P_k	2	$\sigma_\varepsilon^2 + 2\sigma_{PH}^2 + 12\sigma_{RP}^2 + 24\phi_P$
RP_{ik}	2	$\sigma_\varepsilon^2 + 2\sigma_{PH}^2 + 12\sigma_{RP}^2$
$PH_{kj(i)}$	20	$\sigma_\varepsilon^2 + 2\sigma_{PH}^2$
$\varepsilon_{m(ijk)}$	36	σ_ε^2

CHAPTER 12

12.1

Source	df	SS	MS	EMS
R_i	2	337.15	168.58	$\sigma_\varepsilon^2 + 18\sigma_R^2$
S_j	1	16.66	16.66	$\sigma_\varepsilon^2 + 9\sigma_{RS}^2 + 27\phi_S$
RS_{ij}	2	126.78	63.39	$\sigma_\varepsilon^2 + 9\sigma_{RS}^2$
H_k	2	15.59	7.80	$\sigma_\varepsilon^2 + 6\sigma_{RH}^2 + 18\phi_H$
HS_{jk}	2	80.12	40.06	$\sigma_\varepsilon^2 + 3\sigma_{SRH}^2 + 9\phi_{HS}$
RH_{ik}	4	62.85	15.71	$\sigma_2^2 + 6\sigma_{RH}^2$
SRH_{ijk}	4	229.44	57.36	$\sigma_\varepsilon^2 + 3\sigma_{SRH}^2$
$\varepsilon_{m(ijk)}$	36	200.00	5.56	σ_ε^2
Totals	53	1068.59		

Replications (blocks) and replications by all other factors are significant at the 5-percent level; however, these are not the factors of chief interest.

12.3

Source	df	SS	EMS
R_i	1	As before	$\sigma_\varepsilon^2 + 3\sigma_{RM}^2 + 12\sigma_R^2$
T_j	1	with R in	$\sigma_\varepsilon^2 + 3\sigma_{RM}^2 + 6\sigma_M^2 + 6\sigma_{RT}^2 + 12\phi_T$
RT_{ij}	1	place of S	$\sigma_\varepsilon^2 + 3\sigma_{RM}^2 + 6\sigma_{RT}^2$
$M_{k(j)}$	2		$\sigma_\varepsilon^2 + 3\sigma_{RM}^2 + 6\sigma_M^2$
$RM_{ik(j)}$	2		$\sigma_\varepsilon^2 + 3\sigma_{RM}^2$
$\varepsilon_{m(ijk)}$	16		σ_ε^2
Total	23		

No direct test on types, but results look highly significant. No other significant effects.

12.5 Compare models.

12.7 Complete: $Y_{ijkm} = \mu + R_k + M_i + RM_{ik} + H_{j(i)} + RH_{kj(i)} + \varepsilon_{m(ijk)}$
Reduced: $Y_{ijkm} = \mu + R_k + M_i + H_{j(i)} + \varepsilon_{m(ijk)}$
Take fewer readings per head.

12.9 Complete: $Y_{ijkmq} = \mu + R_q + M_i + RM_{iq} + G_j + RG_{jq} + MG_{ij} + RMG_{ijq} + T_{k(j)} + RT_{qk(j)} + MT_{ik(j)} + RMT_{iqk(j)} + \varepsilon_{m(ijkq)}$
Reduced: Omit all interactions with R.

CHAPTER 13

13.1

	Source	df	EMS
Whole plot	R_i	1	$\sigma_\varepsilon^2 + 6\sigma_{RS}^2 + 18\sigma_R^2$
	H_j	2	$\sigma_\varepsilon^2 + 2\sigma_{RHS}^2 + 4\sigma_{HS}^2 + 6\sigma_{RH}^2 + 12\phi_H$
	RH_{ij}	2	$\sigma_\varepsilon^2 + 2\sigma_{RHS}^2 + 6\sigma_{RH}^2$
Split plot	S_k	2	$\sigma_\varepsilon^2 + 6\sigma_{RS}^2 + 12\sigma_S^2$
	RS_{ik}	2	$\sigma_\varepsilon^2 + 6\sigma_{RS}^2$
	HS_{jk}	4	$\sigma_\varepsilon^2 + 2\sigma_{RHS}^2 + 4\sigma_{HS}^2$
	RHS_{ijk}	4	$\sigma_\varepsilon^2 + 2\sigma_{RHS}^2$
	$\varepsilon_{m(ijk)}$	18	σ_ε^2
	Total	35	

13.3 $F'_{1,v} = 24.3$ and $v = 1530$; significant at the 1-percent level.

13.5

	Source	df	EMS
Whole plot	R_i	3	$\sigma_\varepsilon^2 + 8\sigma_{RA}^2 + 40\sigma_R^2$
	S_j	1	$\sigma_\varepsilon^2 + 4\sigma_{RSA}^2 + 16\sigma_{SA}^2 + 20\sigma_{RS}^2 + 80\phi_S$
	RS_{ij}	3	$\sigma_\varepsilon^2 + 4\sigma_{RSA}^2 + 20\sigma_{RS}^2$
	J_k	1	$\sigma_\varepsilon^2 + 4\sigma_{RJA}^2 + 16\sigma_{JA}^2 + 20\sigma_{RJ}^2 + 80\phi_J$
	RJ_{ik}	3	$\sigma_\varepsilon^2 + 4\sigma_{RJA}^2 + 20\sigma_{RJ}^2$
	SJ_{jk}	1	$\sigma_\varepsilon^2 + 2\sigma_{RSJA}^2 + 10\sigma_{RSJ}^2 + 8\sigma_{SJA}^2 + 40\phi_{SJ}$
	RSJ_{ijk}	3	$\sigma_\varepsilon^2 + 2\sigma_{RSJA}^2 + 10\sigma_{RSJ}^2$
Split plot	A_m	4	$\sigma_\varepsilon^2 + 8\sigma_{RA}^2 + 32\sigma_A^2$
	RA_{im}	12	$\sigma_\varepsilon^2 + 8\sigma_{RA}^2$
	SA_{jm}	4	$\sigma_\varepsilon^2 + 4\sigma_{RSA}^2 + 16\sigma_{SA}^2$
	RSA_{ijm}	12	$\sigma_\varepsilon^2 + 4\sigma_{RSA}^2$
	JA_{km}	4	$\sigma_\varepsilon^2 + 4\sigma_{RJA}^2 + 16\sigma_{JA}^2$
	RJA_{ikm}	12	$\sigma_\varepsilon^2 + 4\sigma_{RJA}^2$
	SJA_{jkm}	4	$\sigma_\varepsilon^2 + 2\sigma_{RSJA}^2 + 8\sigma_{SJA}^2$
	$RSJA_{ijkm}$	12	$\sigma_\varepsilon^2 + 2\sigma_{RSJA}^2$
	$\varepsilon_{q(ijkm)}$	80	σ_ε^2

13.7

	Source	df	MS	EMS
Whole	R_i	2	60.46	$\sigma_\varepsilon^2 + 5\sigma_{RO}^2 + 20\sigma_R^2$
plot	D_j	4	276.19	$\sigma_\varepsilon^2 + \sigma_{RDO}^2 + 3\sigma_{DO}^2 + 4\sigma_{RD}^2 + 12\phi_D$
	RD_{ij}	8	40.63	$\sigma_\varepsilon^2 + \sigma_{RDO}^2 + 4\sigma_{RD}^2$
Split	O_k	3	16.73	$\sigma_\varepsilon^2 + 5\sigma_{RO}^2 + 15\sigma_O^2$
plot	RO_{ik}	6	57.44	$\sigma_\varepsilon^2 + 5\sigma_{RO}^2$
	DO_{jk}	12	12.78	$\sigma_\varepsilon^2 + \sigma_{RDO}^2 + 3\sigma_{DO}^2$
	RDO_{ijk}	24	46.83	$\sigma_\varepsilon^2 + \sigma_{RDO}^2$
	Total	59		

Days are significant by an F' test.

13.9 In a nested experiment the levels of B, say, are different within each level of A. In a split-plot experiment the same levels of B (in the split) are used at each level of A.

13.11 Laboratory means:

1	3	2
11.22	11.25	12.53

(Laboratories 1 and 3 underlined together; laboratory 2 underlined separately.)

Mix means:

B	A	C
10.29	11.58	13.13

(Each mix underlined separately.)

Laboratory × mix means all different in decreasing order: 165, 155, 145 with B, A, and C.

13.13

Source	EMS
R_i	$\sigma_\varepsilon^2 + 9\sigma_{RL}^2 + 27\sigma_R^2$
L_j	$\sigma_\varepsilon^2 + 9\sigma_{RL}^2 + 36\sigma_L^2$
RL_{ij}	$\sigma_\varepsilon^2 + 9\sigma_{RL}^2$
T_k	$\sigma_\varepsilon^2 + 3\sigma_{RLT}^2 + 12\sigma_{LT}^2 + 9\sigma_{RT}^2 + 36\phi_T$
RT_{ik}	$\sigma_\varepsilon^2 + 3\sigma_{RLT}^2 + 9\sigma_{RT}^2$
LT_{jk}	$\sigma_\varepsilon^2 + 3\sigma_{RLT}^2 + 12\sigma_{LT}^2$
RLT_{ijk}	$\sigma_\varepsilon^2 + 3\sigma_{RLT}^2$
M_m	$\sigma_\varepsilon^2 + 3\sigma_{RLM}^2 + 12\sigma_{LM}^2 + 9\sigma_{RM}^2 + 36\phi_M$
RM_{im}	$\sigma_\varepsilon^2 + 3\sigma_{RLM}^2 + 9\sigma_{RM}^2$
LM_{jm}	$\sigma_\varepsilon^2 + 3\sigma_{RLM}^2 + 12\sigma_{LM}^2$
RLM_{ijm}	$\sigma_\varepsilon^2 + 3\sigma_{RLM}^2$
TM_{km}	$\sigma_\varepsilon^2 + \sigma_{RLTM}^2 + 4\sigma_{LTM}^2 + 3\sigma_{RTM}^2 + 12\phi_{TM}$
RTM_{ikm}	$\sigma_\varepsilon^2 + \sigma_{RLTM}^2 + 3\sigma_{RTM}^2$
LTM_{jkm}	$\sigma_\varepsilon^2 + \sigma_{RLTM}^2 + 4\sigma_{LTM}^2$
$RTLM_{ijkm}$	σ_ε^2

CHAPTER 14

14.1 Confounding AC (each value is sum of four readings) gives

Block 1	Block 2
$(1) = 2$	$a = -5$
$b = 15$	$ab = 13$
$ac = -17$	$c = -12$
$abc = -7$	$bc = -2$
-7	-6

and $SS_{\text{block}} = SS_{AC} = 0.13$.

14.3 $SS_{\text{blocks}} = 29{,}153.5 = SS_{AB} + SS_{ACD} + SS_{BCD}$.

14.5 The principal block contains: (1), *ab, ad, bd, c, abc, acd, bcd.*

14.7 One scheme is to confound *ABC*, *CDE*, and *ABDE*. The principal block contains (1), *ab, de, abde, ace, bce, acd, bcd.*

Source	df
Main effects	5
Two-way interaction	10
Three-way interaction	8
Four-way interaction	4
Five-way interaction	1
Blocks (or *ABC*, *CDE*, *ABDE*)	3
Total	31

14.9 One scheme is to confound $ABCD^2$. The principal block contains 0000, 1110, 1102, 1012, 2111, 2122, 0221, 0210, 0120, 0101, 1211, 1200, 2220, 0022, 0011, 0202, 0112, 1121, 1020, 1001, 1222, 2010, 2201, 2100, 2002, 2021, 2212.

14.13 Confound a six-way interaction and get principal block. Seven-way no better as *AB* is pseudo-factor of temperature.

14.15

Source		df	
Heats	*A*	1	7 (*A* through *ABC*)
	B	1	
	AB	1	
	C	1	
	AC	1	
	BC	1	
	ABC	1	
T		1	
TH		7	
P		2	
HP		14	
TP		2	
THP		14	
Total		47	

14.17 If $TABC$ is confounded (1 df),

Source	df	SS	MS
Heats	7	59.81	8.54*
Treatments	1	28.52	28.52**
HT	6	10.96	1.83
Positions	2	1.54	0.77
HP	14	23.16	1.65
TP	2	2.54	1.27
THP	14	30.84	2.20
Furnaces or $TABC$	1	4.69	4.69
Totals	47	162.06	

* One asterisk indicates significance at the 5-percent level; two, the 1-percent level.

CHAPTER 15

15.1 Aliases are $A = C$, $B = ABC$, $AB = BC$. SS_{effects} for A (or C) = 105.06, for B (or ABC) = 33.06, for AB (or BC) = 0.56.

15.3 Aliases are

$$A = BC = AB^2C^2$$
$$B = AB^2C = AC$$
$$C = ABC^2 = AB$$
$$AB^2 = AC^2 = BC^2$$

15.5 $A = BD$, $B = AD$, $C = ABCD$, $D = AB$, $AC = BCD$, $BC = ACD$, $CD = ABC$.

15.7

$$A = AB^2C^2D = BCD^2$$
$$B = AB^2CD^2 = ACD^2$$
$$C = ABC^2D^2 = ABD^2$$
$$D = ABC = ABCD$$
$$AB = ABC^2D = CD^2$$
$$AB^2 = AC^2D = BC^2D$$
$$AC = AB^2CD = BD^2$$
$$AC^2 = AB^2D = BC^2D^2$$
$$AD = AB^2C^2 = BCD$$
$$BC = AB^2C^2D^2 = AD^2$$
$$BC^2 = AB^2D^2 = AC^2D^2$$
$$BD = AB^2C = ACD$$
$$CD = ABC^2 = ABD$$

15.9

Source	df
Main effects with 5- or 6-way	7
2-way interaction with 4- or 6-way	21
3-way interaction, and so on as error	35
Total	63

15.11

Source	df	SS		MS
Heats (or 2-way)	6	51.92		8.65*
Treatments (or part of H)	1	12.04		12.04*
Positions (or 3-way)	2	0.58		0.29
PH (or 3-way)	12	29.83	33.42	2.39
PT (or 2-way)	2	3.59		
Totals	23	97.96		

* One asterisk indicates significance at the 5-percent level.

CHAPTER 16

16.1 Before covariance: $F_{6,100} = 50$ and significant.
After covariance: $F_{6,99} = 16.5$ and still significant.

16.3

Source	df	$\sum x^2$	$\sum xy$	$\sum y^2$	Adjusted $\sum y^2$	df	MS
Between lots	3	500	−620	790			
Error	16	1182	428	406	251.02	15	16.73
Totals	19	1682	−192	1196	1174.08	18	
					923.06	3	307.69*

16.5

End of cycle	Estimate of s	Significance
2, block 1	0.60	None
2, block 2	0.64	All except AB
3, block 1	0.60	
3, block 2	0.61	All

16.7

Source	df	MS
Machines M	1	0.0136
Lumber grade L	1	0.0003
ML	1	0.0066
Replications R	1	0.0001
MR	1	0.0028
LR	1	0.0021
MLR	1	0.0015

Index

Index

E.

F.

G.

H.

I.

J.

K.

L.

M.

N.